www.KnowledgePublications.com

SAVE HARMLESS AGREEMENT

Because use of the information, instructions and materials discussed and shown in this book, document, electronic publication or other form of media is beyond our control, the purchaser or user agrees, without reservation to save Knowledge Publications Corporation, its agents, distributors, resellers, consultants, promoters, advisers or employees harmless from any and all claims, demands, actions, debts, liabilities, judgments, costs and attorney's fees arising out of, claimed on account of, or in any manner predicated upon loss of or damage to property, and injuries to or the death of any and all persons whatsoever, occurring in connection with or in any way incidental to or arising out of the purchase, sale, viewing, transfer of information and/or use of any and all property or information or in any manner caused or contributed to by the purchaser or the user or the viewer, their agents, servants, pets or employees, while in, upon, or about the sale or viewing or transfer of knowledge or use site on which the property is sold, offered for sale, or where the property or materials are used or viewed, discussed, communicated, or while going to or departing from such areas.

Laboratory work, scientific experiment, working with hydrogen, high temperatures, combustion gases as well as general chemistry with acids, bases and reactions and/or pressure vessels can be EXTREMELY DANGEROUS to use and possess or to be in the general vicinity of. To experiment with such methods and materials should be done ONLY by qualified and knowledgeable persons well versed and equipped to fabricate, handle, use and or store such materials. Inexperienced persons should first enlist the help of an experienced chemist, scientist or engineer before any activity thereof with such chemicals, methods and knowledge discussed in this media and other material distributed by KnowledgePublications Corporation or its agents. Be sure you know the laws, regulations and codes, local, county, state and federal, regarding the experimentation, construction and or use and storage of any equipment and or chemicals BEFORE you start. Safety must be practiced at all times. Users accept full responsibility and any and all liabilities associated in any way with the purchase and or use and viewing and communications of knowledge, information, methods and materials in this media.

FEDERAL EMERGENCY MANAGEMENT AGENCY RR-28 / Reprinted July 1989

Hydrogen Generator Gas for Vehicles and Engines
Volume 3

Construction of a Simplified Wood Gas Generator for Fueling Internal Combustion Engines in a Petroleum Emergency

Final Report

FEDERAL EMERGENCY MANAGEMENT AGENCY

Construction of a Simplified Wood Gas Generator for Fueling Internal Combustion Engines in a Petroleum Emergency

Final Report

by:
H. LaFontaine, Biomass Energy Foundation, Inc.
Miami, Florida
and
F. P. Zimmerman, Oak Ridge National Laboratory, Energy Division

FEMA Interagency Agreement Number: EMW-84-E-1737
Work Unit: 3521D

for:
Federal Emergency Management Agency
Washington, D.C. 20472

"This report has been reviewed in the Federal Emergency Management Agency and approved for publication. Approval does not signify that the contents necessarily reflect the views and policies of the Federal Emergency Management Agency."

Date Published: March 1989

APPROVED FOR PUBLIC RELEASE: DISTRIBUTION UNLIMITED

Prepared by:
Oak Ridge National Laboratory
Oak Ridge, Tennessee 37831-6285
for the U.S. Department of Energy

SECURITY CLASSIFICATION OF THIS PAGE

REPORT DOCUMENTATION PAGE

1a. REPORT SECURITY CLASSIFICATION: Unclassified	1b. RESTRICTIVE MARKINGS: None
2a. SECURITY CLASSIFICATION AUTHORITY	3. DISTRIBUTION / AVAILABILITY OF REPORT: Unlimited
2b. DECLASSIFICATION / DOWNGRADING SCHEDULE: N/A	
4. PERFORMING ORGANIZATION REPORT NUMBER(S): ORNL-6404	5. MONITORING ORGANIZATION REPORT NUMBER(S): N/A
6a. NAME OF PERFORMING ORGANIZATION: Oak Ridge National Lab. / 6b. OFFICE SYMBOL (If applicable)	7a. NAME OF MONITORING ORGANIZATION
6c. ADDRESS (City, State, and ZIP Code): P. O. Box 2008, Oak Ridge, TN 37831-6190	7b. ADDRESS (City, State, and ZIP Code)
8a. NAME OF FUNDING / SPONSORING ORGANIZATION: Federal Emergency Mgmt. Agncy / 8b. OFFICE SYMBOL (If applicable)	9. PROCUREMENT INSTRUMENT IDENTIFICATION NUMBER: Interagency Agreement No. EMW-84-E-1737
8c. ADDRESS (City, State, and ZIP Code): 500 C Street, SW, Washington, DC 20472	10. SOURCE OF FUNDING NUMBERS: PROGRAM ELEMENT NO. / PROJECT NO. / TASK NO. / WORK UNIT ACCESSION NO. 3521D

11. TITLE (Include Security Classification) Construction of a Simplified Wood Gas Generator for Fueling Internal Combustion Engines in a Petroleum Emergency

12. PERSONAL AUTHOR(S)
Harry LaFontaine, Gregory P. Zimmerman

13a. TYPE OF REPORT	13b. TIME COVERED	14. DATE OF REPORT (Year, Month, Day)	15. PAGE COUNT
Final	FROM ______ TO ______	March 1989	85

16. SUPPLEMENTARY NOTATION

17. COSATI CODES			18. SUBJECT TERMS (Continue on reverse if necessary and identify by block number)
FIELD	GROUP	SUB-GROUP	Emergency Technology, Wood Gasification, Wood Gas, Producer Gas, Downdraft Gasifier, Stratified Gasifier, Petroleum Crisis, Alternate Fuels, Internal Combustion Engines

19. ABSTRACT (Continue on reverse if necessary and identify by block number)
This report is one in a series of emergency technology assessments sponsored by the Federal Emergency Management Agency (FEMA). The purpose of this report is to develop detailed, illustrated instructions for the fabrication, installation, and operation of a biomass gasifier unit (that is, a "producer gas" generator, also called a "wood gas" generator) which is capable of providing emergency fuel for vehicles, such as tractors and trucks, in the event that normal petroleum sources were severely disrupted for an extended period of time. These instructions are prepared in the format of a manual for use by any mechanic who is reasonably proficient in metal fabrication or engine repair.

This report attempts to preserve the knowledge about wood gasification as put into practical use during World War II. Detailed, step-by-step fabrication procedures are presented for a simplified version of the World War II, Imbert wood gas generator. This simple, stratified,

(contd)

20. DISTRIBUTION / AVAILABILITY OF ABSTRACT: ☐ UNCLASSIFIED/UNLIMITED ☐ SAME AS RPT. ☐ DTIC USERS	21. ABSTRACT SECURITY CLASSIFICATION	
22a. NAME OF RESPONSIBLE INDIVIDUAL	22b. TELEPHONE (Include Area Code)	22c. OFFICE SYMBOL

DD FORM 1473, 84 MAR — 83 APR edition may be used until exhausted. All other editions are obsolete. — SECURITY CLASSIFICATION OF THIS PAGE

19. ABSTRACT (contd)

downdraft gasifier unit can be constructed from materials which would be widely available in the United States in a prolonged petroleum crisis. For example, the body of the unit consists of a galvanized metal garbage can atop a small metal drum; common plumbing fittings are used throughout; and a large, stainless steel mixing bowl is used for the grate. The entire compact unit was mounted onto the front of a farm tractor and successfully field tested, using wood chips as the only fuel. Photographic documentation of the actual assembly of the unit as well as its operation is included.

CONTENTS

EXECUTIVE SUMMARY

This report is one in a series of emergency technology assessments sponsored by the Federal Emergency Management Agency (FEMA). The purpose of this report is to develop detailed, illustrated instructions for the fabrication, installation, and operation of a biomass gasifier unit (i.e., a "producer gas" generator, also called a "wood gas" generator) which is capable of providing emergency fuel for vehicles, such as tractors and trucks, should normal petroleum sources be severely disrupted for an extended period of time. These instructions have been prepared as a manual for use by any mechanic who is reasonably proficient in metal fabrication or engine repair.

Fuel gas, produced by the reduction of coal and peat, was used for heating as early as 1840 in Europe and by 1884 had been adapted to fuel engines in England. Prior to 1940, gas generator units were a familiar, but not extensively utilized, technology. However, petroleum shortages during World War II led to widespread gas generator applications in the transportation industries of Western Europe. (Charcoal-burning taxis, a related application, were still common in Korea as late as 1970.) The United States, never faced with such prolonged or severe oil shortages, has lagged far behind Europe and the Orient in familiarity with and application of this technology. However, a catastrophic event could disrupt the supply of petroleum in this country so severely that this technology might be critical in meeting the energy needs of some essential economic activities, such as the production and distribution of food.

In occupied Denmark during World War II, 95% of all mobile farm machinery, tractors, trucks, stationary engines, and fishing and ferry boats were powered by wood gas generator units. Even in neutral Sweden, 40% of all motor traffic operated on gas derived from wood or charcoal. All over Europe, Asia, and Australia, millions of gas generators were in operation between 1940 and 1946. Because of the wood gasifier's somewhat low efficiency, the inconvenience of operation, and the potential health risks from toxic fumes, most of such units were abandoned when oil again became available in 1945. Except for the technology of producing alternate fuels, such as methane or alcohol, the only solution for operating existing internal combustion engines, when oil and petroleum products are not available, has been these simple, inexpensive gasifier units.

This report attempts to preserve the knowledge about wood gasification that was put into practical use during World War II. In this report, detailed step-by-step procedures are presented for constructing a simplified version of the World War II wood gas generator; this simple, stratified, downdraft gasifier unit (shown schematically in Fig. S-1) can be constructed from materials which would be widely available in the United States in a prolonged petroleum crisis. For example, the body of the unit consists of a galvanized metal garbage can atop a small metal drum; common plumbing fittings are used throughout; and a large, stainless steel mixing bowl is used for the grate. A prototype gasifier unit was fabricated from these instructions (see Fig. S-2); this unit was then mounted onto the front of a farm tractor and successfully field tested, using wood chips as the only fuel (see Fig. S-3). Photographic documentation of the actual assembly of the unit, as well as its operational field test, is included in the body of this report.

The use of wood gas generators need not be limited to transportation applications. Stationary engines can also be fueled by wood gasifiers to run electric generators, pumps, and industrial equipment. In fact, the use of wood gas as a fuel is not even restricted to gasoline engines; if a small amount of diesel fuel is used for ignition, a properly adjusted diesel engine can be operated primarily on wood gas introduced through the intake manifold.

S.1. PRINCIPLES OF SOLID FUEL GASIFICATION

All internal combustion engines actually run on vapor, not liquid. The liquid fuels used by gasoline engines are vaporized before they enter the combustion chamber above the pistons. In diesel engines, the fuel is sprayed into the combustion chamber as fine droplets which burn as they vaporize. The purpose of a gasifier, then, is to transform solid fuels into gaseous ones and to keep the gas free of harmful constituents. A gas generator unit is simultaneously an energy converter and a filter. In these twin tasks lie its advantages and its difficulties.

In a sense, gasification is a form of incomplete combustion–heat from the burning solid fuel creates gases which are unable to burn completely because of the insufficient amounts of oxygen from the available supply of air. The same chemical laws which govern combustion processes also apply to gasification. There are many solid biomass fuels suitable for gasification–from wood and paper to peat, lignite, and coal, including coke derived from coal. All of these solid fuels are composed primarily of carbon with varying amounts of hydrogen, oxygen, and impurities, such as sulphur, ash, and moisture. Thus, the aim of gasification is the almost complete transformation of these constituents into gaseous form so that only the ashes and inert materials remain. In creating wood gas for fueling internal combustion engines, it is important that the gas not only be properly produced, but also preserved and not consumed until it is introduced into the engine where it may be appropriately burned.

Gasification is a physiochemical process in which chemical transformations occur along with the conversion of energy. The chemical reactions and thermochemical conversions which occur inside a wood gas generator are too long and too complicated to be covered here; however, such knowledge is not necessary for constructing and operating a wood gasifier. By weight, gas (wood gas) produced in a gasifier unit contains approximately 20% hydrogen (H_2), 20% carbon monoxide (CO), and small amounts of methane, all of which are combustible, plus 50 to 60% nitrogen (N_2). The nitrogen is not combustible; however, it does occupy volume and dilutes the wood gas as it enters and burns in an engine. As the wood gas burns, the products of combustion are carbon dioxide (CO_2) and water vapor (H_2O).

One of the by-products of wood gasification is carbon monoxide, a poisonous gas. The toxic hazards associated with breathing this gas should be avoided during refueling operations or prolonged idling, particularly in inadequately ventilated areas. Except for the obvious fire hazard resulting from the combustion processes inside the unit, carbon monoxide poisoning is the major potential hazard during normal operation of these simplified gasifier units.

S.2. THE STRATIFIED, DOWNDRAFT GASIFIER

Until the early 1980s, wood gasifiers all over the world (including the World War II designs) operated on the principle that both the fuel hopper and the combustion unit be absolutely airtight; the hopper was sealed with a top or lid which had to be opened every time wood was added. Smoke and gas vented into the atmosphere while wood was being loaded; the operator had to be careful not to breathe the unpleasant smoke and toxic fumes.

Over the last few years, a new gasifier design has been developed through cooperative efforts among researchers at the Solar Energy Research Institute in Colorado, the University of California in Davis, the Open University in London, the Buck Rogers Company in Kansas, and the Biomass Energy Foundation, Inc., in Florida. This simplified design employs a balanced, negative-pressure concept in which the old type of sealed fuel hopper is no longer necessary. A closure is only used to preserve the fuel when the engine is stopped. This new technology has several popular names, including "stratified, downdraft gasification" and "open top gasification." Several years of laboratory and field testing have indicated that such simple, inexpensive gasifiers can be built from existing hardware and will perform very well as emergency units.

A schematic diagram of the stratified, downdraft gasifier is shown in Fig. S-1. During operation of this gasifier, air passes uniformly downward through four zones, hence the name "stratified:"

1. The uppermost zone contains unreacted fuel through which air and oxygen enter. This region serves the same function as the fuel hopper in the older, World War II designs.

2. In the second zone, the wood fuel reacts with oxygen during pyrolysis. Most of the volatile components of the fuel are burned in this zone and provide heat for continued pyrolysis reactions. At the bottom of this zone, all of the available oxygen from the air should be completely reacted. The open top design ensures uniform access of air to the pyrolysis region.

3. The third zone is made up of charcoal from the second zone. Hot combustion gases from the pyrolysis region react with the charcoal to convert the carbon dioxide and water vapor into carbon monoxide and hydrogen.

4. The inert char and ash, which constitute the fourth zone, are normally too cool to cause further reactions; however, because the fourth zone is available to absorb heat or oxygen as conditions change, it serves both as a buffer and as a charcoal storage region. Below this zone is the grate. The presence of char and ash serves to protect the grate from excessive temperatures.

The stratified, downdraft design has a number of advantages over the World War II gasifier designs. The open top permits fuel to be fed more easily and allows easy access. The cylindrical shape is easy to fabricate and permits continuous flow of fuel. No special fuel shape or pretreatment is necessary; any blocky fuel can be used.

The foremost question about the operation of the stratified, downdraft gasifier concerns char and ash removal. As the charcoal reacts with the combustion gases, it eventually reaches a very low density and breaks up into a dust containing all of the ash as well as a percentage of the original carbon. This dust may be partially carried away by the gas and might eventually begin to plug the gasifier. Hence, it must be removed by shaking or agitation. When the stratified gasifier unit is used to power vehicles, it is automatically shaken by the vehicle's motion.

An important issue in the design of the stratified, downdraft gasifier is the prevention of fuel bridging and channeling. High-grade biomass fuels, such as wood blocks or chips, will flow down through the gasifier because of gravity and downdraft air flow. However, other fuels (such as shredded chips, sawdust, and bark) can form a bridge, which will obstruct continuous flow and cause very high temperatures. Bridging can be prevented by stirring, shaking, or by agitating the grate or by having it agitated by the vehicle's movement. For prolonged idling, a hand-operated shaker has been included in the design in this report.

A prototype unit of the stratified, downdraft gasifier design (see Figs. S-2 and S-3) has been fabricated according to the instructions in this report; however, it has not been widely tested at this time. The reader is urged to use his ingenuity and initiative in constructing his own wood gas generator. As long as the principle of airtightness in the combustion regions, in the connecting piping, and in the filter units is followed, the form, shape, and method of assembly is not important.

The wood gasifier design presented in this report has as its origin the proven technology used in World War II during actual shortages of gasoline and diesel fuel. It should be acknowledged that there are alternate technologies (such as methane production or use of alcohol fuels) for keeping internal combustion engines in operation during a prolonged petroleum crisis; the wood gasifier unit described in this report represents only one solution to the problem.

ORNL-DWG 87-14553

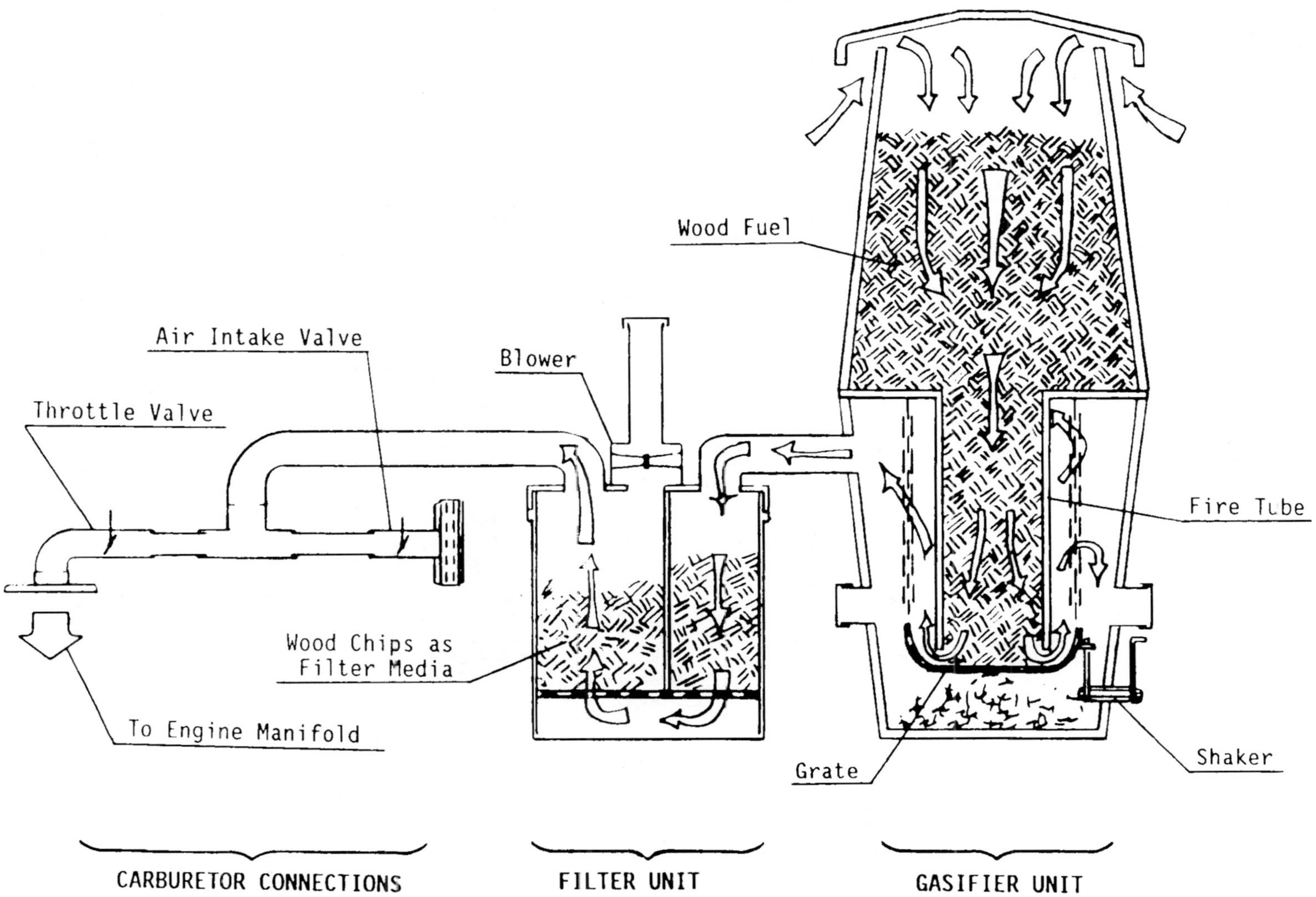

Fig. S-1. Schematic view of the stratified, downdraft gasifier.

ORNL Photo 5321-86

Fig. S-2. The prototype wood gas generator unit mounted onto a tractor.

ORNL-Photo 5325-86

Fig. S-3. Wood gas generator unit in operation during field testing.

CONVERSION FACTORS FOR SI UNITS

English units have been retained in the body of this report. The report refers to commercially available materials and sizes which are commonly expressed in English units. The conversion factors for SI units are given below:

To convert from	To	Multiply by
cubic feet (ft^3)	cubic meters (m^3)	0.0283
cubic yards (yd^3)	cubic meters (m^3)	0.7646
Fahrenheit degrees (°F)	Kelvin degrees (°K)	(see Note 1)
foot (ft)	meter (m)	0.3048
gallon (gal)	cubic meters (m^3)	3.785×10^{-3}
horsepower (hp)	watt (W)	745.7
inch (in.)	meter (m)	0.0254
pound (lb)	kilogram (kg)	0.4536
quart (qt.)	cubic meters (m^3)	9.464×10^{-4}

Note 1: To convert temperatures, use the following equation,

$$K = 273 + 0.5556 \times (F - 32) \;,$$

where

F is the temperature in Fahrenheit degrees, and

K is the temperature in Kelvin degrees.

LIST OF FIGURES

LIST OF FIGURES (continued)

LIST OF TABLES

CONSTRUCTION OF A SIMPLIFIED WOOD GAS GENERATOR FOR FUELING INTERNAL COMBUSTION ENGINES IN A PETROLEUM EMERGENCY

H. LaFontaine
G. P. Zimmerman

ABSTRACT

This report is one in a series of emergency technology assessments sponsored by the Federal Emergency Management Agency (FEMA). The purpose of this report is to develop detailed, illustrated instructions for the fabrication, installation, and operation of a biomass gasifier unit (i.e., a "producer gas" generator, also called a "wood gas" generator) that is capable of providing emergency fuel for vehicles, such as tractors and trucks, in the event that normal petroleum sources were severely disrupted for an extended period of time. These instructions have been prepared as a manual for use by any mechanic who is reasonably proficient in metal fabrication or engine repair.

This report attempts to preserve the knowledge about wood gasification that was put into practical use during World War II. Detailed, step-by-step fabrication procedures are presented for a simplified version of the World War II, Imbert wood gas generator. This simple, stratified, downdraft gasifier unit can be constructed from materials that would be widely available in the United States in a prolonged petroleum crisis. For example, the body of the unit consists of a galvanized metal garbage can atop a small metal drum; common plumbing fittings throughout; and a large, stainless steel mixing bowl for the grate. The entire compact unit was mounted onto the front of a farm tractor and successfully field tested, using wood chips as the only fuel. Photographic documentation of the actual assembly of the unit as well as its operation is included.

---

1. WHAT IS A WOOD GAS GENERATOR AND HOW DOES IT WORK?

This report is one in a series of emergency technology assessments sponsored by the Federal Emergency Management Agency (FEMA). The purpose of this report is to develop detailed, illustrated instructions for the fabrication, installation, and operation of a biomass gasifier unit (i.e., a "producer gas" generator, also called a "wood gas" generator) that is capable of providing emergency fuel for vehicles, such as tractors and trucks, in the event that normal petroleum sources were severely disrupted for an extended period of time. These instructions have been prepared as a manual for use by any mechanic who is reasonably proficient in metal fabrication or engine repair.

1.1 INTRODUCTION

Fuel gas, produced by the reduction of coal and peat, was used for heating, as early as 1840 in Europe, and by 1884 it had been adapted to fuel engines in England. Before 1940, gas generator units were a familiar, but not extensively utilized, technology. However, petroleum shortages during World War II led to widespread gas generator applications in the transportation industries of Western Europe. (Charcoal-burning taxis, a related application, were still common in Korea as late as 1970.) The United States, never faced with such prolonged or severe oil shortages, has lagged far behind Europe and the Orient in familiarity with and application of this technology; however, a catastrophe could so severely disrupt the supply of petroleum in this country that this technology might be critical in meeting the energy needs of some essential economic activities, such as the production and distribution of food.

This report attempts to preserve the knowledge about wood gasification as put into practical use during World War II. Detailed, step-by-step procedures are presented in this report for constructing a simplified version of the World War II, Imbert wood gas generator. This simple, stratified, downdraft gasifier unit can be constructed from materials that would be widely available in the United States in a prolonged petroleum crisis. For example, the body of the unit consists of a galvanized metal garbage can atop a small metal drum; common plumbing fittings throughout; and a large, stainless steel mixing bowl for the grate. A prototype gasifier unit was fabricated from these instructions. This unit was then mounted onto the front of a gasoline-engine farm tractor and successfully field tested, using wood chips as the only fuel; see Fig. 1-1 (all figures and tables are presented at the end of their respective sections). Photographic documentation of the actual assembly of the unit, as well as its operational field test, is included in this report.

The use of wood gas generators need not be limited to transportation applications. Stationary engines can also be fueled by wood gasifiers to run electric generators, pumps, and industrial equipment. In fact, the use of wood gas as a fuel is not even restricted to gasoline engines; if a small amount of diesel fuel is used for ignition, a properly adjusted diesel engine can be operated primarily on wood gas introduced through the intake manifold. However, this report is concerned with the operation of four-cycle gasoline engines rated from 10 to 150 horsepower. If more information is needed about operating gasifiers on other fuels (such as coal, charcoal, peat, sawdust or seaweed), a list of relevant literature is contained in the Bibliography at the end of this report.

The goal of this report is to furnish information for building a homemade wood gas generator made out of ordinary, available hardware, in order to get tractors, trucks, and other vehicles operating without delay, if a severe liquid fuel emergency should arise. Section 1 describes gasification principles and wood gas generators, in general, and gives some historical background about their operation and effectiveness. Section 2 contains detailed step-by-step instructions for constructing your own wood gas generator unit; illustrations and photographs are included to prevent confusion. Section 3 contains information on operating, maintaining, and trouble-shooting your wood gas generator; also included are some very important guidelines on safety when using your gasifier system.

The wood gasifier design presented in this report has as its origin the proven technology used in World War II during actual shortages of gasoline and diesel fuel. It should be acknowledged that there are alternate technologies (such as methane production or use of alcohol fuels) for keeping internal combustion engines in operation during a prolonged petroleum crisis; the wood gasifier unit described in this report represents only one solution to the problem.

1.2 PRINCIPLES OF SOLID FUEL GASIFICATION

All internal combustion engines actually run on vapor, not liquid. The liquid fuels used in gasoline engines are vaporized before they enter the combustion chamber above the pistons. In diesel engines, the fuel is sprayed into the combustion chamber as fine droplets which burn as they vaporize. The purpose of a gasifier, then, is to transform solid fuels into gaseous ones and to keep the gas free of harmful constituents. A gas generator unit is, simultaneously, an energy converter and a filter. In these twin tasks lie its advantages and its difficulties.

The first question many people ask about gasifiers is, "Where does the combustible gas come from?" Light a wooden match; hold it in a horizontal position; and notice that while the wood becomes charcoal, it is not actually burning but is releasing a gas that begins to burn brightly a short distance away from the matchstick. Notice the gap between the matchstick and the luminous flame; this gap contains the wood gas which starts burning only when properly mixed with air (which contains oxygen). By weight, this gas (wood gas) from the charring wood contains approximately 20% hydrogen (H_2), 20% carbon monoxide (CO), and small amounts of methane, all of which are combustible, plus 50 to 60% nitrogen (N_2). The nitrogen is not combustible; however, it does occupy volume and dilutes the wood gas as it enters and burns in an engine. As the wood gas burns, the products of combustion are carbon dioxide (CO_2) and water vapor (H_2O).

The same chemical laws which govern combustion processes also apply to gasification. The solid, biomass fuels suitable for gasification cover a wide range, from wood and paper to peat, lignite, and coal, including coke derived from coal. All of these solid fuels are composed primarily of carbon with varying amounts of hydrogen, oxygen, and impurities, such as sulphur, ash, and moisture. Thus, the aim of gasification is the almost complete transformation of these constituents into gaseous form so that only the ashes and inert materials remain.

In a sense, gasification is a form of incomplete combustion; heat from the burning solid fuel creates gases which are unable to burn completely, due to insufficient amounts of oxygen from the available supply of air. In the matchstick example above, as the wood was burned and pyrolyzed into charcoal, wood gas was created, but the gas was also consumed by combustion (since there was an enormous supply of air in the room). In creating wood gas for fueling internal combustion engines, it is important that the gas not only be properly produced, but also preserved and not consumed until it is introduced into the engine where it may be appropriately burned.

Gasification is a physiochemical process in which chemical transformations occur along with the conversion of energy. The chemical reactions and thermochemical conversions which occur inside a wood gas generator are too long and too complicated to be covered here. Such knowledge is not necessary for constructing and operating a wood gasifier. Books with such information are listed in the Reference Section (see, for example, Reed 1979, Vol. II; or Reed and Das 1988).

1.3 BACKGROUND INFORMATION

The use of wood to provide heat is as old as mankind; but by burning the wood we only utilize about one-third of its energy. Two-thirds is lost into the environment with the smoke. Gasification is a method of collecting the smoke and its combustible components. Making a combustible gas from coal and wood began around 1790 in Europe. Such manufactured gas was used for street lights and was piped into houses for heating, lighting, and cooking. Factories used it for steam boilers, and farmers operated their machinery on wood gas and coal gas. After the discovery of large petroleum reserves in Pennsylvania in 1859, the entire world changed to oil–a cheaper and more convenient fuel. Thousands of gas works all over the world were eventually dismantled.

Wood gas generators are not technological marvels that can totally eliminate our current dependence on oil, reduce the impacts of an energy crunch, or produce long-term economic relief from high fossil fuel prices, but they are a proven emergency solution when such fuels become unobtainable in case of war, civil upheaval, or natural disaster. In fact, many people can recall a widespread use of wood gas generators during World War II, when petroleum products were not available for the civilian populations in many countries. Naturally, the people most affected by oil and petroleum scarcity made the greatest advancements in wood gas generator technology.

In occupied Denmark during World War II, 95% of all mobile farm machinery, tractors, trucks, stationary engines, fishing and ferry boats were powered by wood gas generators. Even in neutral Sweden, 40% of all motor traffic operated on gas derived from wood or charcoal (Reed and Jantzen 1979). All over Europe, Asia, and Australia, millions of gas generators were in operation between 1940 and 1946. Because of the wood gasifier's somewhat low efficiency, the inconvenience of operation, and the potential health risks from toxic fumes, most of such units were abandoned when oil again became available in 1945. Except for the technology of producing alternate fuels, such as methane or alcohol, the only solution for operating existing internal combustion engines, when oil and petroleum products are not available, has been these simple, inexpensive gasifier units.

1.3.1 The World War II, Imbert Gasifier

The basic operation of two gasifiers is described in this and the following section. Their operating advantages and disadvantages will also be discussed. This information is included for the technically interested reader only; it is intended to give the reader more insight into the subtleties of the operating principles of the wood gas generator described in this manual. Those readers who are anxious to begin construction of their own wood gas generator may skip the material below and proceed directly to Sect. 2 without any loss of continuity.

The constricted hearth, downdraft gasifier shown in Fig. 1-2 is sometimes called the "Imbert" gasifier after its inventor, Jacques Imbert; although, it has been commercially manufactured under various names. Such units were mass produced during World War II by many European automotive companies, including General Motors, Ford, and Mercedes-Benz. These units cost about $1500 (1985 evaluation) each. However, after World War II began in 1939, it took six to eight months before factory-made gasifiers were generally available. Thousands of Europeans were saved from certain starvation by home-built, simple gasifier units made from washing machine tubs, old water heaters, and metal gas or oxygen cylinders. Surprisingly, the operation of these units was nearly as efficient as the factory-made units; however, the homemade units lasted for only about 20,000 miles with many repairs, while the factory-made units operated, with few repairs, up to 100,000 miles.

In Fig. 1-2, the upper cylindrical portion of the gasifier unit is simply a storage bin or hopper for wood chips or other biomass fuel. During operation, this chamber is filled every few hours as needed. The spring-loaded, airtight cover must be opened to refill the fuel hopper; it must remain closed and sealed during gasifier operation. The spring permits the cover to function as a safety valve because it will pop open in case of any excessive internal gas pressure.

About one-third of the way up from the bottom of the gasifier unit, there is a set of radially directed air nozzles; these allow air to be injected into the wood as it moves downward to be gasified. In a gas generator for vehicle use, the downstroke of the engine's pistons creates the suction force which moves the air into and through the gasifier unit; during startup of the gasifier, a blower is used to create the proper airflow. The gas is introduced into the engine and consumed a few seconds after it is made. This gasification method is called "producer gas generation," because no storage system is used; only that amount of gas demanded by the engine is produced. When the engine is shut off, the production of gas stops.

During normal operation, the incoming air burns and pyrolyzes some of the wood, most of the tars and oils, and some of the charcoal that fills the constricted area below the nozzles. Most of the fuel mass is converted to gas within this combustion zone. The Imbert gasifier is, in many ways, self-adjusting. If there is insufficient charcoal at the air nozzles, more wood is burned and pyrolyzed to make more charcoal. If too much charcoal forms, then the charcoal level rises above the nozzles, and the incoming air burns the charcoal. Thus, the combustion zone is maintained very close to the nozzles.

Below this combustion zone, the resulting hot combustion gases–carbon dioxide (CO_2) and water vapor (H_2O)–pass into the hot charcoal where they are chemically reduced to combustible fuel gases: carbon monoxide (CO) and hydrogen (H_2). The hearth constriction causes all gases to pass through the reaction zone, thus giving maximum mixing and minimum heat loss. The highest temperatures are reached in this region.

Fine char and ash dust can eventually clog the charcoal bed and will reduce the gas flow unless the dust is removed. The charcoal is supported by a movable grate which can be shaken at intervals. Ash buildup below the grate can be removed during cleaning operations. Usually, wood contains less than 1% ash (by weight). However, as the charcoal is consumed, it eventually collapses to form a powdery charcoal/ash mixture which may represent 2 to 10% (by weight) of the total fuel mass.

The cooling unit required for the Imbert gasifier consists of a water-filled precipitating tank and an automotive radiator-type gas cooler. The precipitating tank removes all unacceptable tars and most of the fine ash from the gas flow, while the radiator further cools the gas. A second filter unit, containing a fine-mesh filtration material, is used to remove the last traces of any ash or dust that may have survived passage through the cooling unit. Once out of the filter unit, the wood gas is mixed with air in the vehicle's carburetor and is then introduced directly into the engine's intake manifold.

The World War II, Imbert gasifier requires wood with a low moisture content (less than 20% by weight) and a uniform, blocky fuel in order to allow easy gravity feed through the constricted hearth. Twigs, sticks, and bark shreds cannot be used. The constriction at the hearth and the protruding air nozzles present obstructions to the passage of the fuel and may create bridging and channeling followed by poor quality gas output, as unpyrolyzed fuel falls into the reaction zone. The vehicle units of the World War II era had ample vibration to jar the carefully sized wood blocks through the gasifier. In fact, an entire industry emerged for preparing wood for use in vehicles at that time (Reed and Jantzen 1979). However, the constricted hearth design seriously limits the range of wood fuel shapes that can be successfully gasified without expensive cubing or pelletizing pretreatment. It is this limitation that makes the Imbert gasifier less flexible for emergency use.

In summary, the World War II Imbert gasifier design has stood the test of time and has successfully been mass produced. It is relatively inexpensive, uses simple construction materials, is easy to fabricate, and can be operated by motorists with a minimum amount of training.

1.3.2 The Stratified, Downdraft Gasifier

Until the early 1980s, wood gasifiers all over the world (including the World War II designs) operated on the principle that both the fuel hopper and the combustion unit be airtight; the hopper was sealed with a top or lid that had to be opened every time wood was added. Smoke and gas vented into the atmosphere while new wood was being loaded; the operator had to be careful not to breathe the unpleasant smoke and toxic fumes.

Over the last few years, a new gasifier design has been developed through cooperative efforts among researchers at the Solar Energy Research Institute in Colorado, the University of California in Davis, the Open University in London, the Buck Rogers Company in Kansas, and the Biomass Energy Foundation, Inc., in Florida (Reed and Das 1988). This simplified design employs a balanced, negative-pressure concept in which the old type of sealed fuel hopper is no longer necessary. A closure is only used to preserve the fuel when the engine is stopped. This new technology has several popular names, including "stratified, downdraft gasification" and "open top gasification." Two years of laboratory and field testing have indicated that such simple, inexpensive gasifiers can be built from existing hardware and will perform very well as emergency units.

A schematic diagram of the stratified, downdraft gasifier is shown in Fig. 1-3. During operation of this gasifier, air passes uniformly downward through four zones, hence the name "stratified:"

1. The uppermost zone contains unreacted fuel through which air and oxygen enter. This region serves the same function as the fuel hopper in the Imbert design.

2. In the second zone, the wood fuel reacts with oxygen during pyrolysis. Most of the volatile components of the fuel are burned in this zone and provide heat for continued pyrolysis reactions. At the bottom of this zone, all of the available oxygen from the air has completely reacted. The open top design ensures uniform access of air to the pyrolysis region.

3. The third zone is made up of charcoal from the second zone. Hot combustion gases from the pyrolysis region react with the charcoal to convert the carbon dioxide and water vapor into carbon monoxide and hydrogen.

4. The inert char and ash, which constitute the fourth zone, are normally too cool to cause further reactions; however, since the fourth zone is available to absorb heat or oxygen as conditions change, it serves both as a buffer and as a charcoal storage region. Below this zone is the grate. The presence of char and ash serves to protect the grate from excessive temperatures.

The stratified, downdraft design has a number of advantages over the World War II, Imbert gasifier. The open top permits fuel to be fed more easily and allows easy access. The cylindrical shape is easy to fabricate and permits continuous flow of fuel. No special fuel shape or pretreatment is necessary; any blocky fuel can be used.

The foremost question about the operation of the stratified, downdraft gasifier concerns char and ash removal. As the charcoal reacts with the combustion gases, it eventually reaches a very low density and breaks up into a dust containing all of the ash as well as a percentage of the original carbon. This dust may be partially carried away by the gas; however, it might eventually begin to plug the gasifier, and so it must be removed by shaking or agitation. Both the Imbert gasifiers and the stratified concept have a provision for shaking the grate; when they are used to power vehicles, they are automatically shaken by the vehicle's motion.

An important issue in the design of the stratified, downdraft gasifier is the prevention of fuel bridging and channeling. High-grade biomass fuels such as wood blocks or chips will flow down through the gasifier under the influence of gravity, and downdraft air flow. However, other fuels (such as shredded wood, sawdust, and bark) can form a bridge that will prevent continuous flow and cause very high temperatures. Obviously, it is desirable to use these widely available biomass residues. Bridging can be prevented by stirring, shaking, or by agitating the grate or by having it agitated by the vehicle's movement. For prolonged idling, a hand-operated shaker has been included in the design.

A prototype design of the stratified, downdraft gasifier design has been developed. The detailed but simple design is described and illustrated in Sect. 2; however, it has not been widely tested at this time. The reader is urged to use his ingenuity and initiative in constructing his own wood gas generator. As long as the principle of airtightness in the combustion regions, in the connecting piping, and in the filter units is followed, the form, shape, and method of assembly is not important.

ORNL-Photo 5325-86

Fig. 1-1. Wood gas generator unit in operation during field testing.

ORNL-DWG 87-14552

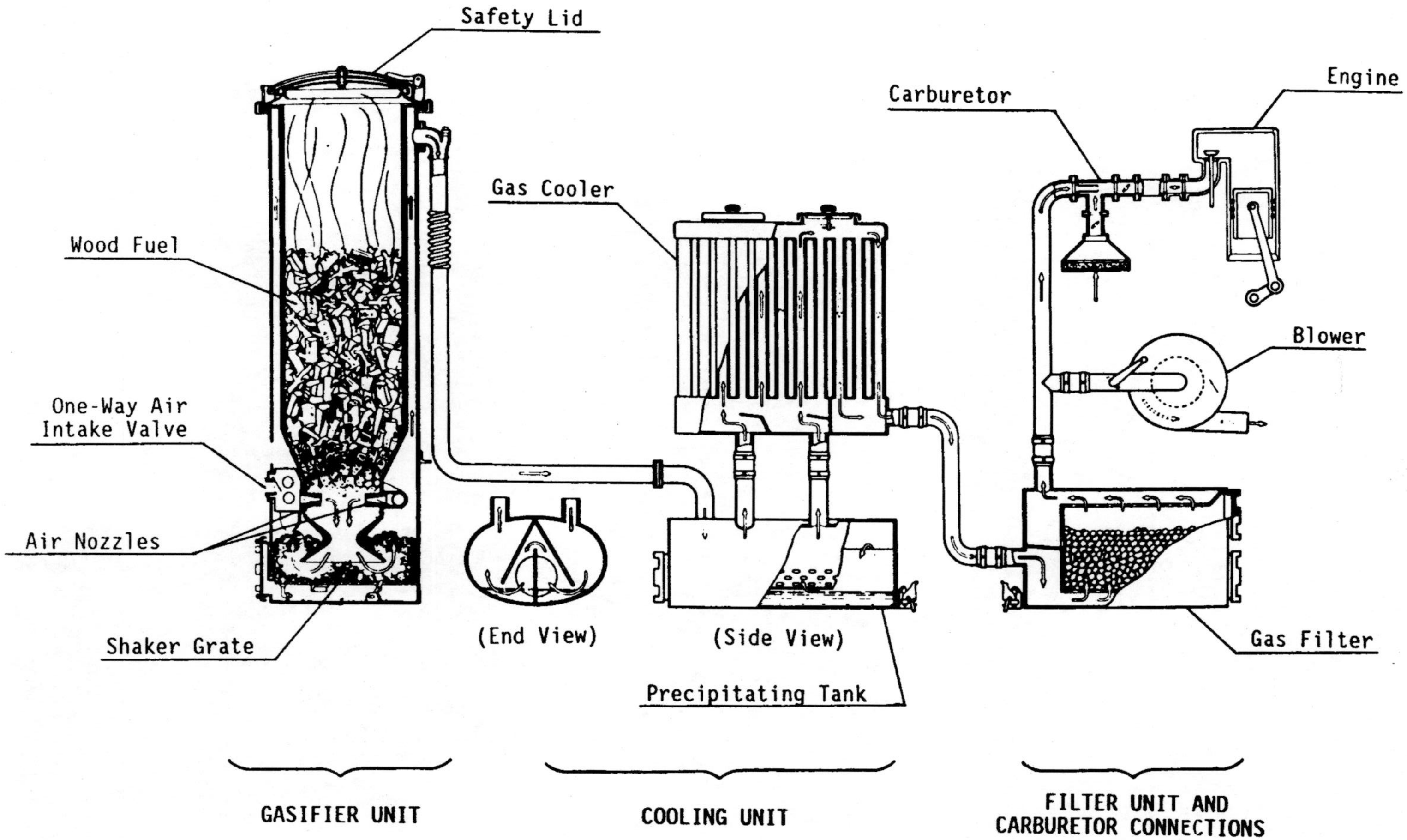

Fig. 1-2. Schematic view of the World War II, Imbert gasifier.

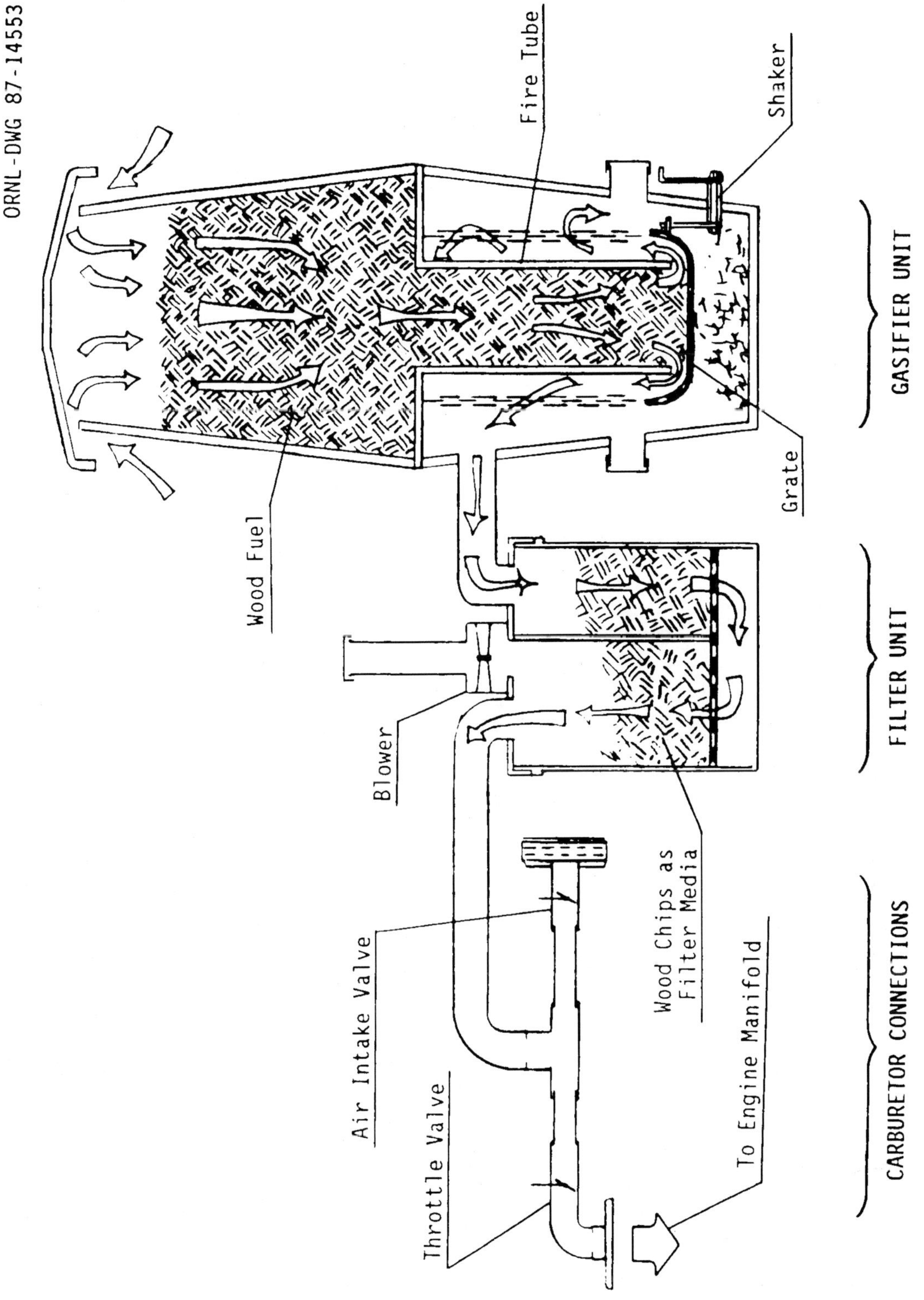

Fig. 1-3. Schematic view of the stratified, downdraft gasifier.

2. BUILDING YOUR OWN WOOD GAS GENERATOR

The following fabrication instructions, parts lists, and illustrations describe the prototype gasifier unit shown schematically in Fig. 1-3. These instructions are simple and easy to follow. The dimensions in the following plans are given in inches rather than in millimeters to make construction easier for those who might be unfamiliar with the metric system and to allow the builder to take advantage of available, alternate construction materials. It will be obvious to the experienced engineer, mechanic, or builder that most of the dimensions (for example, plate thicknesses and cleanout diameters) are not critical to the acceptable performance of the finished gasifier unit.

The prototype gasifier unit described in the following text was actually constructed and field tested on a gasoline-engine farm tractor (a 35-hp, John Deere 1010 Special); see Fig. 2-1. The unit operated very well, and on par with the European, World War II designs, but it has not had the test of time nor the millions of operating hours like the older Imbert design. This new stratified design was developed for the construction of simple, inexpensive emergency wood gas generator units. The prototype design below should be considered to be the absolute minimum in regard to materials, piping and filter arrangement, and carburetor system connections.

The gasifier unit, as described below, is designed to maintain proper cooling, even at moderate vehicle speeds. If this unit is to be used on stationary engines or on slow-moving vehicles, a gas cooler and a secondary filter must be placed in the piping system between the generator unit and the carburetor. The ideal temperature for the wood gas at the inlet to the carburetor manifold would be 70°F, with acceptable peaks of 140 to 160°F. For every 10 degrees above 70°F, an estimated 1% horsepower is lost. Cooler gas has higher density and, therefore, contains more combustible components per unit volume.

The millions of wood gasifiers built during World War II proved that shape, form, and construction material had little or no effect on the performance of the unit. Judicious substitution or the use of scavenged parts is, therefore, quite acceptable. What <u>is</u> important is that:

1. the fire tube dimensions (inside diameter and length) must be correctly selected to match the rated horsepower of particular engine which is to be fueled,

2. airtightness of the gas generator unit and all connecting piping must be maintained at all times, and

3. unnecessary friction should be eliminated in all of the air and gas passages by avoiding sharp bends in the piping and by using piping sizes which are not too small.

2.1. BUILDING THE GAS GENERATOR UNIT AND THE FUEL HOPPER

Figure 2-2 shows an exploded view of the gas generator unit and the fuel hopper; the list of materials is given in Table 2-1 (all figures and tables mentioned in Sect. 2 are presented at the end of Sect. 2). Only the dimensions of the fire tube (Item 1A) must be reasonably close; all other dimensions and materials can be substituted as long as complete airtightness is maintained. In the following instructions, all item numbers refer both to Fig. 2-2 and to Table 2-1.

The prototype unit described in this report was constructed for use with a 35-hp gasoline engine; the unit has a fire tube diameter of 6 in. (as determined from Table 2-2). A gas generator unit containing a fire tube up to 9-in. diameter (i.e., a gasifier unit for fueling engines up to about 65 hp) can be constructed from the following instructions. If your engine requires a fire tube diameter of 10 in. or more, use a 55-gal drum for the gas unit and another 55-gal drum for the fuel hopper.

The following fabrication procedure is very general and can be applied to the construction of gas generator units of any size; however, the specific dimensions which are given in the parts list and in the instructions below are for this particular prototype unit. All accompanying photographs were taken during the actual assembly of the prototype unit.

The fabrication procedure is as follows:

1. Using the displacement or horsepower rating of the engine to be fueled by the gasifier unit, determine the dimensions (inside diameter and length) of the fire tube (Item 1A) from Table 2-2. Fabricate a cylindrical tube or cut a length of correctly sized pipe to match the dimensions from Table 2-2. (For the prototype gasifier unit illustrated in this report, a 6-in.-diam firetube was used; its length was 19 in.)

2. The circular top plate (Item 2A) should be cut to a diameter equal to the outside diameter of the gasifier housing drum (Item 3A) at its top. A circular hole should then be cut in the center of the top plate; the diameter of this hole must be equal to the outside diameter of the fire tube. The fire tube (Item 1A) should then be welded at a right angle to the top plate (Item 2A) as shown in Fig. 2-3.

3. The grate (Item 4A) should be made from a stainless steel mixing bowl or colander. Approximately 125 holes with diameters of 1/2 in. should be drilled in the bottom and up the sides of the mixing bowl; see Fig. 2-4. A U-bolt (Item 5A) should be welded horizontally to the side of the grate, 2 in. from its bottom. This U-bolt will be interlocked with the shaker mechanism (Item 12A) in a later step.

4. The support chains (Item 6A) are to be attached to the grate in three evenly spaced holes drilled under the lip of the mixing bowl or colander; see Fig. 2-5. These chains are to be connected to the top plate (Item 2A) with eyebolts (Item 7A), as shown in Fig. 2-6. Each eyebolt should have two nuts, one on each side of the top plate, so that the eyebolts can be adjusted to the proper length. When assembled, the bottom of the firetube should be 1.25 in. above the bottom of the mixing bowl.

5. A hole equal to the outside diameter of the ash cleanout port (Item 8A) should be cut into the side of the gasifier housing drum (Item 3A); the bottom edge of this hole

should be about 1/2 in. from the bottom of the drum. Because of the thin wall thickness of oil drums and garbage cans, welding is not recommended; brazing such parts to the drums or cans will ensure both strength and airtightness (see Fig. 2-7).

6. Two holes, equal to the outside diameters of the ignition ports (Item 10A), are to be cut with their centers at a distance from the top of the housing drum (Item 3A) equal to the firetube length less 7 in. (19 in. less 7 in. equals 12 in. for this prototype unit); the holes should be placed opposite each other as shown in Fig. 2-2. The ignition ports should be attached to the wall of the housing drum by brazing.

7. When the ash cleanout port (Item 8A) and the ignition ports (Item 10A) have been attached to the wall of the gasifier housing drum (Item 3A), they should then be closed with pipe caps, Items 9A and 11A respectively. The threads of the pipe caps should be first coated with high temperature silicone (Item 27A) to ensure airtightness. An optional steel crossbar welded to the pipe cap will reduce the effort required to open these caps later.

8. The shaker assembly (Item 12A) is shown in Fig. 2-8. The 1/2-in. pipe (Item 1AA) should be brazed into the side of the housing drum (Item 3A), 1.5 inches from the bottom of the drum; the length of this pipe which protrudes into the drum must be chosen so that the upright bar (Item 2AA) is in line with the U-bolt (Item 5A) on the grate. Likewise, the length of the upright bar must be selected so as to connect into the U-bolt.

9. Weld the upright bar (Item 2AA) to the head of the bolt (Item 3AA). The threaded end of the bolt should be ground down or flattened on one side, as shown in Fig. 2-9, to positively interlock with a slot to be drilled and filed in the handle (Item 4AA). The handle can be formed or bent into any desired or convenient shape.

10. A hole should be drilled in the pipe cap (Item 7AA) so that there is a close fit between this hole and the bolt (Item 3AA). The close fit will help to ensure airtightness.

11. Before assembling the shaker, as shown in Fig. 2-8, coat the bolt (Item 3AA) with a small amount of grease. Before inserting the bolt, fill the pipe (Item 1AA) with high temperature silicone (Item 27A) to ensure airtightness. Tighten the nuts (Item 6AA) so that the position of the handle (Item 4AA) is maintained by friction, yet is capable of being turned and agitated during cleanout or stationary operation.

12. Fabricate the supports (Item 13A) for the gasifier unit housing drum (Item 3A) out of rectangular, iron bar stock. The shape and height of the support flanges must be determined by the frame of the vehicle to which the gasifier is to be mounted. The supports can either be bolted to the bottom and side with the 1/4-in. bolts (Item 14A) or can be brazed directly to the drum; see Fig. 2-10. Remember to seal all bolt holes for airtightness.

13. Completely cover the bottom of the housing drum (Item 3A) with 1/2 in. of hydraulic cement (Item 28A). The cement should also be applied to the inside of the drum for

about 5 in. up the inside walls near the bottom. All edges should be rounded for easy ash removal.

14. The fuel hopper (Item 15A) is to be made from a second container with its bottom up as shown in Fig. 2-11. Remove the bottom, leaving a 1/4-in. lip around the circumference.

15. A garden hose (Item 17A) should be cut to a length equal to the circumference of the fuel hopper (Item 15A) and should then be slit along its entire length. It should be placed over the edge of the fuel hopper from which the bottom was removed. This will prevent injury to the operator when adding wood fuel to the unit. To insure close fit of the garbage can lid (Item 16A), a piece of weather stripping (Item 18A) should be attached under the lid where it makes contact with the fuel hopper.

16. Cut four support bars (Item 19A) to lengths 2.5 in. longer than the height of the fuel hopper (Item 15A). Drill a 3/8-in. hole in each end of all four support bars; these holes should be centered 3/4 in. from the ends. Bend 2 in. of each end of these support bars over at a right angle; then, mount them evenly spaced around the fuel hopper (Item 15A) with 1/4-in. bolts (Item 20A). One of the bends on each support bar should be as close to the lower edge of the fuel hopper as possible.

17. Cut four metal triangular standoffs (Item 21A) and braze, weld, or rivet them flat against the edge of the garbage can lid (Item 16A) as shown in Fig. 2-12; they must be aligned with the four support bars (Item 19A) attached to the fuel hopper. During operation, the garbage can lid must have a minimum 3/4-in. opening for air passage; the standoffs should provide this clearance when they are engaged into the holes in the top edges of the support bars (Item 19A); see Fig. 2-13.

18. Two eye hooks (Item 22A) should be attached to opposite sides of the garbage can lid (Item 16A). Two screen door springs (Item 23A) should be attached to the garbage can handles and used under tension to keep the top lid (Item 16A) either open or closed.

19. Cut the oil drum lock ring (Item 24A) to the exact circumference of the top plate (Item 2A) so that it will fit snugly around the gasifier unit housing drum (Item 3A).

20. Cut four 2 by 2 by 1/4-in. tabs (Item 25A); then, braze these tabs to the lock ring (Item 24A), evenly spaced and in alignment with the support bars (Item 19A) on the fuel hopper. Drill a 3/8-in. hole in each tab to align with the holes in the fuel hopper support bars (Item 19A). The lock ring is shown in Fig. 2-14.

21. The connecting pipe (Item 29A) between the gasifier unit and the filter unit should be attached to the gasifier housing drum (Item 3A) at a point 6 in. below the top of the drum. This pipe must be a minimum of 2-in. in diameter and should be at least 6 ft long for cooling purposes. At least one of the ends of this pipe must be removable for cleaning and maintenance. On this prototype unit, an airtight electrical conduit connector was used; this connection is visible in Fig. 2-1. Many similar plumbing devices

are available and can be used if they are suitable for operation at 400°F and above. The pipe can also be welded or brazed directly to the housing drum.

22. When assembling the gasifier unit, the upright bar (Item 2AA) on the shaker assembly must be placed inside the U-bolt (Item 5A) on the grate.

23. The lock ring will then clamp the gasifier unit housing drum (Item 3A) and the top plate (Item 2A) together. The fuel hopper support bars (Item 19A) must be attached to the tabs (Item 25A) on the lock ring with bolts (Item 26A). High temperature silicone (Item 27A) should be applied to all edges to make an airtight connection. The lock ring connections are shown in the lower portion of Fig. 2-13.

2.2 BUILDING THE PRIMARY FILTER UNIT

Figures 2-15 and 2-16 show exploded views of the primary filter unit; the list of materials is given in Table 2-3 (all figures and tables mentioned in Sect. 2 are presented at the end of Sect. 2). In the following instructions, all item numbers refer to either Fig. 2-15 or 2-16 and to Table 2-3.

The prototype primary filter unit was made from a 5-gal paint can. That size seems to be sufficient for gasifiers with fire tubes up to 10 in. in diameter. If a fire tube diameter of more than 10 in. is required, then a 20-gal garbage can or a 30-gal oil drum should be used. The filter unit could be fabricated in any shape or form as long as airtightness and unobstructed flow of gas are provided. If a 5-gal container is used, it must be clean and free of any chemical residue. The top edge must be straight and without any indentations. If an alternate container can be found or fabricated, a larger diameter will permit longer operation between cleanings.

The piping (Item 29A in Figs. 2-2 and 2-15) which connects the gas generator unit to the primary filter should be considered to be a necessary part of the cooling system and should never have an inside diameter less than 2 in. A flexible automotive exhaust pipe was used on the prototype filter unit described below; it was shaped into a semicircular arc so that increased length would achieve a greater cooling effect.

The fabrication procedure for the filter unit follows:

1. A hole equal to the outside diameter of the drain tube (Item 13B in Fig. 2-15) should be cut into the side of the filter container (Item 1B); the bottom edge of this hole should be about 1/2 in. from the inside bottom of the container.

2. The drain tube (Item 13B) should be inserted into the previously cut hole in the filter container and should be positioned so that its nonthreaded end is near the center of the container and is about 1/2 in. off the bottom. Once this position has been ensured, braze (do not weld) the drain pipe into the side of the filter container. Close the threaded, exterior end of the drain pipe with the pipe cap (Item 14B).

3. Coat the bottom of the filter container (Item 1B) with a 1/2-in. layer of hydraulic cement (Item 28A), taking care not to plug or obstruct the end of the drain tube (Item 13B) with cement (i.e., fill the drain tube with a paper, styrofoam, or other easily removable, but rigid material). The cement should also be applied for about 1.5 in. up

the inside walls of the container near its bottom. Round the edges slightly; the cement is to provide a pathway for any liquid condensate to drain out through the drain tube. The cement must be allowed to harden before proceeding with the fabrication steps below. Remove the filler material from the drain tube when the cement has hardened.

4. A circular bottom plate (Item 2B) should be cut to a diameter 1/2 in. smaller than the inside diameter of the filter container (Item 1B). This will allow for heat expansion and easy removal for cleaning. This bottom plate should be drilled with as many 3/4-in. holes as are practical for the size of the plate. Three evenly spaced 3/8-in. holes should also be drilled around the edge of the bottom plate for the spacer bolts (Item 3B).

5. Fig. 2-16 shows the detail of using three bolts (Item 3B) as spacers for the bottom plate (Item 2B). The length of the bolts should be adjusted to provide a clearance of about 2-in. between the layer of cement in the bottom of the container (Item 1B) and the bottom plate (Item 2B).

6. A rectangular divider plate (Item 4B) should be cut to a width 1/4 in. less than the inside diameter of the filter container (Item 1B) and to a height 2.5 in. less than the inside height of the container. This divider plate should then be welded at a right angle to the centerline of the bottom plate (Item 2B) as shown in Fig. 2-17.

7. Cut a piece of high-temperature hydraulic hose (Item 5B) to a length equal to the circumference of the filter container. It should be slit along its entire length and then placed over the top edge of the filter container (Item 1B) to ensure airtightness.

8. A circular lid (Item 6B) should be cut equal to the outside diameter of the filter container (Item 1B). Three holes should be cut into this lid for the exhaust pipe (Item 29A) from the gasifier unit, the blower (Item 7B), and the filter exhaust pipe (Item 10B) to the engine manifold. Note the arrangement of these holes: the pipe (Item 29A) from the gasifier unit must enter the lid on one side of the divider plate (Item 4B); the blower (Item 7B) and the filter exhaust pipe (Item 10B) to the engine manifold must be located on the other side of the divider plate. This arrangement can be seen in Fig. 2-18.

9. The connecting pipe (Item 29A) between the gasifier unit and the filter unit should be attached to the lid (Item 5B) of the filter container. At least one of the ends of the connecting pipe (Item 29A) must be removable for cleaning and maintenance. On this prototype unit, an airtight electrical conduit connector was used. Many similar plumbing devices are available and can be used if they are suitable for operation at 400°F and above. The pipe can also be welded or brazed directly to the lid.

10. Attach the blower (Item 7B) to the filter container lid (Item 6B). On the prototype gasifier illustrated in this report, a heater blower from a Volkswagen automobile was used. Connections for a vertical extension tube (Item 8B) will have to be fabricated as shown in Fig. 2-19. A closing cap (Item 9B) is required for the blower exhaust tube. A plumbing cap of steel or plastic with a close fit can be used or fabricated to fit. The vertical extension and the closing cap are visible in Fig. 2-1.

11. The gas outlet (Item 10B) to the carbureting unit on the engine should be 1.25 in. minimum diameter. In fabricating this connection, all abrupt bends should be avoided to ensure free flow of gas. Using plumbing elbows is one solution. The gas outlet (Item 10B) can either be welded or brazed to the lid (Item 6B) of the filter container or an airtight, electrical conduit connector can be used.

12. Latching devices (Item 11B) should be welded or brazed to the lid (see Fig. 2-20) and to the sides (see Fig. 2-21) of the filter container. An air-tight connection between the lid and the filter container must be maintained.

13. Cut two lengths of high-temperature hydraulic hose (Item 12B) equal to the height of the divider plate (Item 4B); cut a third length of hose equal to the width of the divider plate. Slit each hose along its entire length. Place the first two hoses on each side of the divider plate, and place the third hose along the top edge of the divider plate as shown in Fig. 2-17.

14. Insert the divider plate (Item 4B) into the filter container (Item 1B), making sure that the hoses (Item 12B) create an airtight seal along all sides. By changing the length of the spacer bolts (Item 3B), adjust the height of the divider plate so that it is exactly flush with the top of the filter container. Make sure that the lid (Item 5B) will seat flatly and tightly against the top edge of the divider plate.

15. Fill the filter container (Item 1B) on both sides of the divider plate with wood chips, the same kind as would be used for fuel in the gasifier unit. After carefully packing and leveling these wood chips, place the lid (Item 6B) on the filter container, and close the latches tightly.

2.3. BUILDING THE CARBURETING UNIT WITH THE AIR AND THROTTLE CONTROLS

Figures 2-22 and 2-23 show exploded views of the carbureting unit; the list of materials is given in Table 2-4 (all figures and tables mentioned in Sect. 2 are presented at the end of Sect. 2). In the following instructions, all item numbers refer both to Figs. 2-22 and 2-23 and to Table 2-4. The following is a simple and easy way to assemble a carburetor to achieve both air mixture and throttle control. It can be mounted to either updraft or downdraft manifolds by simply turning the unit over. Most of the fabrication procedure below is devoted to the assembly of two butterfly valves: one for the throttle valve and one for the air mixture valve. The remainder of the carburetor unit can be assembled from ordinary, threaded plumbing parts.

The inside diameter of the piping used in the carburetor unit must be related to the size of the engine and should never be smaller than the intake opening on the engine manifold. If in doubt on the inside diameter for the pipe and/or hose sizes, always go with a larger diameter. This will reduce friction losses and will give longer operating hours between cleanings.

When the wood gas leaves the filter unit it should normally be below 180°F. About 2 ft from the filter container, an automotive water hose can be connected to the pipe on the carbureting unit. This rubber hose will keep engine vibration from creating air leaks in the filter unit or in the connecting piping. The hose must be a fairly new item; such hoses have a steel spring inside to keep them from collapsing when negative pressure is applied. The spring will soon rust if it has first been subjected to water and then to the hot wood gas enriched with hydrogen.

The fabrication procedure for the assembly of two butterfly valves follows:

1. The manifold adapter (Item 1C in Fig. 2-22) must be fitted with bolts and/or holes for mounting onto the engine's existing intake manifold. Because gasoline engines are produced with so many different types of intake manifolds, ingenuity and common sense must be used to modify the manifold adapter (Item 1C) for each different engine to be operated on wood gas. A gasket (Item 7C) should be cut to match the shape of the engine intake fitting.

2. The butterfly valve (Item 3C) is shown in Figs. 2-24 and 2-25; two such valves are required. A 3/8-in. hole should be drilled through the diameter of each valve body (Item 1CC) at the midpoint of its length.

3. The valve plate (Item 2CC) must be oval in shape with the dimensions given in Table 2-4. An oval valve plate must be used so that, in the closed position, the valve will be about 10° off center. This will ensure that the valve will come to a complete stop in the closed position.

4. The edges of the valve plate (Item 2CC), around the longer diameter of the oval, should be beveled to provide a positive, airtight closure. Two evenly spaced, 1/4-in. holes should be drilled along the shorter diameter of the oval plate.

5. The valve support rod (Item 3CC) should be filed or ground flat on one side as shown in Fig. 2-24; the flat area must begin 1/4 in. from one end and must continue for a distance equal to the inside diameter of the valve body (Item 1CC).

6. Two 3/16-in. holes should be drilled into the flat area of the valve support rod (Item 3CC); these holes must align with the holes in the valve plate (Item 2CC). They must also be tapped (with threads) to accept the valve plate screws (Item 4CC).

7. The butterfly valve (Item 3C) should be assembled by first placing the valve support rod (Item 3CC) through the hole in the valve body (Item 1CC). The valve plate (Item 2CC) should be dropped into one end of the valve body and then inserted into the flat area of the valve support rod. The two screws (Item 4CC) should be used to attach the valve plate to the support rod. Check to see that the assembled valve plate rotates freely and seats completely in the closed position.

8. A nut (Item 6CC) should be welded flat against one side of the throttle arm (Item 5CC) near its end. A 1/8-in. hole should be drilled into the side of the nut and must be threaded to accept the set screw (Item 7CC). At least one hole should be drilled into the throttle arm for attachment of the engine throttle control or air control linkages.

9. Place the nut (Item 6CC) on the throttle arm over the end of the valve support rod (Item 3CC) and use the set screw (Item 7CC) to secure the assembly. The throttle arm can be placed in any convenient orientation. Assembled butterfly valves are shown in Fig. 2-26.

10. The remaining parts of the carburetor assembly should be screwed together as shown in Fig. 2-27. Pipe thread compound should be used to make airtight connections. The assembled carburetor unit should be attached to the engine's intake manifold as shown in Fig. 2-28

11. This prototype gasifier was designed to operate if gasoline were unavailable; but, if dual operation on wood and gasoline is desired, the elbow (Item 2C) could be replaced with a tee, allowing a gasoline carburetor to also be mounted.

12. The arm on the butterfly valve (Item 3C) which is closest to the elbow (Item 2C) is to be connected to the foot- (or, on tractors, hand-) operated accelerator. The other butterfly valve is to be used as the air mixture control valve and can be operated with a manual choke cable. If the engine has an automatic choking device, then a hand-operated choke cable should be installed. Both butterfly valves and their connecting control linkages must operate smoothly with the ability to adjust the valve yet keep it stationary in the selected position during operation. The linkages must close the valves airtight when the engine is off.

13. The air inlet (Item 6C) should be connected by an extension hose or pipe, either iron or plastic, to the existing engine's air filter in order to prevent road dust or agricultural residue from entering the engine.

14. The wood gas inlet (Item 5C) is to be connected to the outlet piping (Item 10B as shown in Fig. 2-15) from the wood gas filter unit. Part of this connection should be a high-temperature rubber or neoprene hose to absorb engine vibration.

ORNL-Photo 5321-86

Fig. 2-1. The prototype wood gas generator unit mounted onto a tractor.

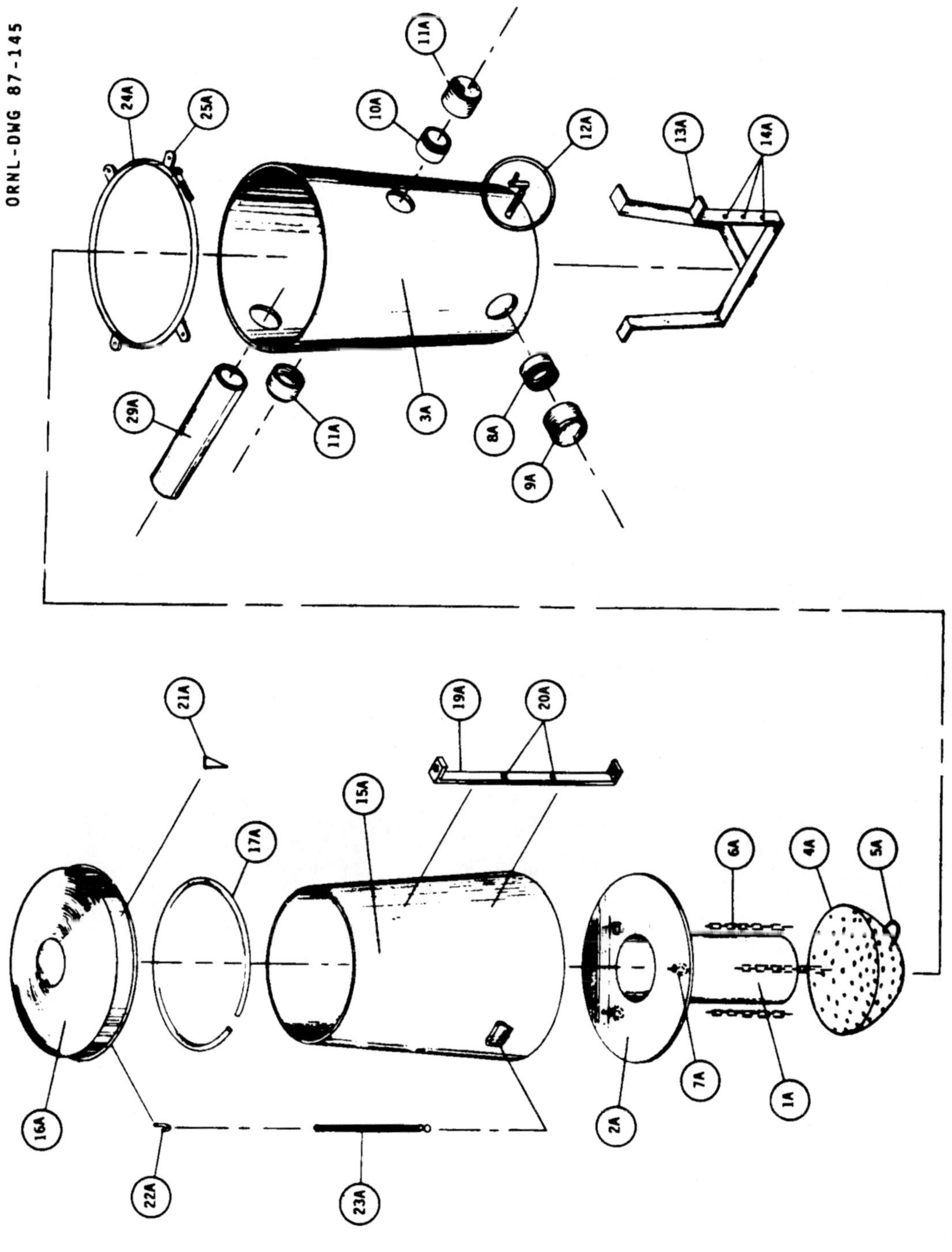

Fig. 2-2. Exploded, schematic diagram of the wood gas generator unit and the fuel hopper.

ORNL-Photo 4533-87

Fig. 2-3. The fire tube and circular top plate of the gasifier unit.

ORNL-Photo 4472-87

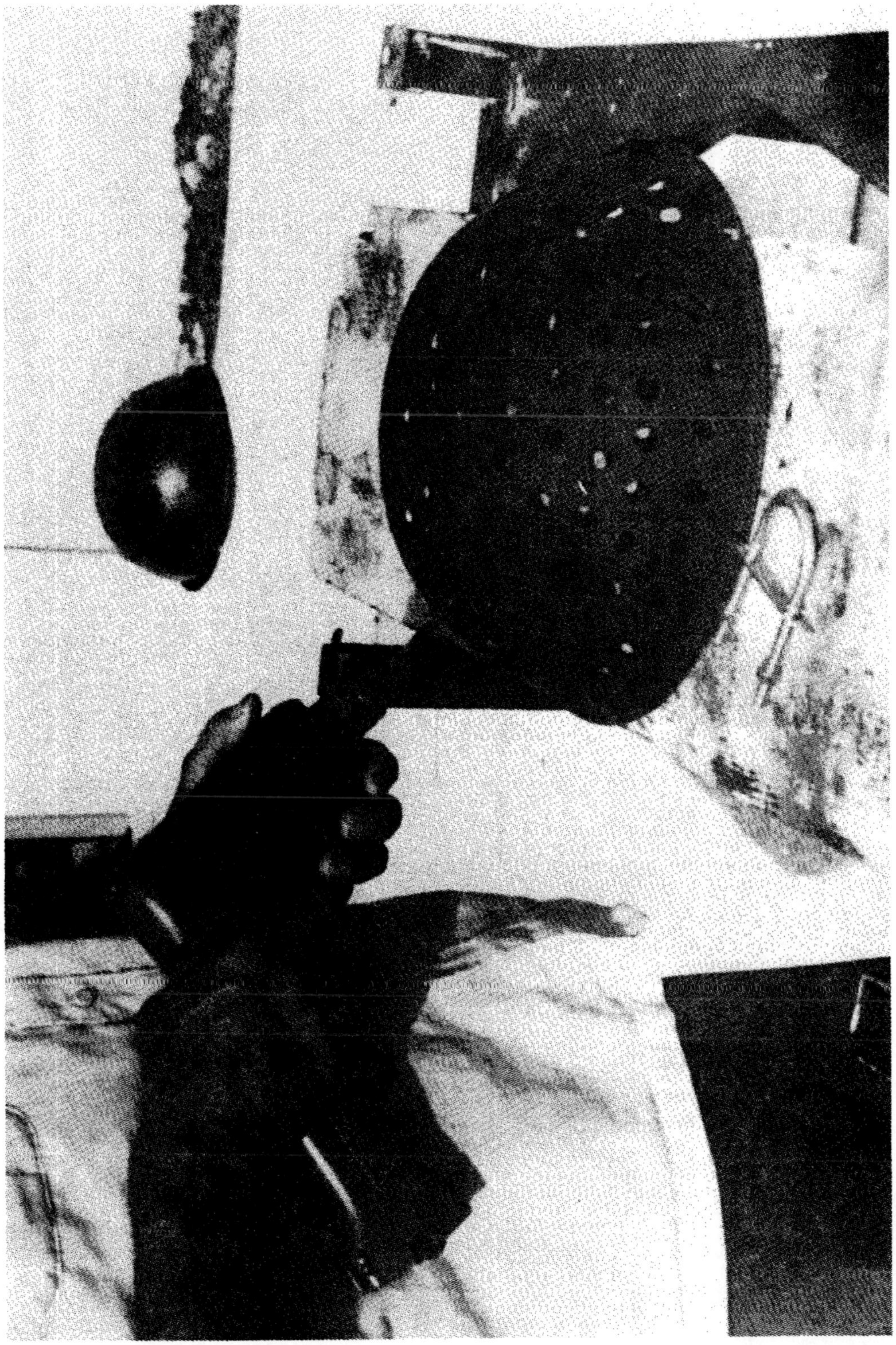

Fig. 2-4. Drilling holes into the stainless steel mixing bowl to be used for the grate. Note the U-bolt in the foreground.

ORNL-Photo 4473-87

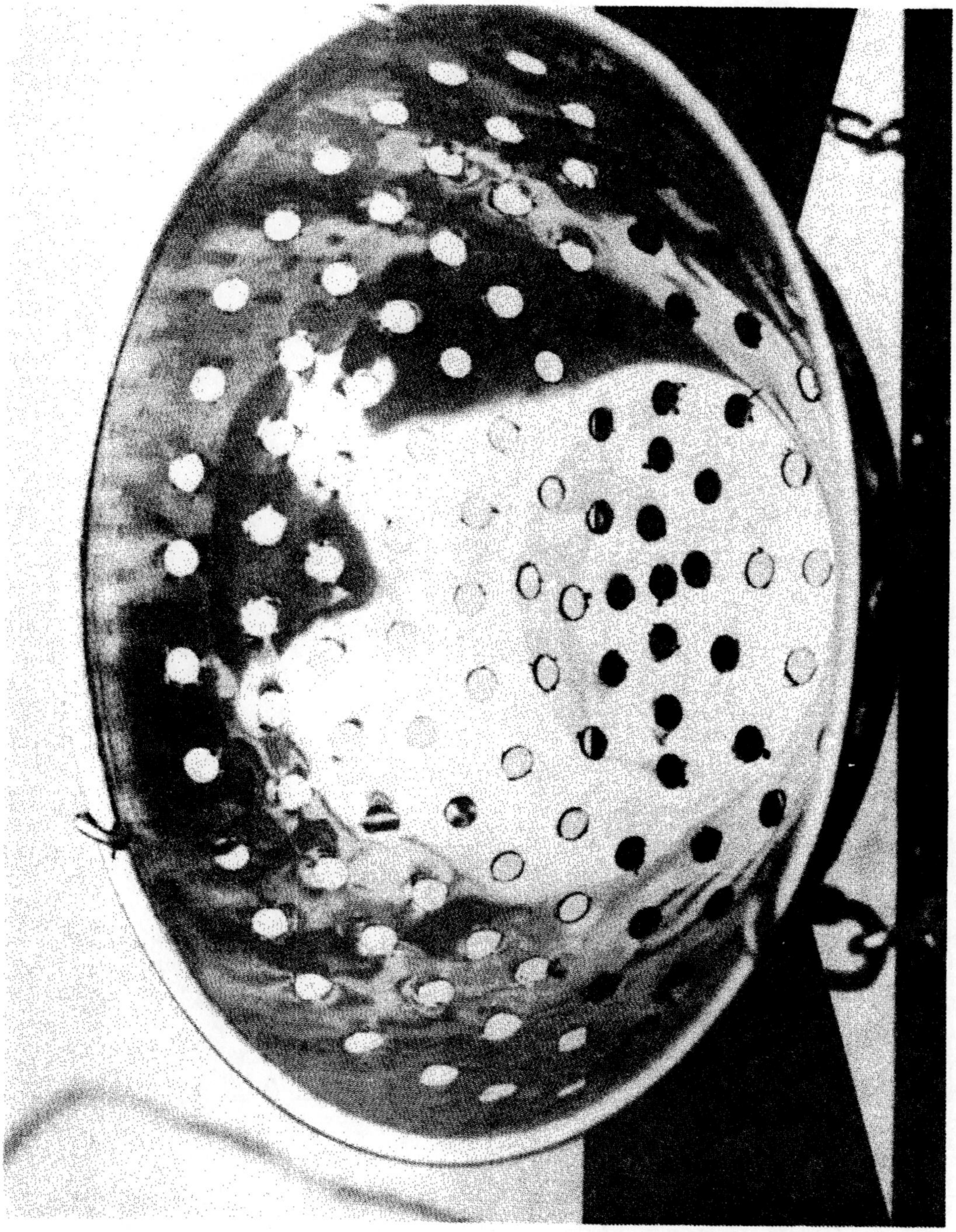

Fig. 2-5. Chains attached to the lip of the stainless steel mixing bowl.

ORNL-Photo 4493-87

Fig. 2-6. Connect the mixing bowl to the top plate with chains. Note that the diesel ignitor "glow plugs" shown in this photograph were included for experimentation only; they were abandoned in the final prototype design.

ORNL-Photo 4529-87

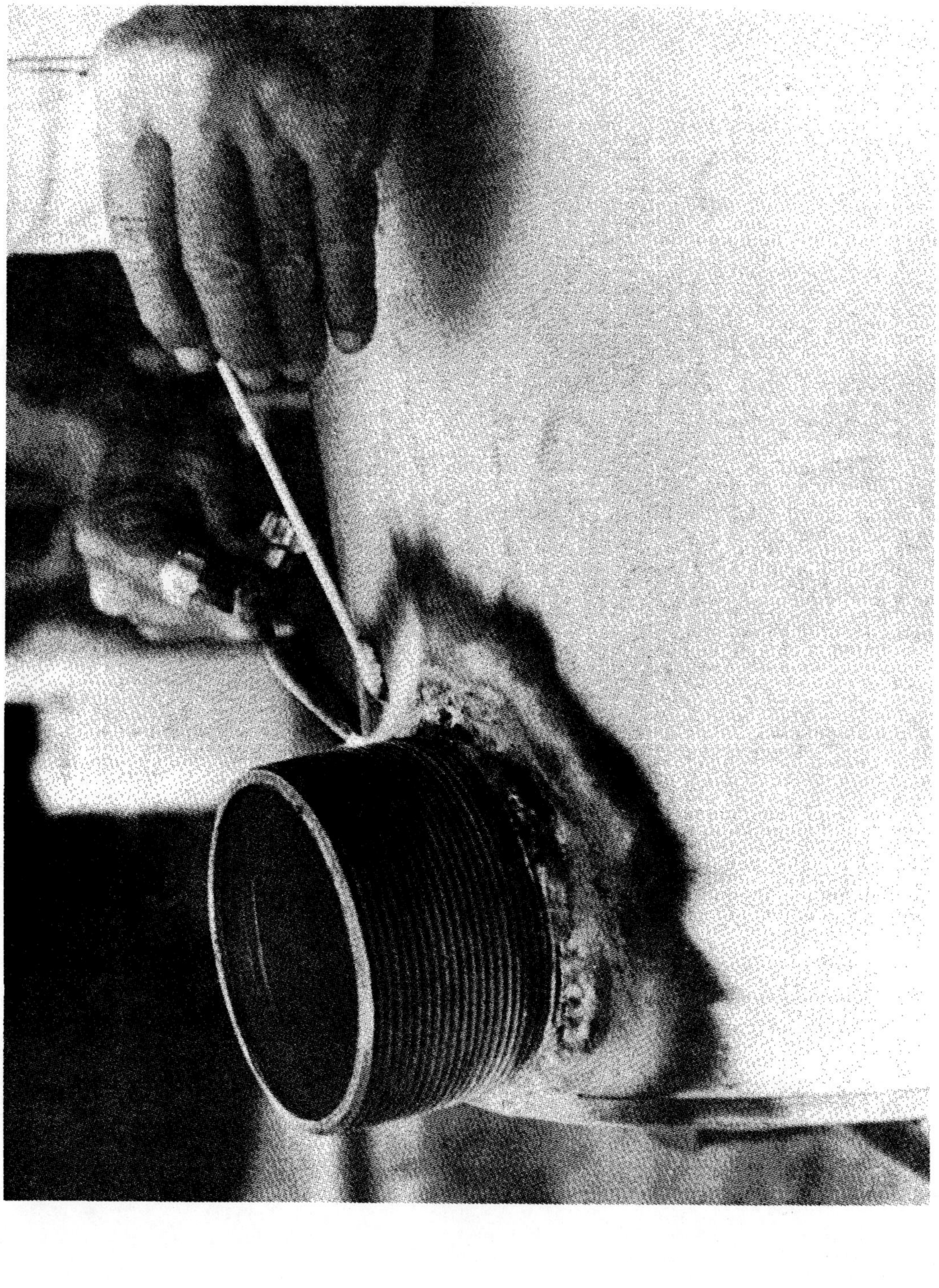

Fig. 2-7. **Braze, <u>do not weld</u>, the plumbing fittings to the thin walled drums.**

ORNL-DWG 87-14555

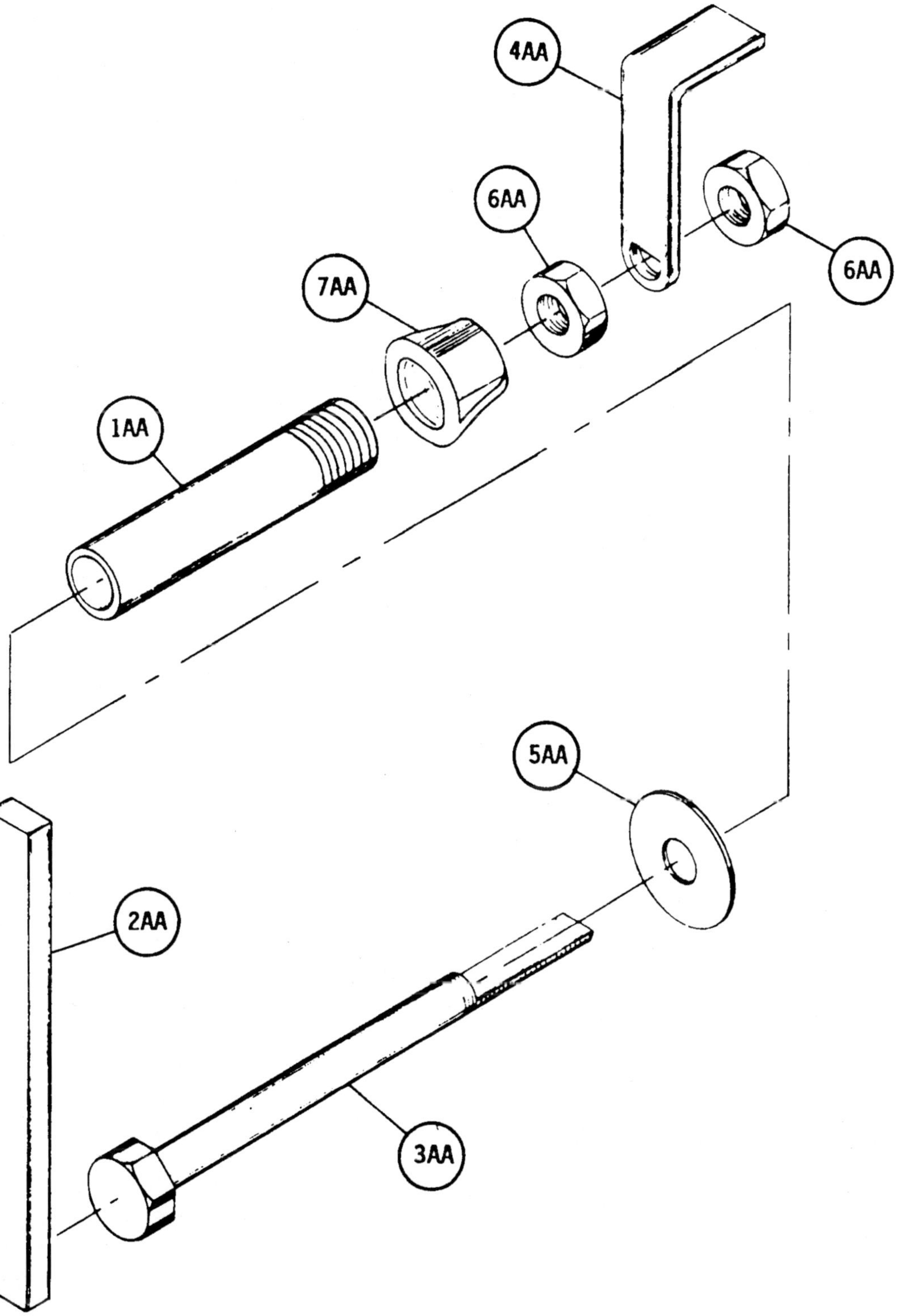

Fig. 2-8. Exploded, schematic diagram of the grate shaking mechanism.

ORNL-Photo 4496-87

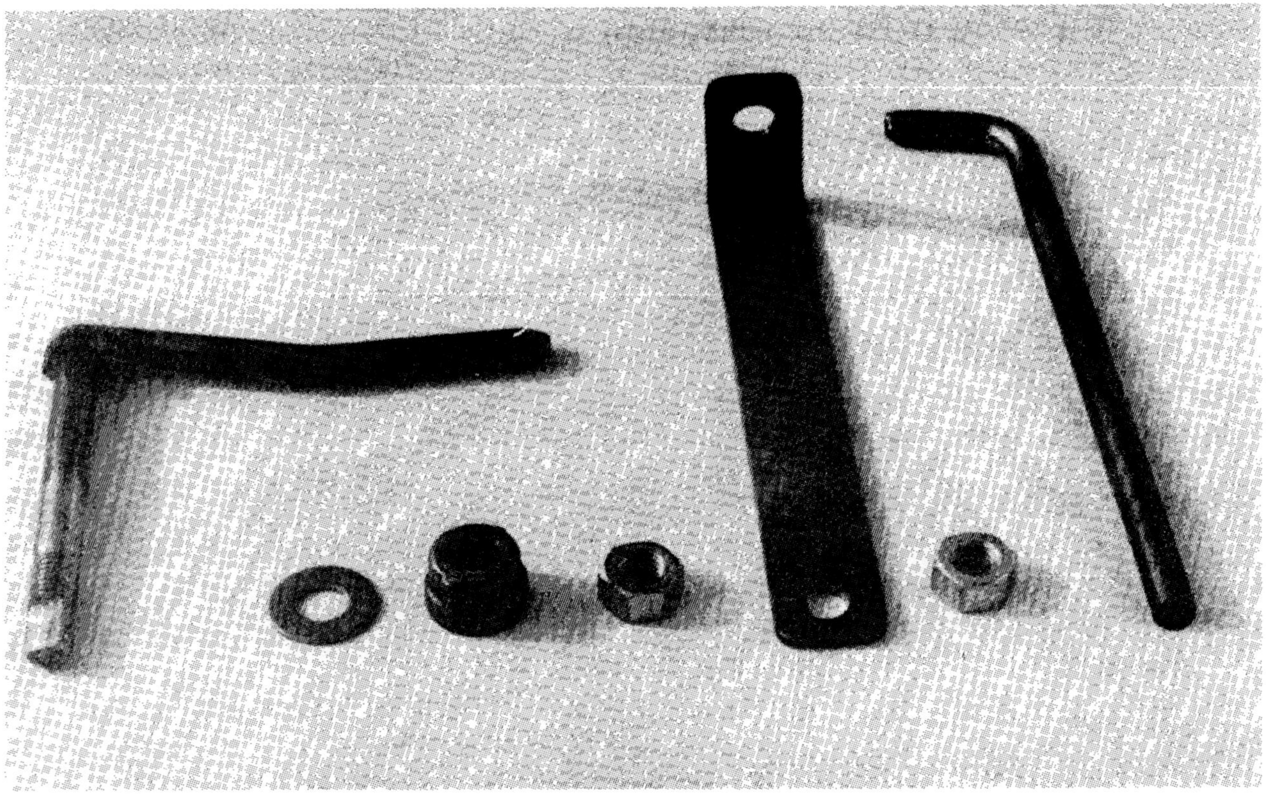

Fig. 2-9. Parts for the shaker assembly. Note the flattened portion of the bolt (at extreme left) which positively locks into the handle (third from right). At the extreme right is a "poker bar" which engages into the hole in the top of the handle to operate the shaker mechanism; the shaker handle will get very hot during normal gasifier operation.

ORNL-Photo 4527-87

Fig. 2-10. The support frame can be brazed or bolted to the side of the gasifier unit. All bolts should be sealed air tight.

ORNL-Photo 4499-87

Fig. 2-11. Containers used in constructing the prototype gasifier unit. At right, a 20-gal garbage can (the fuel hopper) is shown on top of a 30-gal metal drum (the gasifier unit housing). The 5-gal paint can, at left, is used as the filter container.

ORNL-Photo 4508-87

Fig. 2-12. Cover for the fuel hopper. Note the foam weatherstripping (#3) attached to the underside of the lid where it contacts the fuel hopper. Attach four standoffs (#2) to the lid (#1) as shown.

ORNL-Photo 5342-86

Fig. 2-13. Operating configuration of the fuel hopper and its cover.

ORNL-Photo 4506-87

Fig. 2-14. Lock ring and welded tabs. Also pictured inside the lock ring (#1): the ash cleanout cover cap (#2), and the ignition cover cap (#3).

ORNL-DWG 87-14556

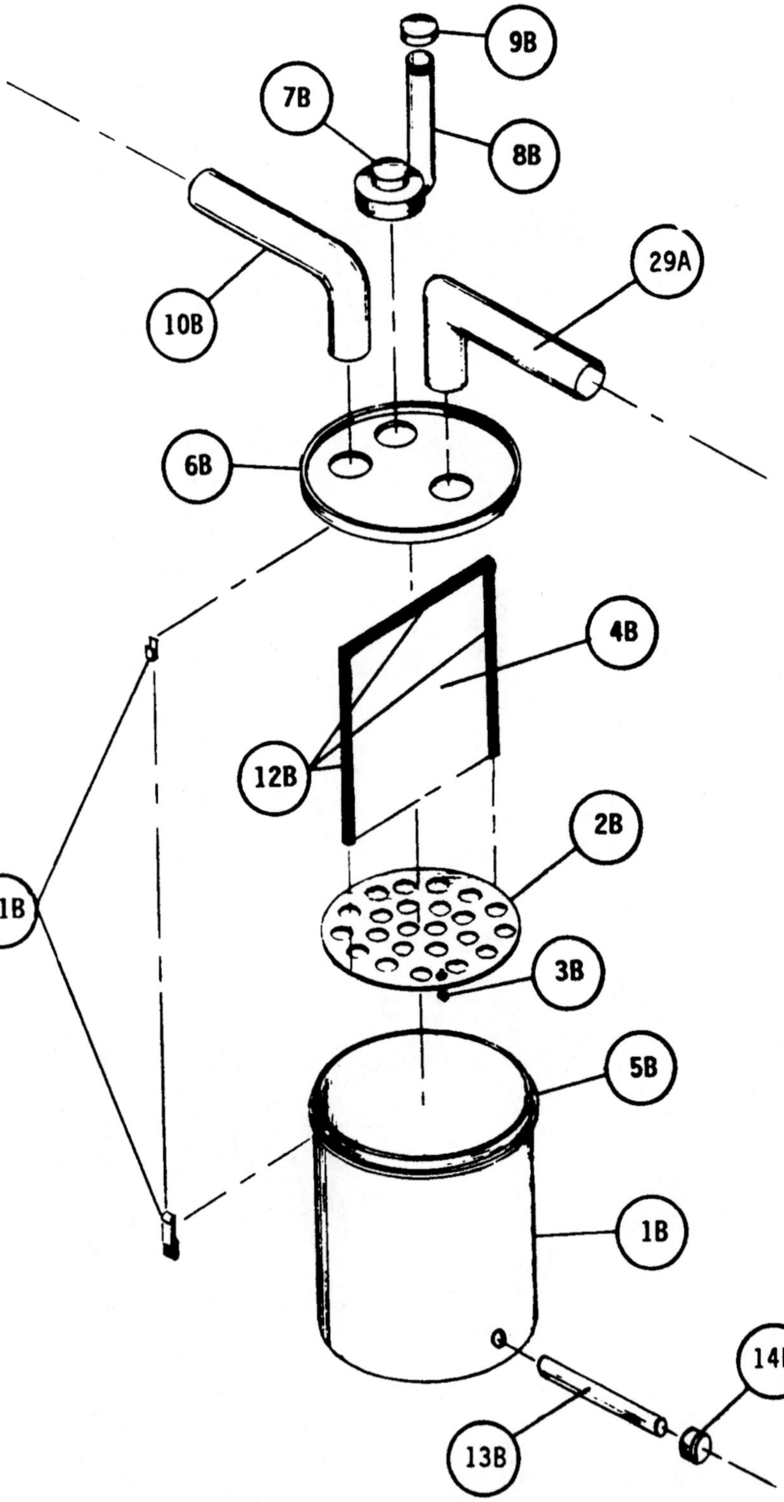

Fig. 2-15. Exploded, schematic diagram of the filter unit.

ORNL-DWG 87-14557

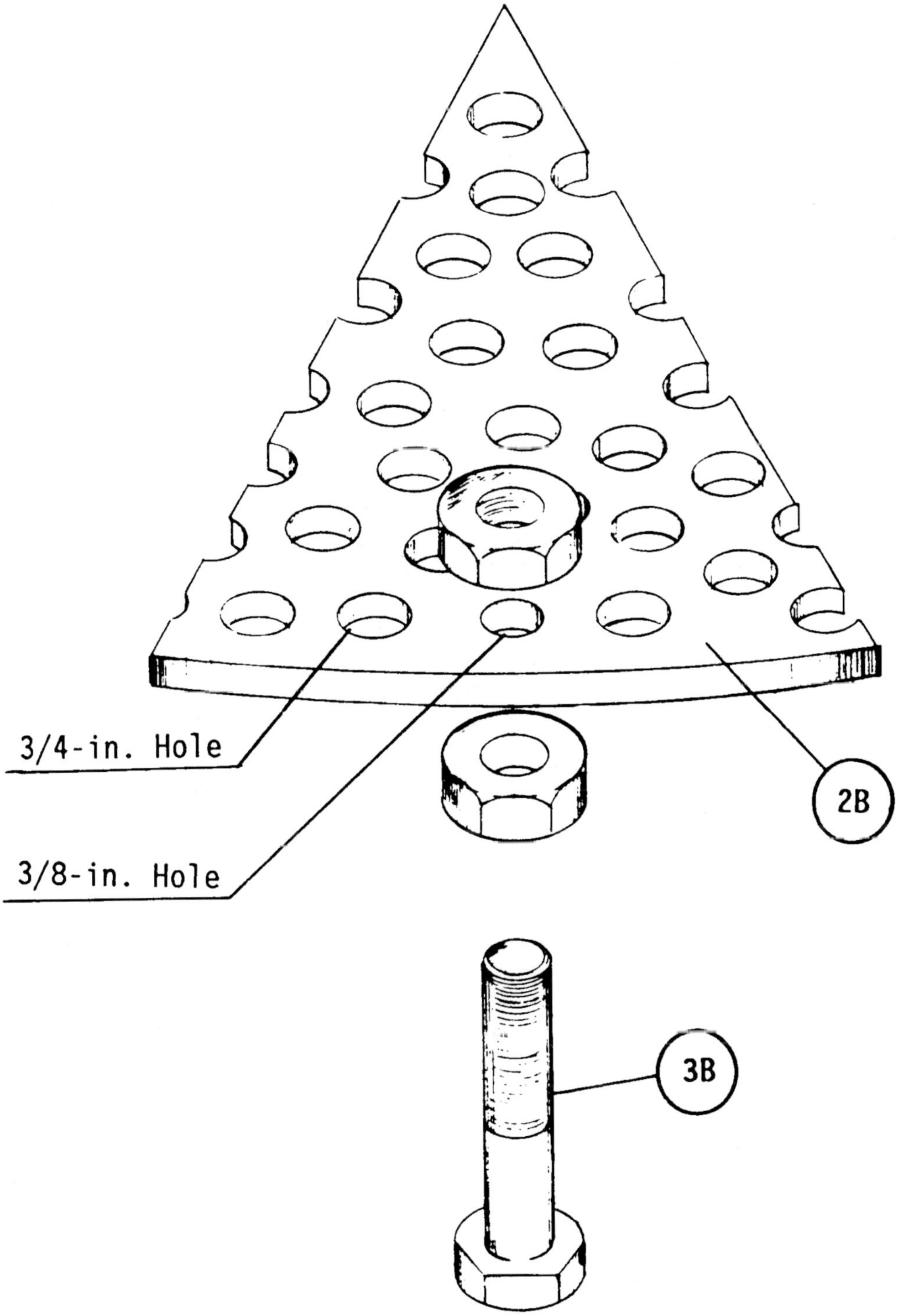

Fig. 2-16. Detail of the standoffs for the bottom plate of the filter unit.

ORNL-Photo 4519-87

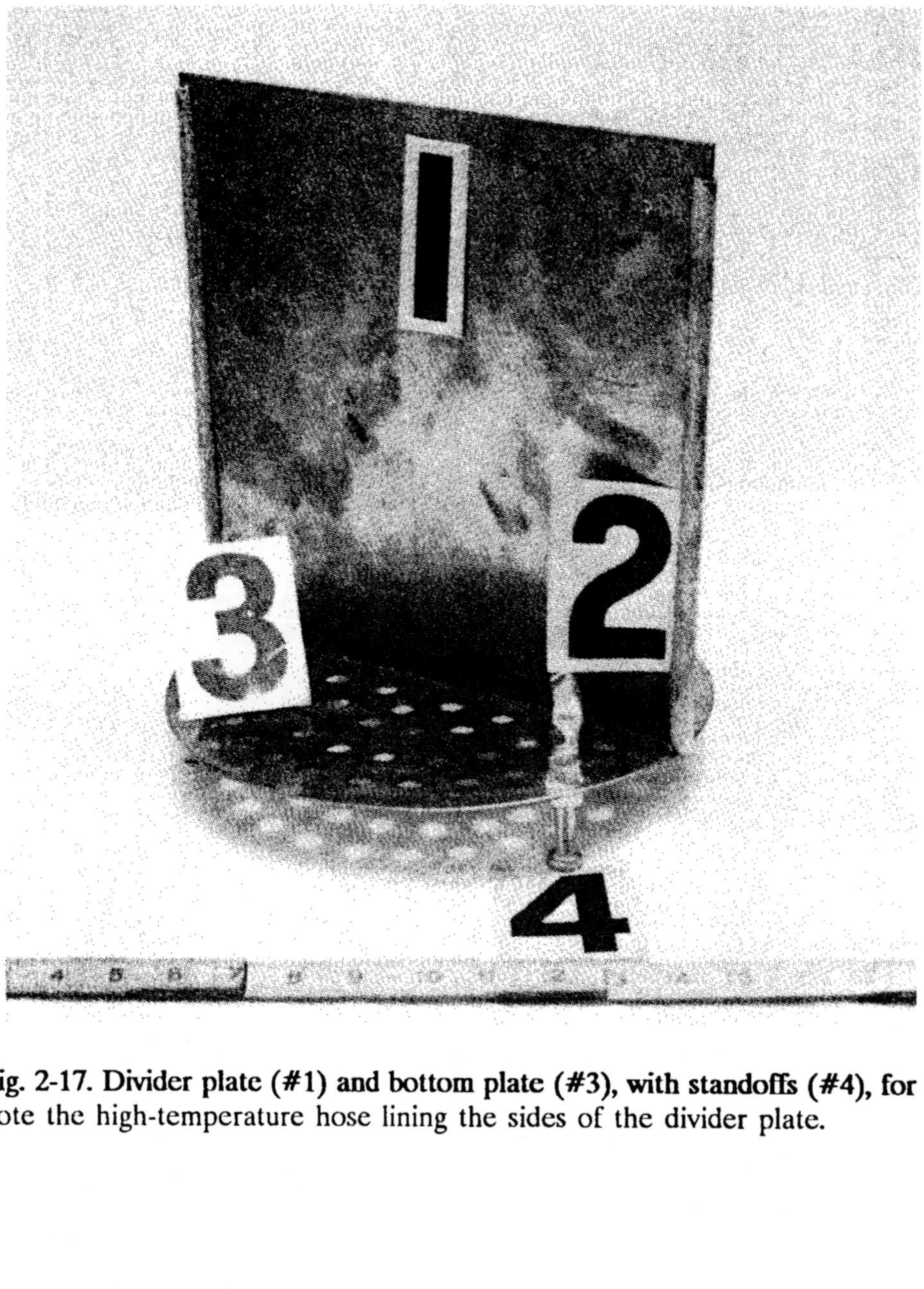

Fig. 2-17. Divider plate (#1) and bottom plate (#3), with standoffs (#4), for the filter unit. Note the high-temperature hose lining the sides of the divider plate.

ORNL-Photo 4525-87

Fig. 2-18. Circular lid (#1) for the filter unit. Note the arrangement of the holes; divider plate would roughly run from 10 o'clock position to 4 o'clock position (assuming 12 o'clock is taken to be at the rear of the photograph). Also shown are the conduit connectors (#2 and #3) and accompanying nuts (#4 and #5) for inside the lid.

ORNL-Photo 4522-87

Fig. 2-19. Blower (#1) with exhaust extension assembly. Note adapter coupling (#2), pipe nipple (#3), and elbow (#4) for vertical exhaust pipe.

ORNL-Photo 4512-87

Fig. 2-20. Assembled and installed blower (#1), extension assembly (#4), and conduit connectors for gas inlet (#2) and outlet (#3) on lid of filter unit. Note hook attachments at edge of lid for latches.

ORNL-Photo 4517-87

Fig. 2-21. Filter container (#1) showing latches (#2) for lid and hose (#3) around top.

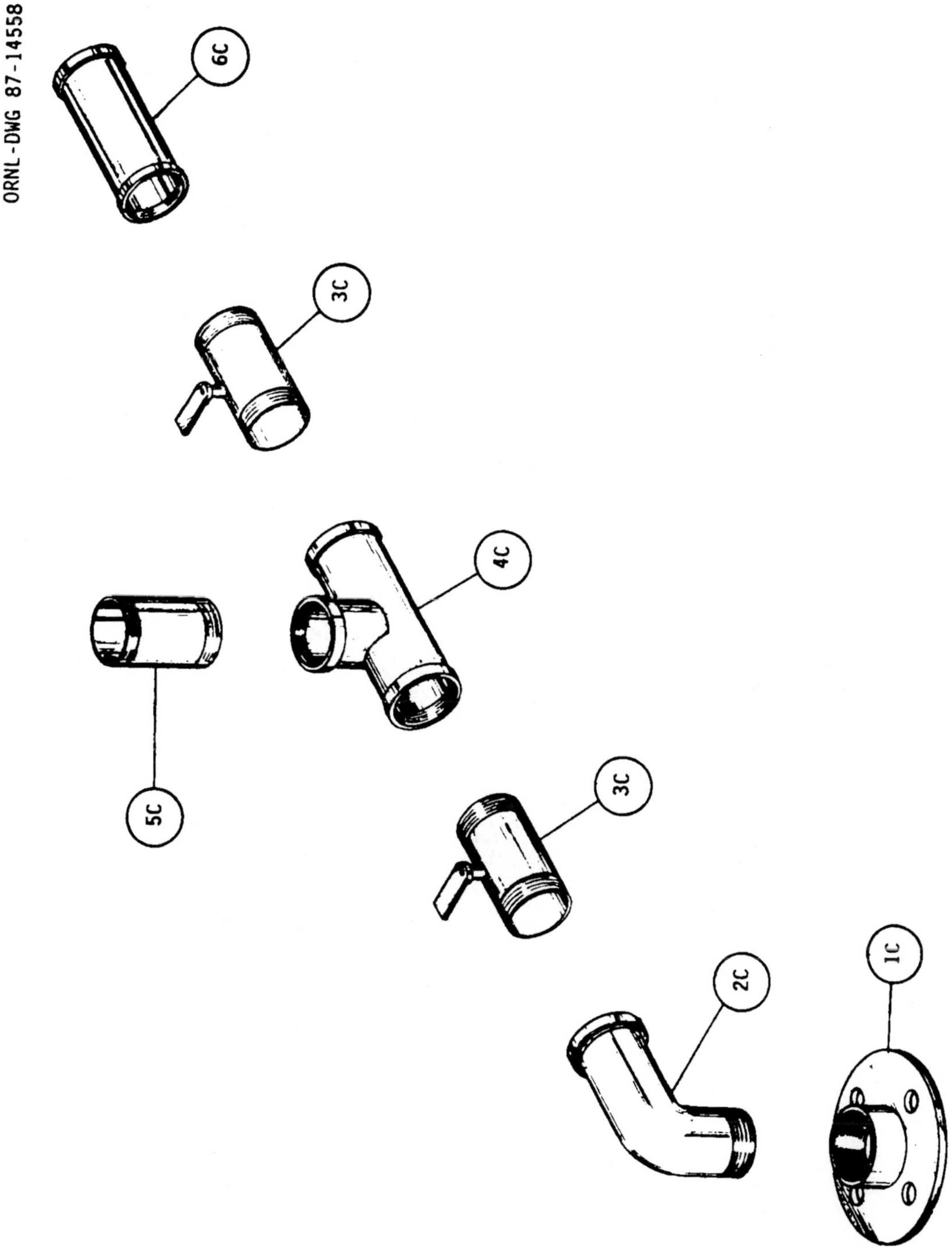

Fig. 2-22. Exploded, schematic diagram of the carbureting unit and control valves.

ORNL-DWG 87-14559

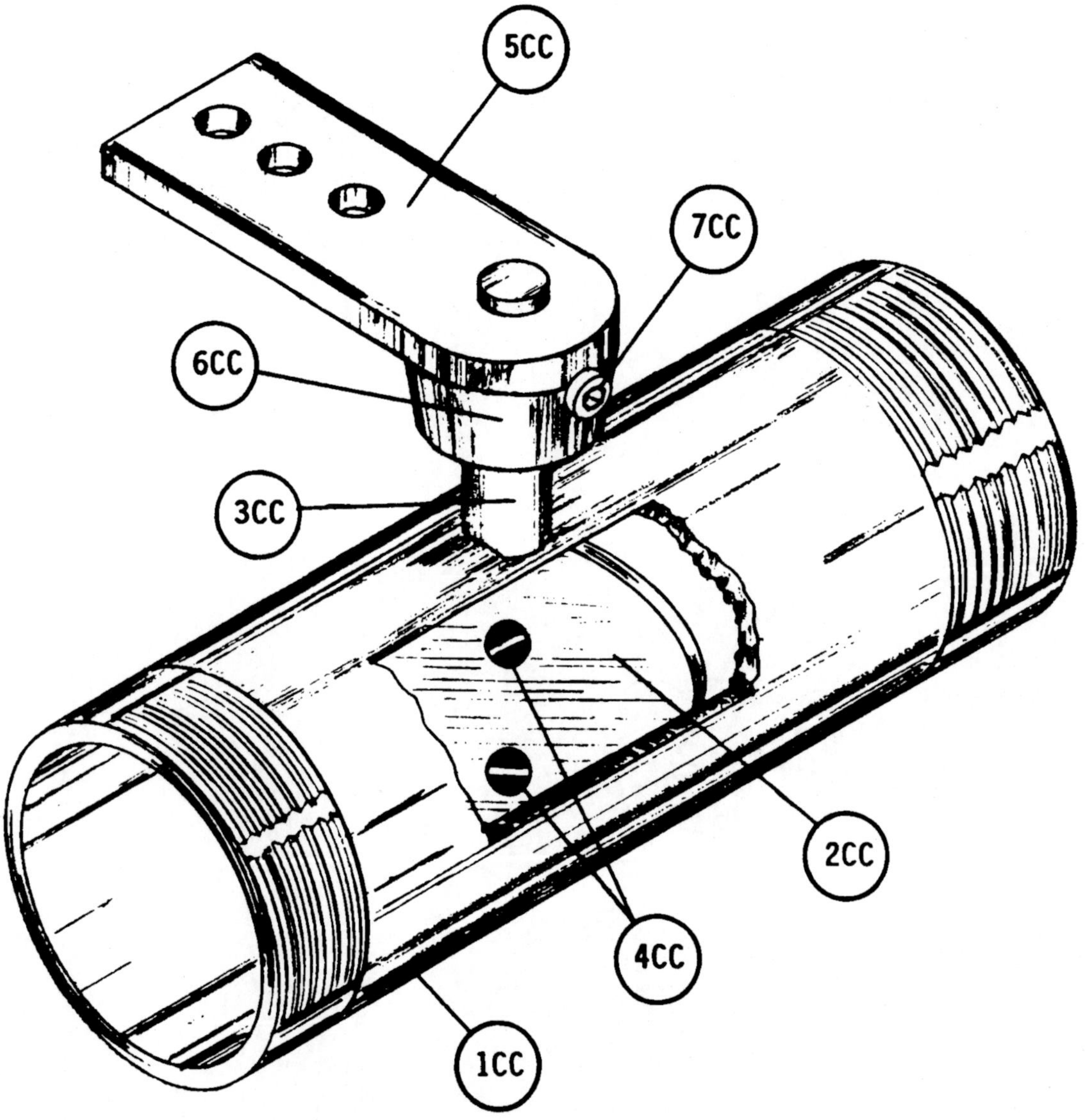

Fig. 2-23. Schematic diagram of a butterfly control valve.

ORNL-Photo 4484-87

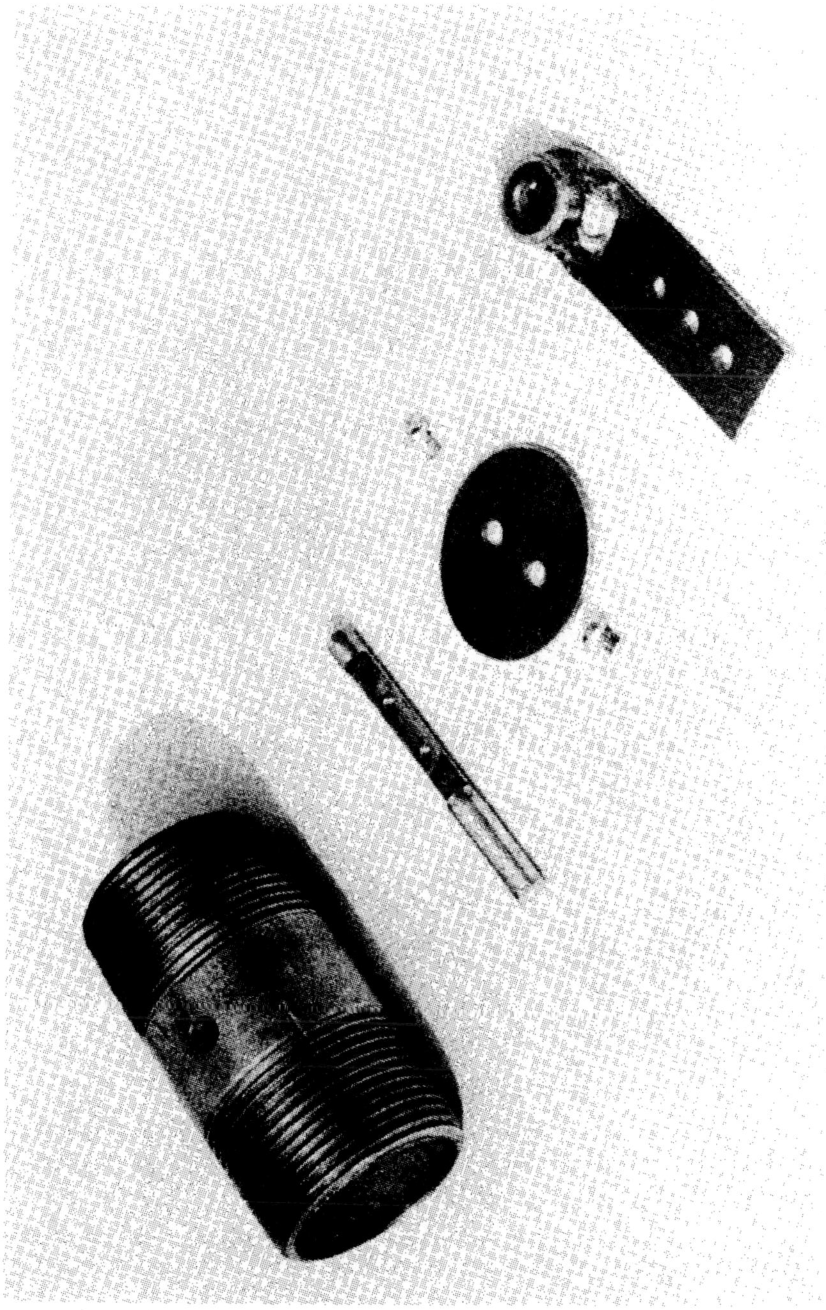

Fig. 2-24. Parts required for the butterfly valve.

ORNL-Photo 4485-87

Fig. 2-25. Butterfly valve assembly. Note that the valve has been assembled outside of the valve body for clarity.

ORNL-Photo 4487-87

Fig. 2-26. Assembled butterfly valves.

ORNL-Photo 4490-87

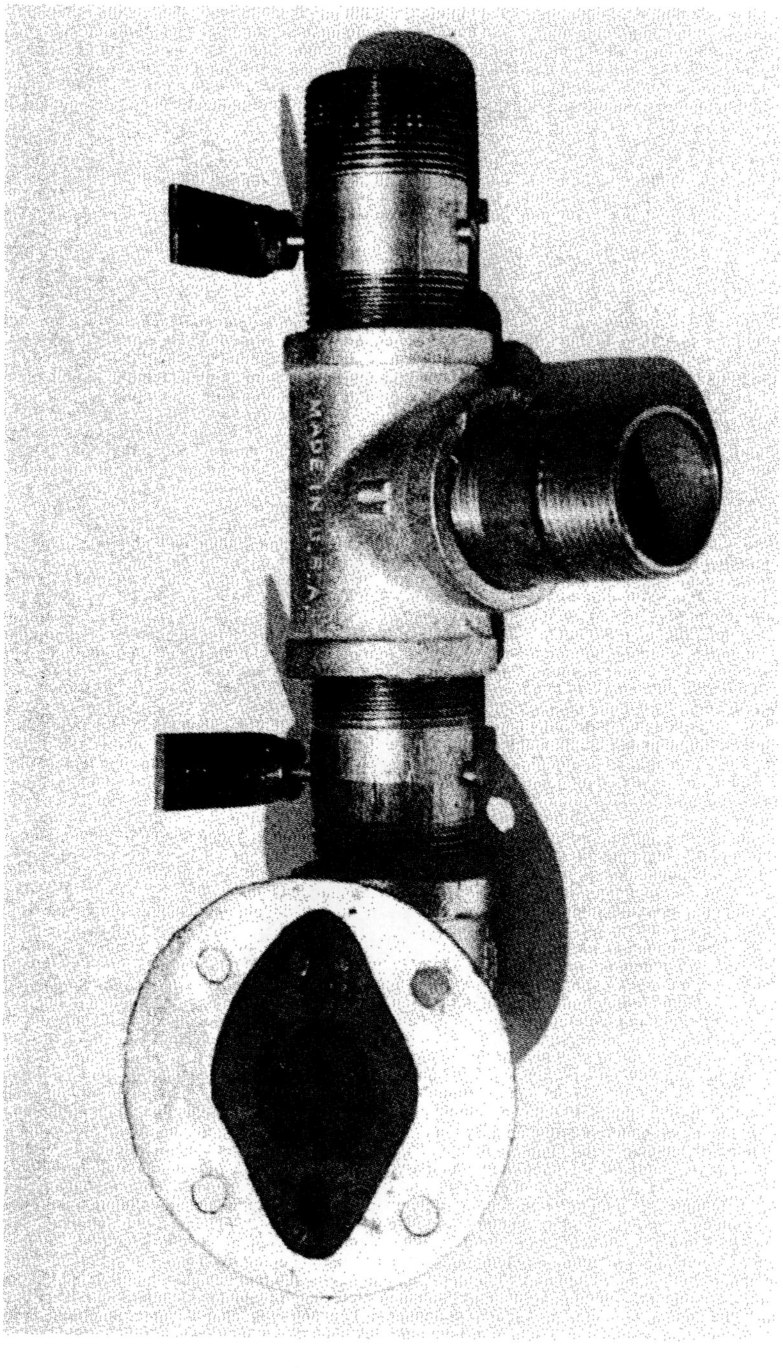

Fig. 2-27. Assembled carburetion unit. Note the gasket on the closet flange.

ORNL-Photo 5317-86

Fig. 2-28. Carburetion unit attached to engine's existing intake manifold. Wood gas enters from the side of the tee; air enters from the right-hand end. The butterfly valve at the right (partially obscured) is connected to the air control (choke) cable; the left valve is connected to the throttle linkage.

Table 2-1. List of materials for the gasifier unit and the wood fuel hopper

Item	Quantity	Description
1A	1	Metal pipe, tube, or other, open-ended metal cylinder; diameter and length from Table 2-2; minimum wall thickness of 1/4 in.
2A	1	Circular metal plate with thickness of 1/8 in.; diameter equal to outside diameter of Item 1A.
3A	1	30-gal metal oil drum or metal container with approximate dimensions of 18 in. (diameter) by 29 in. (height); container must have a bottom.
4A	1	10-quart stainless steel mixing bowl, colander, or other stainless steel bowl with approximately 14-in. diameter and 6-in. depth.
5A	1	2-in. metal U-bolt.
6A	1	3/16-in. metal chain with 1-in. links; 7 ft total length.
7A	3	1/4-in. eyebolts, 3 in. length with two nuts for each eyebolt.
8A	1	4-in. metal pipe nipple.
9A	1	Metal pipe cap for Item 8A.
10A	2	3-in. metal pipe nipple.
11A	2	Metal pipe cap for Item 10A.
12A		Shaker assembly; see Fig. 2-8.
1AA	1	Metal 1/2-in. pipe; 6 in. length.
2AA	1	Iron bar stock; square or round, 1/2 in.; 6 in. length.
3AA	1	1/2-in. bolt; 8 in. long.
4AA	1	Iron bar stock; rectangular, 1/4 by 1 in.; 10 in. length.
5AA	1	1/2-in. flat washer.
6AA	2	1/2-in. nuts.
7AA	1	Metal pipe cap or bushing for Item 1AA.

Table 2-1. (continued)

Item	Quantity	Description
13A	1	Iron bar stock; rectangular, 1/4 by 2 in.; 10 ft length.
14A	25	1/4-in. bolts; 3/4 in. length; with nuts.
15A	1	20-gal metal garbage can or metal container with approximate dimensions of 18 in. (top diameter) by 24 in. (height); bottom is not required.
16A	1	Lid for 20-gal garbage can.
17A	1	Garden hose; 1/2 to 5/8 in. diameter; length equal to circumference of Item 15A.
18A	1	Foam weather stripping with adhesive backing; 1/4 by 1 in.; length equal to circumference of Item 15A.
19A	1	Iron bar stock; rectangular, 1/4 by 2 in.; 10 ft length.
20A	12	1/4-in. bolts; 3/4 in. length; with nuts.
21A	4	Metal triangles; 2 by 2.5 in., 1/8 to 1/4 in. thick.
22A	2	Metal eye hook.
23A	2	Screen door spring, 14 in. length.
24A	1	Lock ring for 30-gal (or larger) oil drum.
25A	4	Metal squares; 2 by 2 in., 1/4 in. thick.
26A	4	3/8-in. bolts; 3 in. length.
27A	1	Tube of high temperature silicone or liquid high temperature gasket material.
28A	1	60-lb. sack of hydraulic or other waterproof cement [such as SEC-PLUG (tm), which is maunfactured by the Atlas Chemical Company, Miami, FL].
29A	1	2-in. pipe, electrical conduit, flexible automobile exhaust pipe, or other metal tubing; 6-ft minimum length. Pipe must be able to withstand temperatures of 400°F.

Table 2-2. Fire tube dimensions

Inside diameter (inches)	Minimum length (inches)	Engine power (horsepower)	Typical engine displacement (cubic inches)
2[a]	16	5	10
4[a]	16	15	30
6	16	30	60
7	18	40	80
8	20	50	100
9	22	65	130
10	24	80	160
11	26	100	200
12	28	120	240
13	30	140	280
14	32	160	320

[a]A fire tube with an inside diameter of less than 6 in. would create bridging problems with wood chips and blocks. If the engine is rated at or below 15 horsepower, use a 6-in. minimum fire tube diameter and create a throat restriction in the bottom of the tube corresponding to the diameter entered in the above table.

<u>NOTES</u>:

For engines with displacement rated in liters, the conversion factor is 1 liter = 61.02 cubic inches.

The horsepower listed above is the SAE net brake horsepower as measured at the rear of the transmission with standard accessories operating. Since the figures vary when a given engine is installed and used for different purposes, such figures are representative rather than exact. The above horsepower ratings are given at the engine's highest operating speed.

Table 2-3. List of materials for the primary filter unit

Item	Quantity	Description
1B	1	5-gal metal can or other metal container with minimum dimensions of 11.5-in. diameter and 13 in. tall.
2B	1	Circular metal plate; diameter equal to 1/2 in. smaller than inside diameter of Item 1B; thickness of 1/8 in.
3B	3	3/8-in. bolts; 3 in. length with two nuts for each bolt.
4B	1	Rectangular metal plate; width equal to 1/4 in. smaller than inside diameter of Item 1B; height equal to 2.5 in. smaller than internal height of Item 1B; 1/8 in. thick.
5B	1	High-temperature hose, 3/8 to 5/8 in. diameter; length equal to circumference of Item 1B.
6B	1	Circular metal plate; diameter equal to outside diameter of Item 1B; thickness of 1/8 in.
7B	1	12-volt blower (automotive heater type); case and fan must be all metal.
8B	1	Metal extension pipe for blower outlet, including elbows and connections for vertical orientation; 1 ft. minimum length.
9B	1	Cap for Item 8B; plastic is acceptable.
10B	1	1.25-in. metal pipe, electrical conduit, automotive exhaust pipe, or other metal tubing; 2 ft minimum length.
11B	3	Metal latch for securely connecting Items 1B and 6B together. Such devices as suitcase or luggage catches, bail-type latches, window sash catches (with strike), or wing-nut latches are acceptable.
12B	1	High-temperature hose, 3/8 to 5/8 in. diameter; length equal to three times the height of Item 4B.
13B	1	Metal 1/2-in. pipe, threaded on one end; 8 in. length.
14B	1	Metal pipe cap for Item 13B.

Table 2-4. List of materials for the carbureting unit

Item	Quantity	Description
1C	1	1.25-in. closet flange.
2C	1	1.25-in. male-to-female 45° pipe elbow.
3C		Butterfly valve; see Fig. 2-23.
1CC	2	1.25-in. pipe nipple or threaded length of pipe, 3-in. length.
2CC	2	Oval metal plate; 1/16 in. thick; short dimension equal to inside diameter of Item 1CC; long dimension equal to 1.02 times the short dimension.
3CC	2	3/8-in. diameter rod; 2.5 in. length.
4CC	4	3/16-in. screws; 3/16 in. length.
5CC	2	Flat bar stock; rectangular, 1/2 by 3 in.; 1/8 in. thick.
6CC	1	7/16-in. nut.
7CC	1	1/8-in. set screw.
4C	1	1.25-in. tee with all female threads.
5C	1	1.25-in. pipe nipple or threaded length of pipe, 3 in. length.
6C	1	1.25-in. pipe or hose.
7C	1	Gasket material; sized to cover Item 1C.
8C	1	Tube of pipe compound or Teflon tape for sealing threaded assemblies.

3. OPERATING AND MAINTAINING YOUR WOOD GAS GENERATOR

3.1 USING WOOD AS A FUEL

Because wood was used extensively as generator fuel during World War II, and since it is plentiful in most parts of the populated United States, it merits particular attention for use as an emergency source of energy. When used in gas generators, about 20 lb of wood have the energy equivalence of one gallon of gasoline.

Wood consists of carbon, oxygen, hydrogen, and a small amount of nitrogen. As a gas generator fuel, wood has several advantages. The ash content is quite low, only 0.5 to 2% (by weight), depending on the species and upon the presence of bark. Wood is free of sulphur, a contaminant that easily forms sulfuric acid which can cause corrosion damage to both the engine and the gas generator. Wood is easily ignited–a definite virtue for the operation of any gas generator unit.

The main disadvantages for wood as a fuel are its bulkiness and its moisture content. As it is a relatively light material, one cubic yard of wood produces only 500 to 600 lb of gas generator fuel. Moisture content is notoriously high in wood fuels, and it must be brought below 20% (by weight) before it can be used in a gas generator unit. By weight, the moisture in green wood runs from 25 to 60%, in air-dried wood from 12 to 15%, and in kiln-dried wood about 8%. Moisture content can be measured quite easily by carefully weighing a specimen of the wood, placing it in an oven at 220°F for thirty minutes, reweighing the specimen, and reheating it until its weight decreases to a constant value. The original moisture content is equivalent to the weight lost.

The prototype unit in this manual (with an 6-in.-diam firetube) operated well on both wood chips (minimum size: 3/4 by 3/4 by 1/4 in.) and blocks (up to 2-in. cubes); see Fig. 3-1 (all figures and tables mentioned in Sect. 3 are presented at the end of Sect. 3). Larger sizes could be used, if the firetube diameter is increased to prevent bridging of the individual pieces of wood; of course, a throat restriction would then have to be added to the bottom of the firetube so as to satisfy the dimensions in Table 2-2 in Sect. 2.

3.2 SPECIAL CONSIDERATIONS AND ENGINE MODIFICATIONS

To start the fire in the gasifier, the blower must be used to create a suction airflow through the wood in the hopper and downward in the firetube. If an especially high horsepower engine is to be fueled by the gasifier unit, then it might be necessary to install two such blowers and run them simultaneously during start-up.

When the wood gas leaves the gasifier unit, all the oxygen pulled down with the air through the firetube has been chemically converted and is contained in carbon monoxide (CO) and water (H_2O). The wood gas is unable to burn without being mixed with the proper amount of additional oxygen. If an air leak develops below the grate area, the hot gas will burn while consuming the available oxygen and will create heat; this will almost certainly destroy the gasifier unit if it is not detected soon. If an air leak develops in the filter unit or in the connecting piping, the gas will become saturated with improper amounts of oxygen and will become too dilute to power the engine. Therefore, airtightness from the gasifier unit to the engine is absolutely essential.

Ideally, as the wood gas enters the engine manifold it should be mixed with air in a ratio of 1:1 or 1.1:1 (air to gas) by volume. The carburetion system described in this report will provide this mixture with a minimum of friction losses in the piping. The throttle control valve and the air control valve must be operable from the driver's seat of the vehicle.

The engine's spark plug gaps should be adjusted to between 0.012 and 0.015 in.; the ignition timing should be adjusted to "early."

3.3 INITIAL START-UP PROCEDURE

Initially, you will need to add charcoal to the grate below the firetube. Subsequent operation will already have the grate full of charcoal which has been left over from the previous operating period.

Fill the firetube with charcoal* to a level 4 in. above the grate. Fill the hopper with air-dried wood; then, proceed with the routine start-up directions below.

3.4 ROUTINE START-UP PROCEDURE

1. Agitate the grate shaker handle for at least twenty seconds to shake down the charcoal from the previous operating period.

2. Open the ash cleanout port and remove the ashes from the generator housing drum. Lubricate the threads of the cleanout port with high-temperature silicone, and close the cover of the cleanout port so that it is airtight.

3. Fill the hopper with wood fuel, and tamp the fuel down lightly. Either leave the lid completely off the fuel hopper, or adjust the opening around the lid to a 3/4-in. (or larger) clearance.

4. Close the carburetor's air control valve and remove the cover from the blower exhaust on top of the filter unit. Start the blower, and let it run for thirty seconds to avoid explosion of residual gas in the system. Then, with the blower still operating, proceed with the next step.

5. Open the ignition port, and ignite a 12- by 12-in. piece of newspaper; with a long stick or wire, push the burning sheet of newspaper into the grate; see Fig. 3-2. Close the ignition port. If no smoke appears at the blower's exhaust port, repeat the start-up

---

*Charcoal produced for outdoor barbecue grills is not well suited for gas generator use. To produce a better grade of charcoal, place a rag soaked in alcohol on the grate, or place 3 to 5 pages of newspaper on the grate, then fill the fire tube to a height of 10 to 12 in. with well-dried wood. Have all the valves closed and let the fire tube act as a chimney until the wood is converted to charcoal.

sequence from Step (e). If repeated attempts fail, new charcoal should be added to the unit as described in Sect. 3.3, above, and the start-up ignition sequence should be repeated.

6. After a few minutes of smoky exhaust, test the gas at the blower exhaust by safely and carefully attempting to ignite it; see Fig. 3-3. When the gas burns consistently well, stop the blower and replace the cover on the blower exhaust.

7. Open the carburetor's air control valve, adjust the engine's accelerator, and start the engine in a normal manner. Let the engine warm up slowly (two to five minutes). If the engine fails to start or dies repeatedly, restart the blower and repeat the ignition sequence from Step (4).

3.5 DRIVING AND NORMAL OPERATION

Shift gears so as to keep the engine speed (rpm) high at all times. Remember that it is the vacuum created by the pistons that provides the force which moves the gas from the gasifier unit into the engine.

Refill the hopper with wood (as shown in Fig. 3-4) before it is completely empty, but avoid refilling just before the end of engine operation. Periodically shake down the ashes from the grate. If your system is equipped with a gas cooler, drain water from the cooler from time to time.

Under operation in dry weather, the gasifier can be operated without the lid on the fuel hopper. However, when the gasifier unit is shut down the hopper must be covered to prevent air from continuing to burn the wood in the hopper. Under wet-weather operation, the cover must be placed on the fuel hopper, and then lifted up and rotated about 2 in. until the triangular pieces line up with the holes in the support bars. The tension of the screen door springs will then hold the lid closed. See Fig. 3-5 for clarification.

3.6 SHUTTING-DOWN THE GASIFIER UNIT

When shutting down the gasifier unit, turn off the ignition switch and open the carburetor's air control valve for ten seconds to relieve any pressure from within the system. Then, completely close the air control valve, and place the cover tightly on the fuel hopper. When restarting after a short stopover, let the engine warm up briefly. After longer stops (up to one hour), tamp down the wood lightly and try to use the blower for restarting without relighting the wood fuel. After very long stops (over two hours) the charcoal must be ignited again.

3.7 ROUTINE MAINTENANCE

Periodically check all nuts on the gasifier unit, the fuel hopper, the filter unit, and the carburetor for snugness; check all penetrations and fittings for airtightness. In addition, perform the following maintenance activities as scheduled:

3.7.1 Daily Maintenance

Open the ash cleanout port of the gasifier housing drum and remove the ashes after shaking the grate for at least thirty seconds. Replace the cover of the port after coating the threads with high-temperature silicone to ensure airtightness. Open the drain tube at the bottom of the filter container and allow any liquid condensate to drain out; remember to close the drain tube when finished.

3.7.2 Weekly Maintenance (or every 15 hours of operation)

Clean out the gasifier housing drum, the fuel hopper, and the filter. Rinse out the piping and connections to and from the filter. Replace the wood chips inside the filter, (The used wood chips from the filter can be dumped into the fuel hopper and burned to produce wood gas.) Use high-temperature silicone on all pipe connections and on the filter lid to ensure airtightness.

3.7.3 Biweekly Maintenance (or every 30 hours of operation)

Make sure that all pipe connections are secure and airtight. Check and tighten all mounting connections to the vehicle chassis. Check for rust on the outside of the gas generator housing drum, especially on the lower region. Coat with high-temperature protective paint as necessary.

3.8 OPERATING PROBLEMS AND TROUBLE-SHOOTING

A discussion of problems and their related causes and cures is contained in the trouble-shooting guide of Table 3-1. Many operational problems can be traced to failure to maintain the airtightness of all piping connections and fittings; the piping should be routinely checked to prevent such problems.

3.9 HAZARDS ASSOCIATED WITH GASIFIER OPERATION

Unfortunately, gas generator operation involves certain problems, such as toxic hazards and fire hazards. These hazards should not be treated lightly; their inclusion here, at the end of this report, does not mean that these hazards are unimportant. The reader should not underestimate the dangerous nature of these hazards.

3.9.1 Toxic Hazards

Many deaths in Europe during World War II were attributed to poisoning from wood gas generators. The danger of "generator gas poisoning" was one of the reasons that such gasifiers were readily abandoned at the end of World War II. It is important to emphasize that "generator gas poisoning" is carbon monoxide (CO) poisoning. Acute "generator gas poisoning" is identical with the symptoms that may develop if a heating stove damper is closed too early, or if a gasoline vehicle is allowed to idle in a poorly ventilated garage. Table 3-2 shows how poisoning symptoms develop according to the concentration of carbon monoxide in breathable air. It is important to note that rather brief exposures to very small concentrations of carbon monoxide result in undesirable physiological effects.

In case of carbon monoxide poisoning, first aid should consist of the following procedures:

1. Move the victim quickly out into the open air or to a room with fresh air and good ventilation. All physical exertion on the part of the victim must be avoided.

2. If the victim is unconscious, every second is valuable. Loosen any tight clothing around the neck. If breathing has stopped, remove foreign objects from the mouth (false teeth, chewing gum, etc.) and immediately give artificial respiration.

3. Keep the victim warm.

4. Always call a physician.

5. In case of mild carbon monoxide poisoning without unconsciousness, the victim should be given oxygen if possible.

3.9.2 Technical Aspects of "Generator Gas Poisoning"

Generator gas poisoning is often caused by technical defects in the functioning of the gas generator unit. When the engine is running, independent of the starting blower, the entire system is under negative pressure created by the engine's pistons; the risk of poisoning through leakage is therefore minimal. However, when the engine is shut off, formation of wood gas continues, causing an increase of pressure inside the generator unit. This pressure increase lasts for approximately 20 minutes after the engine is shut off. For this reason, it is not advisable to stay in the vehicle during this period. Also, the gas generator unit should be allowed to cool for at least 20 minutes before the vehicle is placed in an enclosed garage connected with living quarters. It should be emphasized that the gas formed during the shut-down period has a carbon monoxide content of 23 to 27% and is thus very toxic.

3.9.3 Fire Hazard

The outside of a gas generator housing drum may reach the same temperature as a catalytic converter on today's automobiles. Care should be taken when operating in areas where dry grass or combustible material can come into contact with the housing drum of the gas generator unit. If a gas generator unit is mounted on a personal car, bus, van or truck, a minimum 6-in. clearance must be maintained around the unit. Disposal of ashes must only be attempted after the unit has cooled down (to below 150°F). Such residue must be placed away from any combustible material and preferably be hosed down with water for absolute safety.

ORNL-Photo 5332-86

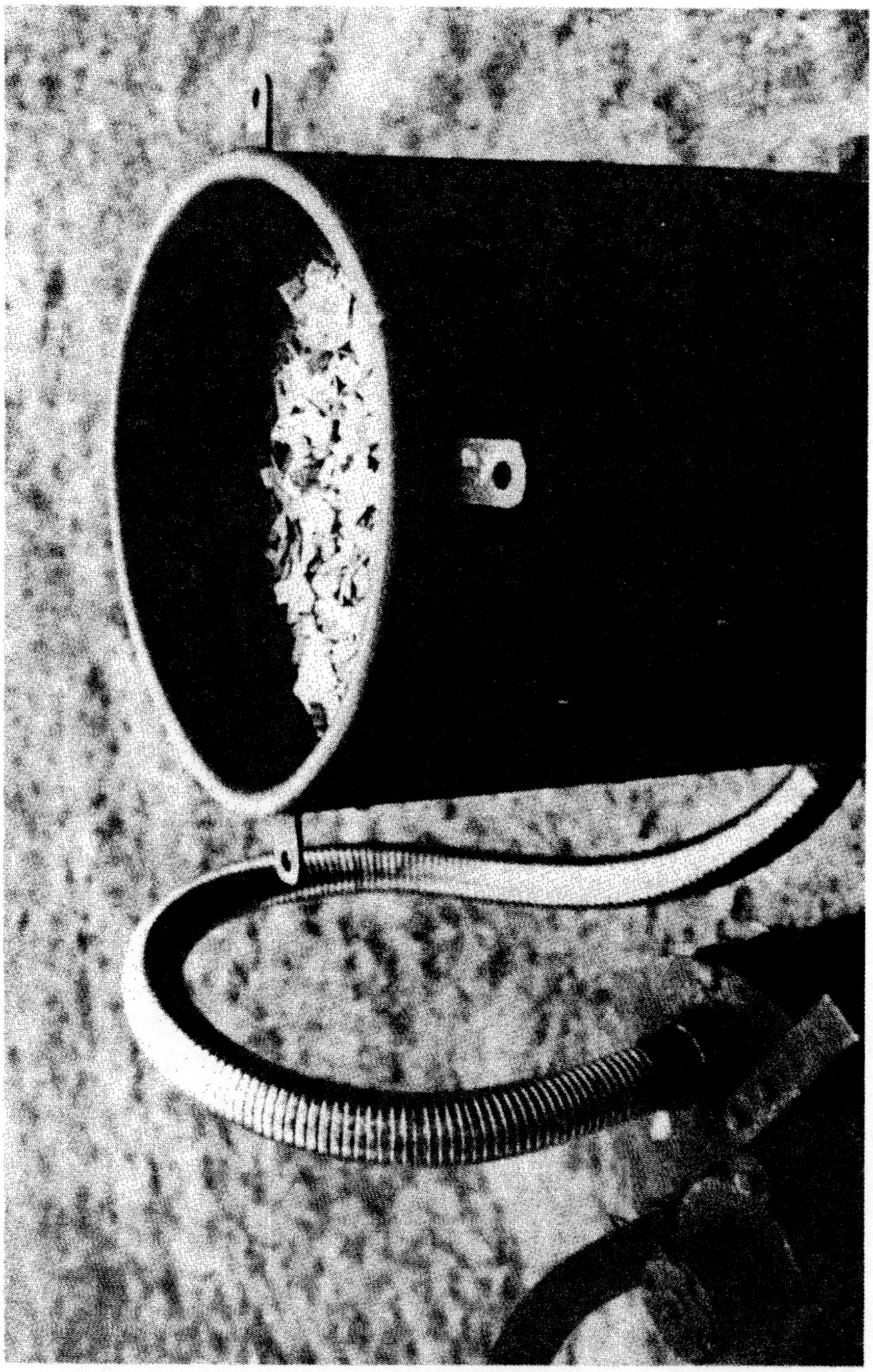

Fig. 3-1. Virtually all varieties of wood chips can be used for fuel. (Minimum size for this 6-in. firetube unit: 3/4 by 3/4 by 1/4 in.; maximum size: 2-in. cubes.)

ORNL-Photo 5313-86

Fig. 3-2. Ignite a single piece of newspaper to start the gasifier unit. Push the flaming newspaper through the ignition port and directly into the grate. (At the right of the photo, note the battery which is operating the blower atop the filter unit.)

ORNL-Photo 5315-86

Fig. 3-3. Igniting the exhaust gas will demonstrate that the gasifier unit is working properly.

ORNL-Photo 5324-86

Fig. 3-4. Refill the fuel hopper before it becomes two-thirds empty.

ORNL-Photo 5314-86

Fig. 3-5. The lid must be used to cover the fuel hopper in wet weather or when shutting the unit down.

Table 3-1. Trouble-shooting your wood gas generator

Trouble	Cause	Remedy
Start up takes too long.	1. Dirty system or clogged pipes.	Clean the gasifier unit and all connecting piping.
	2. Blower is too weak.	Check the blower and test the battery's charge.
	3. Wet or poor quality charcoal.	Check charcoal and replace or refill to proper level.
	4. Wood fuel bridges in the fire tube.	Lightly tamp down the wood fuel in the hopper and fire tube or replace the fuel with smaller-sized chips.
Engine will not start.	1. Insufficient gas.	Use the blower longer during start up.
	2. Wet wood fuel.	Vent steam and smoke through the fire tube and fuel hopper for several minutes.
	3. Incorrect fuel-air mixture.	Regulate the carburetor's air control valve for proper mixing.
Engine starts, but soon dies.	1. Not enough gas has been produced.	Use low RPM while starting engine and do not rev engine for several minutes.
	2. Air channels through fire tube.	Tamp down wood fuel lightly in hopper. Do not crush charcoal above the grate.
Engine loses power under load.	1. Restricted gas flow in piping.	Reduce air mixture valve setting. Check for partial blockage of unit or piping.
	2. Leaks in system.	Check all covers and pipes for air tightness.

Table 3-2. Effects of breathing carbon monoxide

Carbon monoxide content of inhaled air (%)	Physiological effects
0.020	Possible mild frontal headache after two to three hours.
0.040	Frontal headache and nausea after one to two hours; occipital (rear of head) headache after 2.5 to 3.5 hours.
0.080	Headache, dizziness, and nausea in 45 min; collapse and possible unconsciousness in two hours.
0.160	Headache, dizziness, and nausea in 20 min; collapse, unconsciousness and possible death in two hours.
0.320	Headache and dizziness in 5 to 10 min; unconsciousness and danger of death in 30 min.
0.640	Headache and dizziness in 1 to 2 min; unconsciousness and danger of death in 10 to 15 min.
1.280	Immediate physiological effect; unconsciousness and danger of death in 1 to 3 min.

Source: Murakoa, J. S., *Shelter Habitability Studies–The Effects of Oxygen Depletion and Fire Gases on Occupants of Shelter,* NCEL-TR-144, U.S. Naval Civil Engineering Laboratory, Port Hueneme, CA, July 1961.

BIBLIOGRAPHY

Anonymous, "Coast to Coast on Homemade Fuel," **The Mother Earth News** (No. 73) pp. 178-179 (January/February 1982).

Anonymous, "Wood Gas Update," **The Mother Earth News** (No. 71) pp. 164-165 (September/October 1981).

Anonymous, "Mother's Woodburning Truck," **The Mother Earth News** (No. 69) pp. 126-129 (May/June 1981).

Chatterjee, A.K., *State of the Art Report on Pyrolysis of Wood and Agricultural Biomass,* PN-AAK-818, U.S. Department of Agriculture, Newark, NJ, March 1981.

Cruz, I.E., *Producer-Gas Technology for Rural Applications,* published by the Food and Agriculture Organization of the United Nations, New York, NY, 1985.

Donath, E.E., "Vehicle Gas Producers," **Processing Technology (3)** pp. 141-153 (1980).

Eoff, K.M., and D.M. Post, "How to Power a Gasoline Engine with Wood," Technical Reprint No. 1, **The Mother Earth News,** Hendersonville, NC, 1981.

Kaupp, A., and J.R. Goss, *State of the Art Report for Small Scale (to 50kW) Gas Producer-Engine Systems,* USDA Contract No. 53-319R-O-141, University of California, Davis, CA, March 1981.

Miller, B., *State-of-the-Art Survey of Wood Gasification,* EPRI-AP-3101, Fred C. Hart Associates, Inc., Washington, DC, May 1983.

Nygards, N., *Wood Gas Generator for Vehicles* (a translation of the Swedish book), published by Purwaco International, Minneapolis, MN, 1979.

Oliver, E.D., *Technical Evaluation of Wood Gasification,* EPRI-AP-2567, Synthetic Fuels Association, Inc., Palo Alto, CA, August 1982.

Reed, T.B. (ed), *A Survey of Biomass Gasification, Vol. I--Synopsis and Executive Summary,* SERI/TR-33-239 (Vol. I), Solar Energy Research Institute, Golden, CO, July 1979.

Reed, T.B. (ed), *A Survey of Biomass Gasification, Vol. II--Principles of Gasification,* SERI/TR-33-239 (Vol. II), Solar Energy Research Institute, Golden, CO, July 1979.

Reed, T.B. (ed), *A Survey of Biomass Gasification, Vol. III--Current Technology and Research,* SERI/TR-33-239 (Vol. III), Solar Energy Research Institute, Golden, CO, April 1980.

Reed, T.B., and A. Das, *Handbook of Biomass Downdraft Gasifier Engine Systems,* SERI/SP-271-3022, Solar Research Institute, Golden, CO, March 1988.

Reed, T.B., and D. Jantzen, *Generator Gas: The Swedish Experience from 1939-1945* (a translation of the Swedish book, *Gengas*), SERI/SP-33-140, Solar Energy Research Institute, Golden, CO, January 1979.

Skov, N., and M.L. Papworth, *PEGASUS (Petroleum/Gasoline Substitute Systems),* published by the Pegasus Publishing Co., Miami, FL, 1974.

Vietmeyer, N.L., et al., *Producer Gas: Another Fuel for Motor Vehicle Transport,* published by the National Academy Press, Washington, DC, 1983.

Wise, D.L. (ed.), *Fuel Gas from Biomass,* published by the CRC Press, Boca Raton, FL, 1981.

Construction Plans for Other Wood Gas Generator Units can be Purchased from the Following Organizations:

PEGASUS Publishing Co.
1995 Keystone Blvd.
Miami, FL 33181

Missouri Gasification Systems, Inc.
Route 3, Box 198
California, MO 65018

Mother's Plans
The Mother Earth News
P.O. Box 70
Hendersonville, NC 28791

Nunnikhoven Industries
P. O. Box 580
Mediapolis, IA 52637

U.S. GOVERNMENT PRINTING OFFICE: 1989 0—940-366

Hydrogen Generator Gas for Vehicles and Engines Volume 4

Handbook of Biomass Downdraft Gasifier Engine Systems

Thomas B. Reed
and
Agua Das

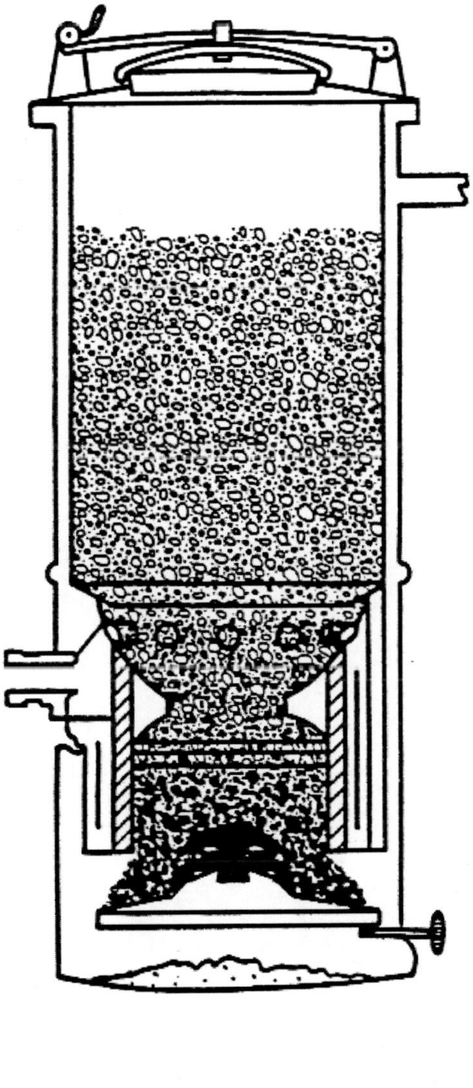

The Biomass Energy Foundation Press

HANDBOOK OF BIOMASS GASIFIER ENGINE SYSTEMS

Second Edition, (Fourth printing)

This handbook was initially prepared by the Solar Energy Research Institute (SERI) under the U.S. Department of Energy Solar Technical Information Program. It was published as SERI/SP-271-3022; DE 88001135, UC Category: 245. Since then SERI has become the National Renewable Energy Laboratory, NREL.

While it was intended as a guide to the design, testing, operation and manufacture of small-scale (less than 200 kW) gasifiers, a great deal of the information will be useful for all levels of biomass gasification.

Since the initial publication in 1988 there has been continuing interest in small scale gasification around the world and the original supply of copies has been exhausted. This second edition has been made to fill the continuing need for information in this interesting field.

Acknowledgments

Since it is impossible for one or two authors to realistically comprehend a subject from all viewpoints, we have solicited input from leading workers in the field. Early versions were sent to a number of investigators, and each was invited to comment on and supplement our effort. We therefore express our heartfelt thanks to the following reviewers for greatly enhancing the quality of the final product:

Dr. Thomas Milne, Solar Energy Research Institute
Dr. Thomas McGowan, Georgia Institute of Technology
Mr. Matthew Mendis, World Bank
Mr. Bill Nostrand, New England Gasification Associates
Dr. Bjorn Kjellstrom, The Beijer Institute, Sweden
Dr. Hubert Stassen, Twente University, The Netherlands
Prof. Ibarra Cruz, University of Manila, The Philippines

We take final responsibility for the contents and omissions, and extend our apologies to those workers whose work we may have unknowingly omitted.

Organization and Use

A gasifier converts solid fuel to gaseous fuel. A gasifier system includes the gasification reactor itself, along with the auxiliary equipment necessary to handle the solids, gases, and effluents going into or coming from the gasifier. The figure below shows the major components of a gasifier system and the chapters in which they are discussed.

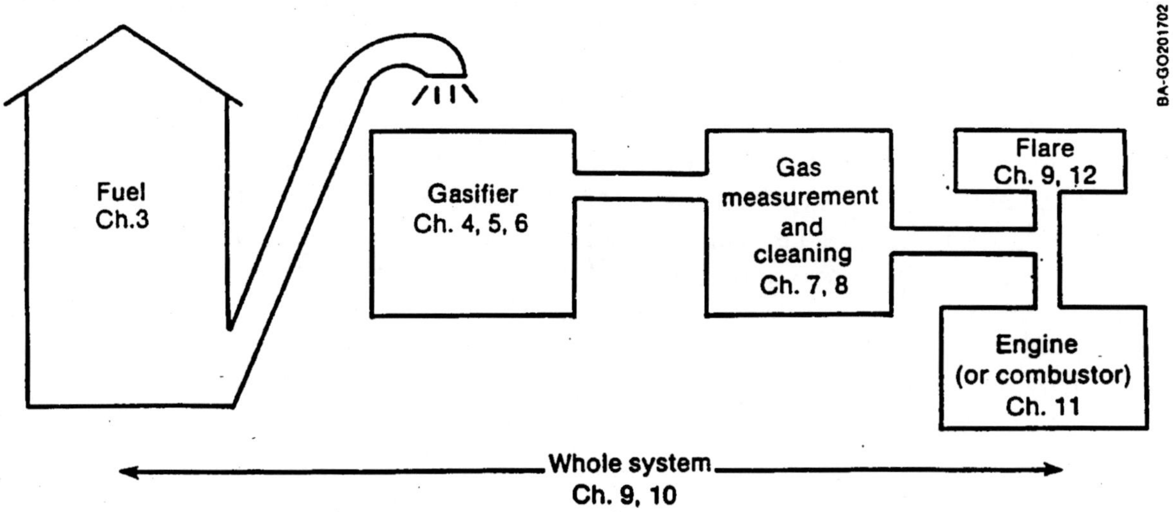

Printed in the United States of America
Available from:
Superintendent of Documents
U.S. Government Printing Office
Washington, DC 20402

National Technical Information Service
U.S. Department of Commerce
5285 Port Royal Road
Springfield, VA 22161

Price: Microfiche A01
Printed Copy A07

Codes are used for pricing all publications. The code is determined by the number of pages in the publication. Information pertaining to the pricing codes can be found in the current issue of the following publications which are generally available in most libraries: *Energy Research Abstracts (ERA)*; *Government Reports Announcements* and Index (*GRA* and I); *Scientific and Technical Abstract Reports (STAR)*; and publication NTIS-PR-360 available from NTIS at the above address.

Contents

Chapter 1

Introduction and Guide to the Literature and Research

1.1 Role of Gasification in Biomass Conversion

This handbook explains how biomass can be converted to a gas in a downdraft gasifier and gives details for designing, testing, operating, and manufacturing gasifiers and gasifier systems, primarily for shaft power generation up to 200 kW. It is intended to help convert gasification from a practical art into a field of engineered design. Although the handbook focuses on downdraft gasification as the only method suitable for small-scale power systems, it also gives extensive detail on biomass fuels, gas testing and cleanup instrumentation, and safety considerations that will be of use to all those who work with gasifiers at whatever scale.

The combustion of biomass in wood stoves and industrial boilers has increased dramatically in some areas, and forest, agricultural, and paper wastes are being used extensively for fuels by some industries. However, more extensive biomass use still waits for the application of improved conversion methods, such as gasification, that match biomass energy to processes currently requiring liquid and gaseous fuels. Examples of such processes include glass, lime, and brick manufacture; power generation; and transportation.

Biomass, like coal, is a solid fuel and thus is inherently less convenient to use than the gaseous or liquid fuels to which we have become accustomed. An overview of various processes now in use or under evaluation for converting biomass to more conventional energy forms such as gas or liquid fuels is shown in Fig. 1-1 (Reed 1978). The figure shows how sunlight is converted to biomass through either traditional activities (e.g., agriculture and silviculture) or new innovative techniques (e.g., as energy plantations, coppicing, and algaeculture) now being developed.

Biomass resources fall into two categories: wet or wettable biomass (molasses, starches, and manures) and dry biomass (woody and agricultural materials and residues). Biological processes require wet biomass and operate at or near room temperature. These processes, shown on the lower left side of Fig. 1-1, include fermentation to produce alcohols and digestion to produce methane.

Thermal processes function best using biomass feedstocks with less than 50% moisture content and are shown on the right side of Fig. 1-1. The simplest thermal process is **combustion**, which yields only heat. **Pyrolysis** uses heat to break down biomass and yields charcoal, wood-oils, tars, and gases.

Gasification processes convert biomass into combustible gases that ideally contain all the energy originally present in the biomass. In practice, gasification can convert 60% to 90% of the energy in the biomass into energy in the gas. Gasification processes can be either **direct** (using air or oxygen to generate heat through exothermic reactions) or **indirect** (transferring heat to the reactor from the outside). The gas can be burned to produce industrial or residential heat, to run engines for mechanical or electrical power, or to make synthetic fuels.

In one sense, biomass gasification is already a well proven technology. Approximately one million downdraft gasifiers were used to operate cars, trucks, boats, trains, and electric generators in Europe during World War II (Egloff 1943), and the history of this experience is outlined in Chapter 2. However, the war's end saw this emergency measure abandoned, as inexpensive gasoline became available (Reed 1985b).

Development of biomass gasification was disrupted in 1946 as the war ended and inexpensive (15¢/gal) gasoline became available. The magnitude of damage inflicted on gasifier technology by this disruption can be seen by the fact that it is difficult for even the "advanced" technology of the 1980s to achieve on tests what was routine operation in the 1940s. The design, research, and manufacturing teams of that decade have all disbanded. We have from the past only that small fraction of knowledge that has been published, whereas the large bulk of firsthand experience in operation design has been lost and forgotten.

Gasification was rediscovered in an era of fuel shortages and higher oil prices, and there are gasifier engine projects under way in more than 20 countries for producing process heat and electrical and mechanical power (Kjellstrom 1983, 1985). In its rebirth, however, the existing technology has uncovered major problems in connection with effluent and gas cleanup and the fuel supply, which were less important during the emergency of World War II. Today, these problems must be solved if biomass gasification is to reemerge as a fuel source. Apparently, it is going to take a few years for the technology of the 1980s to be effectively applied to the accomplishments of the 1940s. Space-age advances in materials and control systems are available for use in today's process designs, so a continuous development effort and lively open exchange should enable us to incorporate latter-day chemical and chemical engineering techniques to build clean, convenient, and reliable systems. A recent workshop on low-energy gasification tabulates research and development needs (Easterling 1985).

The accelerated use of gasification technologies ultimately depends upon their ability to compete with fossil fuels, which in turn depends on unknown factors about resources, economics, and political conditions. At present (1988), gasification and other alternative energy processes are being developed slowly in the United States because of relatively plentiful supplies of low-cost gaseous and liquid fossil fuels. However, political changes could rapidly and dramatically alter this situation, as witnessed during the OPEC oil crises of the seventies. The U.S. Office of Technology Assessment (OTA) recently has issued a report calling for a national capability for emergency implementation of gasifiers (OTA 1984).

1.2 Biomass Energy Potential

Biomass is a renewable fuel that supplies 2% to 3% of U.S. energy needs and an even larger percentage in some other countries (OTA 1980; DOE 1982). OTA projects that biomass could supply from 7% to 20% (6-17 quads*) annually (OTA 1980) from sources such as those shown in Table 1-1 (Reed 1981), if it can be made available in a convenient form and if conversion equipment is accessible. The potential of biomass for world use is equally great (Bioenergy 1985).

*1 quad = 10^{15} Btu

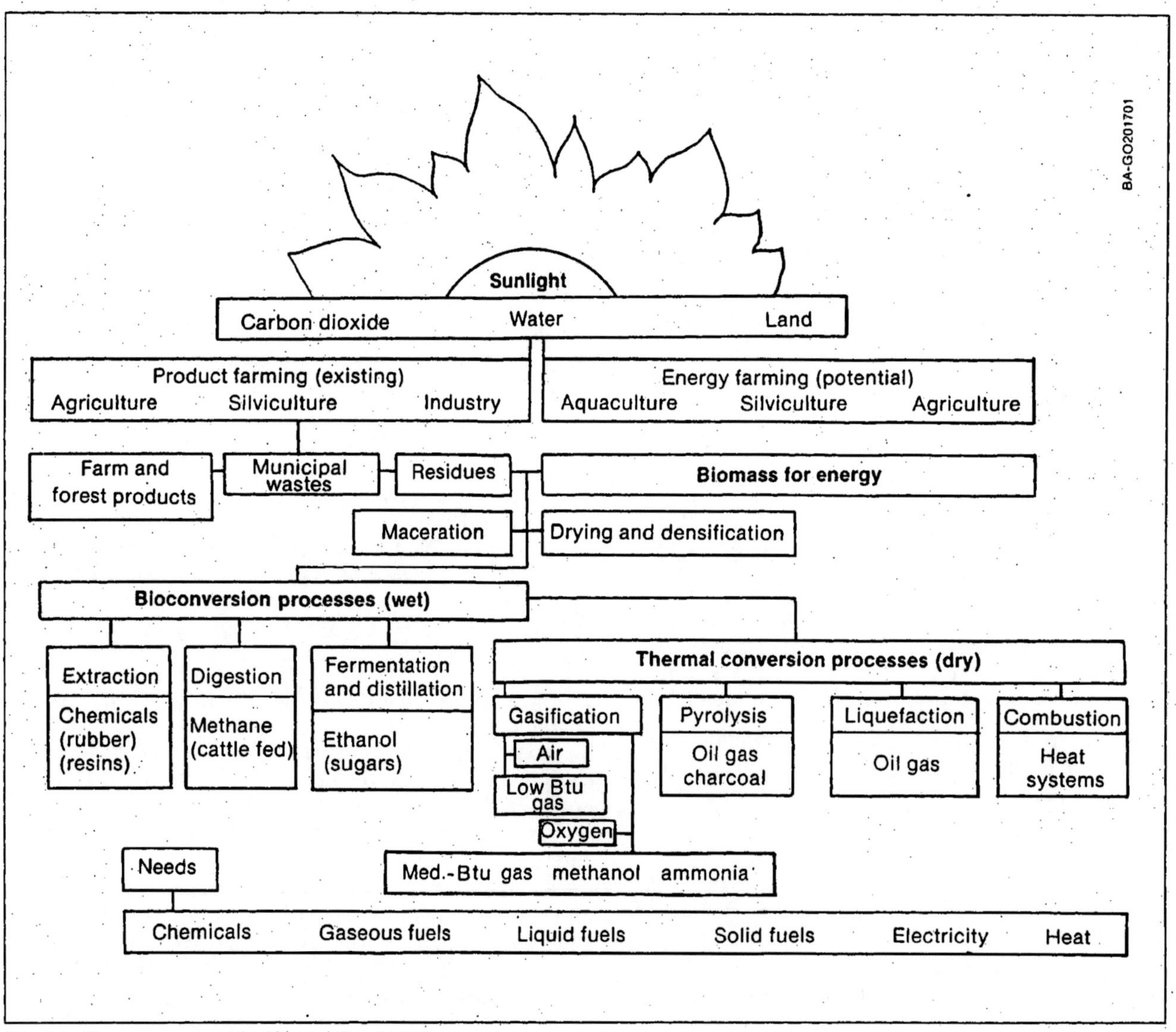

Fig. 1-1. Biomass energy paths (Source: Reed 1978)

Table 1-1. Summary of the Annual Energy Potential of Existing Sources of Biomass in the United States

Resource	10^6 Dry Tons/Year	Quads/Year
Crop residues	278.0	4.15
Animal manures	26.5	0.33
Unused mill residues[a]	24.1	0.41
Logging residues	83.2	1.41
Municipal solid wastes	130.0	1.63
Standing forests	384.0	6.51
Totals	925.8	14.44

[a]Does not include unused bark from wood pulp mills.

Source: Reed 1981, p. 39

Biomass is a renewable energy form with many positive features. The biomass feedstock is often a low-cost byproduct of agriculture or silviculture; it is low in ash and sulfur content, and it does not increase the level of carbon dioxide in the atmosphere and the subsequent greenhouse effect (provided that consumption does not exceed annual production). Care must be taken to ensure that biomass use as fuel is on a renewable basis (Lowdermilk 1975; Reed 1978). Today, many countries (such as China, Korea, Brazil, and South Africa) have active reforestation programs that are helping to increase the total world forest area. With continued diligence, the prospects for making biomass truly renewable will steadily improve.

1.3 Guide to Gasification Literature

1.3.1 Bibliographies

The number of books, articles, and reports on biomass gasification easily exceeds 10,000 (Reed 1985b), with many important studies conducted before 1950. One can easily become discouraged when trying to find the earlier works. Fortunately, much of this early work has been collected; some of it has been summarized, and some of it has been reprinted. We offer here an overview of this body of knowledge in order to help the reader locate required material. In general, the more recent works are still available.

Two major collections of the older papers have been made in the past decade. The U.S. National Academy of Sciences published a bibliography of its extensive collection of early papers in *Producer Gas: Another Fuel for Motor Transport* (NAS 1983). The University of California at Davis acquired an extensive collection of papers while preparing *State of the Art for Small Gas Producer Engine Systems* (Kaupp 1984a). Most of these papers are also in the possession of A. Kaupp at GATE in Germany and also are on file at SERI. A very recent publication from India, *State of Art Report on Biomass Gasification*, (Parikh 1985) contains more than 1200 abstracts of articles on gasification as well as an assessment of its viability and an excellent list of more than 1000 writers and workers in the field. Unfortunately, massive bibliographies of undifferentiated material can confuse the reader or give an impression of a level of understanding that does not exist for gasification. We hope this manual will help the reader to put this material into perspective.

1.3.2 Books

There was a great deal of research and commercialization directed toward coal and biomass gasification between 1850 and 1950. However, cheap and plentiful gas and oil prevented the commercial development of the technology except in times of emergency. The reader is referred especially to a number of excellent historical books. *Modern Gas Producers* (Rambush 1923) gives an account of experiences with updraft and coal gasifiers. *Generator Gas* (Gengas 1950) and its sequel, *Wood Gas Generator for Vehicles* (Nygards 1979), give the reader a complete coverage of all aspects of downdraft gasifiers during World War II. *Gas Producers and Blast Furnaces* (Gumz 1950) looks at the thermodynamics and kinetics of coal and wood gasification. The article by Schläpfer and Tobler, "Theoretical and Practical Studies of Operation of Motorcars on Wood Gas," (Schläpfer 1937) is the best practical and scientific discussion of small gasifiers to appear during that period.

A more general survey of biomass thermal conversion was published during 1979-80 in the SERI three-volume *Survey of Biomass Gasification* (Reed 1981). This work subsequently was published commercially as *Principles of Biomass Gasification* (Reed 1981). The work *Producer Gas: Another Fuel for Motor Transport* (NAS 1983) contains an excellent historical perspective as well as a projection of coming developments. A monumental work, *Small-Scale Gas Producer Engine Systems*, is available in the United States and Germany (Kaupp 1984a). In addition to other considerations, this work contains an in-depth treatment of the use of forest and agricultural residues.

Finally, several private groups have published or republished gasifier plans or gasifier books and pamphlets (TIPI 1986; Skov 1974; Mother 1982; Nunnikhoven 1984; Nygards 1979).

1.3.3 Gasification Proceedings

Current gasification work generally is reported at conferences and then appears in the published proceedings. The U.S. Department of Energy (DOE) (PNL 1982; Easterling 1985) the U.S. Department of Agriculture (USDA), the Forest Products Research Society (FPRS 1983), the U.S. Environmental Protection Agency (EPA), and the Institute of Gas Technology (IGT) all have had continuing interest in various forms of gasification and have sponsored conferences dealing with this field. These publications contain many articles of interest, and the proceedings often span many years of research. The Electric Power Research Institute (EPRI) has commissioned two studies on the use of producer gas (Miller 1983; Schroeder 1985). Government interest in gasification has tended to focus on large-scale systems.

Biomass gasification is perceived by the foreign aid agencies of the developed countries (such as the U.S. Agency for International Development [U.S. AID]) as a major potential energy source for many parts of the developing world. The Beijer Institute of Sweden has organized two international conferences for these donor agencies and published three volumes of recent studies of gasification relevant to the problems of developing countries (Kjellstrom 1983, 1985).

South Africa is uniquely situated relative to producer gas research because it is highly developed technically and produces much of its fuel by gasification. However, it also has a native population of 20 million whose needs match those of less developed countries. A major world conference in timber utilization in May 1985 included week-long sessions on both wood gasification and charcoal manufacture (NTRI 1985).

The European Economic Community (EEC) has shown a great deal of interest in biomass energy in all forms and has been very active in gasification during the last five years (CEC 1980, 1982; Bridgwater 1984; Bioenergy 1985). The EEC has focused on the high-tech aspects of gasification (such as oxygen gasification), but has also funded work in small-scale gasifiers as part of its perceived responsibility toward "associated" developing countries (Beenackers and van Swaaij 1982; Carré 1985; Bridgwater 1984; NTRI 1985; Manurung and Beenackers 1985).

1.3.4 Commercial Information

Another source of gasifier information is provided by companies developing commercial gasifier systems. These groups write advertising brochures as often as they write scientific articles, and it is sometimes difficult to separate actual from projected performance. Their publications should be read critically but usually contain important (if optimistic) information.

1.3.5 Producer Gas Research

Much research into air gasification is being conducted at various universities around the world. However, it is difficult to trace this work if it is occurring either unfunded or on a small scale. The work of Goss and his students at the University of California at Davis deserves special mention because it has spanned a decade and includes both experimental and theoretical studies (Goss 1979). Twente University in the Netherlands has had a large program in gasification for many years (Groeneveld 1980a,b; Aarsen 1985; Buekens 1985). The University of Florida at Gainesville has a very active research group in producer gas (IGT 1984). In addition, excellent gasification work is proceeding in Canada, Europe, Brazil, the Philippines, New Zealand, and other parts of the world, primarily at the university level.

1.3.6 Producer Gas R&D Funding

U.S. AID has had a strong interest in producer gas technology because it offers a means for reducing the dependency of developing nations on imported fuels and has supported a number of projects around the world. The Producer Gas Roundtable of Stockholm, Sweden, is an oversight organization supported by various international development agencies to promote information exchange on gasification, to and between developing countries. It has sponsored two major international conferences (Kjellstrom 1983, 1985).

A moderate level of funding ($2 million to $5 million/yr) has been maintained since 1975 by DOE for "advanced concept" gasification and pyrolysis processes. Most of the work is aimed at large industrial processes and is supported in government laboratories, industrial firms, and universities. Progress in these programs is reported at the meetings of DOE's Thermochemical Conversion Contractors (PNL 1986), as well as at other meetings. DOE recently sponsored a meeting to examine the potential and problems of low energy gasification (Easterling 1985) but is currently focusing on direct liquefaction of wood. The status of many of the government research and development projects and commercial gasifiers projects was summarized in *Survey of Biomass Gasification* (Reed 1981).

EPRI (Schroeder 1985) has evaluated the potential of gasifiers for making electricity. The Forest Service of the USDA holds annual meetings at which gasifiers are discussed (FPRS 1983).

Reports on government programs are maintained by the Office of Scientific and Technical Information (OSTI) where they can be obtained in either microfiche or printed copies. They are sometimes difficult to obtain after the original supply of reports is exhausted. Copies of these reports are also available in GPO depository libraries. There are at least two such libraries—one public and one university—in each state.

1.3.7 Federal Emergency Management Agency (FEMA) Gasifier Work

The downdraft gasifier reached its highest development during the emergency of World War II. FEMA has taken interest in small-scale gasifiers because they could function during a period of breakdown in our oil supply due to atomic attack or other disruption of conventional fuels.

With this in mind, FEMA contracted with H. LaFontaine of the Biomass Energy Foundation to build a prototype gasifier that could be made with readily available parts and to write a "craftsman manual" description of gasifier construction and operation (LaFontaine 1987). The gasifier has passed the test, and the manual is now in the process of being published by FEMA.

Chapter 2
History, Current Developments, and Future Directions

2.1 Historical Development

2.1.1 Early Development of Gasification

Gasification was discovered independently in both France and England in 1798, and by 1850 the technology had been developed to the point that it was possible to light much of London with manufactured gas or "town gas" from coal (Singer 1958; Kaupp 1984a). Manufactured gas soon crossed the Atlantic to the United States and, by 1920, most American towns and cities supplied gas to the residents for cooking and lighting through the local "gasworks."

In 1930, the first natural gas pipeline was built to transport natural gas to Denver from the oil fields of Texas. As pipelines crisscrossed the country, very low-cost natural gas displaced manufactured gas, and the once-widespread industry soon was forgotten. "Town gas" continued to be used in England until the 1970s, but the plants were dismantled following the discovery of North Sea oil. Today, a few plants are still operating in the third world.

2.1.2 Vehicle Gasifiers

Starting about the time of World War I, small gasifiers were developed around charcoal and biomass feedstocks to operate vehicles, boats, trains, and small electric generators (Rambush 1923). Between the two world wars, development was pursued mostly by amateur enthusiasts because gasoline was relatively inexpensive and simpler to use than biomass. In 1939 the German blockade halted all oil transport to Europe. Military use of gasoline received top priority, and the civilian populations had to fend for themselves for transport fuels. Approximately one million gasifiers were used to operate vehicles worldwide during the war years. The subsequent development of wood producer gas units is a testament to human ingenuity in the face of adversity. Extended accounts make fascinating reading and inform the reader of both the promise and difficulties of using producer gas. (Egloff 1941, 1943; Gengas 1950; NAS 1983; Kaupp 1984a).

At the beginning of World War II, there was a great deal of interest in all forms of alternative fuels (Egloff 1941, 1943). By 1943, 90% of the vehicles in Sweden were powered by gasifiers. By the end of the war, there were more than 700,000 wood-gas generators powering trucks, cars, and buses in Europe and probably more than a million worldwide (Egloff 1943). However, these impressive numbers included only six wood-fueled vehicles in the United States and two in Canada, where low-cost gasoline continued to be available throughout the war. Many articles were written on gasification during that time (see Chapter 1). Some photographs of gasifiers fitted to vehicles of that era are shown in Fig. 2-1. Most gasifiers were simply "belted on" and

BA-GO201703

Fig. 2-1. Vehicle gasifiers before 1950 (Source: NAS 1983)

regarded as only temporary modifications for wartime conditions. However, a few car makers went so far as to modify the body work for gasifier installation. Soon after the war, low-cost gasoline became available again, and most users went back to burning gasoline because of its convenience.

2.2 Current Development Activities

After the OPEC oil embargo of 1973, there was renewed interest in all forms of alternative energy, including gas produced from coal and biomass. Most of the early work supported by the United States and foreign energy establishments focused on large-scale coal-fed gasifiers that were intended to produce synthetic natural gas as a fuel. There was little interest in biomass or biomass gasification (PNL 1986), except for groups concerned with uses in less developed countries (NAS 1983; Kjellstrom 1981, 1983, 1985) and private individuals (Skov 1974; Mother 1982; TIPI 1986).

Recently, there has been increased interest in biomass as a renewable energy source. In the last few years, a number of individuals and groups have built versions of small downdraft gasifiers and have operated them as demonstration units. A few of the gasifier-powered vehicles from this effort are shown in Fig. 2-2, and today one can obtain shop plans for constructing gasifiers (Nunnikhoven 1984; Mother 1982; Skov 1974). Unfortunately, no body of information is available to help either the latter-day hobbyists or their counterparts involved in full-time research to evaluate critical factors such as gasifier operation, gas quality, gas-cleanup systems, engine operation, and engine wear.

Interest in small-scale gasifiers is strong among organizations that deal with less developed countries such as the World Bank, the U.S. Agency for International Development, and the equivalent organizations in European countries. The Producer Gas Roundtable (of the Beijer Institute in Stockholm) has published a number of books on gasification and drawn together technical expertise from around the world. In addition, this group has hosted several conferences on producer gas for less developed countries (Kjellstrom 1981, 1983, 1985).

Producer gas from charcoal has been developed commercially in the Philippines (Kjellstrom 1983), where more than 1000 units have operated. Producer gas is generated for industrial heat by more than 30 large units operating in Brazil (Makray 1984).

2.3 Future Development Directions

Predicting the needs and direction of development in our modern world is very dangerous, because we don't know how future conditions will change and what our response will be. Since the first OPEC embargo in 1973, we have oscillated between a concern with energy supplies and business as normal. Therefore, we can't predict which direction we are likely to go, but we can at least list the possible options and factors that affect the choice.

In normal times, development is driven by **economic considerations**, and some of the economic factors influencing use of gasification are listed in Chapter 13. In times of **emergency**, our priorities change drastically and quite different developments occur.

Small gasifiers were developed very rapidly during the **emergency** of World War II and just as rapidly disappeared when liquid fuels were available. Transporation is a very high priority, and the U.S. Department of Defense currently has a program to disseminate information on small gasifiers in case of national emergency. However, for economic reasons, no work on gasifiers for vehicles is in progress in the United States. During the late 1970s, we imported more than 40% of our oil. We reserved much of our liquid fuel for transport, and there was no government call to develop gasifiers in the United States. (However, Sweden-Volvo manufactured and stored 10,000 units for emergency use.)

In the private sector of the United States during the last 10 years, there has been a corresponding development of biomass gasifiers for **heat applications** at the scale found in lumber and paper mills. There has been interest in power generation at a small scale in the United States stimulated by attractive power buy back rates in some states under the Public Utilities Regulatory Policy Act (PURPA) discussed in Chapter 13.

A very active area of development for small gasifiers is to generate power in **developing countries**, which have biomass resources and cannot easily afford liquid fuels. They do not have an electrical distribution grid so power systems of 10 to 1000 kW are very attractive. Thus, the scale of operation has an important influence on what is developed in this case.

Finally, new developments in gasifiers may extend their use to other new areas. One of our authors (Das) has developed a small gasifier suitable for firing a foundry. The other author (Reed) is developing small batch-type gasifiers for cooking and lighting applications in third world countries.

Fig. 2.2. Vehicle gasifiers after OPEC (Source: NAS 1983)

Chapter 3
Gasifier Fuels

3.1 Introduction

Biomass fuels occur in a multitude of physical forms. The often-heard manufacturer's claim that a particular gasifier can gasify "any biomass fuel" is a naive statement, and each form can be expected to have unique problems until proven otherwise. This physical disparity accounts in part for the large number of gasifier designs available today. The gasifiers used widely during World War II used specially prepared 1x2x2 cm^3 hardwood blocks. However, such blocks could represent only a tiny fraction of the biomass materials available for gasification. Some gasifiers currently are undergoing design evolutions that will enable them to use a wider range of fuels; nevertheless, fuel properties are very important in determining satisfactory operating conditions. Therefore, these multifeedstock gasifiers will be able to use only a limited range of biomass with controlled specifications, and anyone installing such a gasifier should have tests run on the fuel to be used before deciding upon a purchase. The ability to specify fuel parameters is very important, and we discuss them in this chapter. Fortunately, a wide variety of tests are available for biomass and charcoal gasifiers that can be useful to those interested in gasification.

Green wood can contain up to 50% water by weight, so its properties vary widely with moisture content. The chemical composition of biomass (expressed on a dry, ash-free basis) is more constant than that of the various coals (bituminous, anthracite, lignite) as shown in Fig. 3-1. Furthermore, more than 80% of the biomass is volatile. Coal is typically only 20% volatile; the remaining 80% is unreactive coke, which is more difficult to gasify than charcoal. Biomass generally has very low sulfur and ash content compared to coal. However, unlike coal, biomass comes in a wide variety of physical forms, making it necessary to tailor the shapes of the gasifier, fuel-drying equipment, feed systems, and ash-removal equipment to each form. Therefore, the resulting gasifier design must be very fuel-specific.

3.2 Biomass Fuel Analysis

3.2.1 Proximate and Ultimate Analysis

Two types of analyses, **proximate** and **ultimate**, are useful for defining the physical, chemical, and fuel properties of a particular biomass feedstock. These analyses were initially developed for coal and are widely available from commercial laboratories. They are described in detail in the publications of the American Society for Testing Materials (ASTM), shown in Table 3-1. The equipment necessary for performing elemental analysis is shown in Table 3-2. The **proximate analysis**

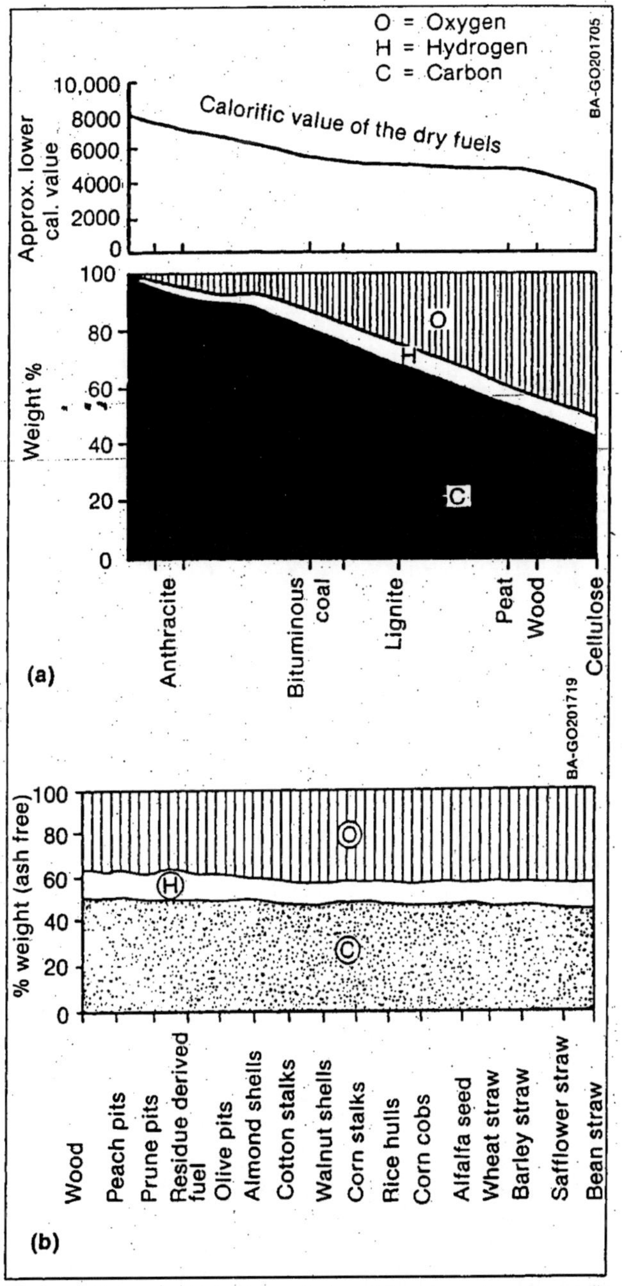

Fig. 3-1. Elemental (ultimate) analysis of (a) coals and wood and (b) biomass fuels (Sources: Skov 1974, p. 35. (©1974. Used with permission of Biomass Energy Foundation, Inc.) and Kaupp 1984a, Fig. 96)

Table 3-1. ASTM Standards Methods for Proximate and Ultimate Analysis of Wood Feedstocks

Method	Test No.
Proximate Analysis	
Moisture	E871
Volatile matter	E872
Ash	D1102
Ultimate Analysis	
C	E777
H	E777
O	E870
N	E778
S	E775
Gross Heating Value	E711

Table 3-2. Elemental Analyzer Equipment

Instrument	Oxidant	Capability	Detection[a]
Carlo Erba 1104	oxygen	C,H,N,O	FID & TC
Chemical Data Systems (CDS 1200)	oxygen	C,H,N,O,S and functional groups	FID & TC
Hewlett-Packard HP-185	MnO_2 added	C,H,N	FID & TC
Perkin Elmer 240	oxygen	C,H,N,O,S	TC

[a]FID—Flame ionization detector
TC—Thermal conductivity
Source: Reed 1981

is relatively simple and can be performed with a drying oven, a laboratory furnace, and a balance. The **ultimate analysis** involves more advanced chemical techniques. Both analyses can be performed in commercial laboratories for $25 to $100.

The proximate analysis determines the moisture (M), volatile matter (VM), ash (A), and (by difference) fixed carbon content (C) of a fuel, using standard ASTM tests. Moisture is analyzed by the weight loss observed at 110°C. The volatile matter is driven off in a closed crucible by slow heating to 950°C, and the sample is weighed again. The high heating rates encountered within an actual gasifier yield a higher volatile content and a lower fixed carbon content than the slow rate used in the ASTM measurement, but char yield from the gasifier is expected to be proportional to char yield from the ASTM test.

The proximate analyses for selected biomass feedstocks and other solids are shown in Table 3-3. Note that more than 70% of most biomass material is volatile under the conditions of the test. The proximate analysis generally includes moisture content measured on a wet basis, MCW, where

$$MCW = (\text{wet weight} - \text{dry weight})/\text{wet weight}. \quad (3\text{-}1)$$

Sometimes, moisture content is reported on a dry-weight basis, MCD, where

$$MCD = (\text{wet weight} - \text{dry weight})/\text{dry weight}. \quad (3\text{-}2)$$

Values given in one form can be converted to the other as shown in Fig. 3-2 according to the relationships:

$$MCD = MCW/(1 - MCW), \text{ and} \quad (3\text{-}3)$$

$$MCW = MCD/(1 + MCD). \quad (3\text{-}4)$$

Moisture contents for typical biomass fuels are shown in Table 3-4. The effect of moisture content on heat recovery and combustion efficiency is shown in Table 3-5. Recoverable heat drops dramatically with increased moisture since the heat of vaporization of the water is not normally recovered during combustion (see Table 4-1).

Since biomass varies in its properties from day to day and from load to load, it is common to report analyses on a dry basis, and sometimes on a moisture- and ash-free (MAF) basis. It is then a simple matter to calculate other specific conditions from this value.

The ultimate analysis gives the chemical composition and the higher heating value of the fuels. The chemical analysis usually lists the carbon, hydrogen, oxygen, nitrogen, sulfur, and ash content of the dry fuel on a weight percentage basis. Ultimate analyses for a number of biomass and other solid fuels are given in Table 3-6 and for various chars in Table 3-7.

Note in Table 3-6 that biomass is typically very low in both nitrogen and sulfur content relative to fossil fuels. However, selected biomass feedstocks may have much higher values. The sulfur and nitrogen contents of selected biomass fuels are shown in Tables 3-8 and 3-9.

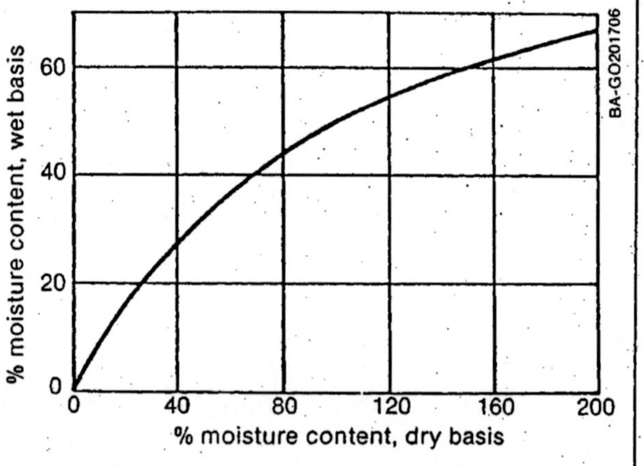

Fig. 3-2. Wet basis-dry basis moisture content comparison (Source: McGowan 1980, Fig. 1-1)

The ash content of biomass is typically much less than that of coals, but some forms have a high ash content, as shown in Table 3-3. This can lead to ash melting (known as "slagging"), which can cause severe problems in some gasifiers. A standard ASTM method is available for measuring the slagging temperature for ash (Table 3-1).

The higher heating value of the fuel is determined by reacting the fuel with oxygen in a bomb calorimeter and measuring the heat released to a known quantity of water. The heat released during this procedure represents the maximum amount of energy that can be obtained from combusting the fuel and is a necessary value for calculating the efficiency of gasification. The high heating value (HHV) is measured in this test, since liquid water is produced; however, the low heating value (LHV) is more relevant to the amount of energy produced, and this can be calculated from the HHV value shown in Table 4-1.

The heat of combustion is determined by the composition of the biomass and in fact can be calculated with considerable accuracy from

Table 3-3. Proximate Analysis Data for Selected Solid Fuels and Biomass Materials (Dry Basis, Weight Percent)

	Volatile Matter (VM)	Fixed Carbon (FC)	Ash	Reference
Coals				
Pittsburgh seam coal	33.9	55.8	10.3	(1)
Wyoming Elkol coal	44.4	51.4	4.2	(1)
Lignite	43.0	46.6	10.4	(1)
Oven Dry Woods				
Western hemlock	84.8	15.0	0.2	(2)
Douglas fir	86.2	13.7	0.1	(2)
White fir	84.4	13.1	0.5	(2)
Ponderosa pine	87.0	12.8	0.2	(2)
Redwood	83.5	16.1	0.4	(2)
Cedar	77.0	21.0	2.0	(2)
Oven Dry Barks				
Western hemlock	74.3	24.0	1.7	(2)
Douglas fir	70.6	27.2	2.2	(2)
White fir	73.4	24.0	2.6	(2)
Ponderosa pine	73.4	25.9	0.7	(2)
Redwood	71.3	27.9	0.8	(2)
Cedar	86.7	13.1	0.2	(2)
Mill Woodwaste Samples				
-4 mesh redwood shavings	76.2	23.5	0.3	(3)
-4 mesh Alabama oakchips	74.7	21.9	3.3	(3)
Municipal Refuse and Major Components				
National average waste	65.9	9.1	25.0	(4)
Newspaper (9.4% of average waste)	86.3	12.2	1.5	(4)
Paper boxes (23.4%)	81.7	12.9	5.4	(4)
Magazine paper (6.8%)	69.2	7.3	23.4	(4)
Brown paper (5.6%)	89.1	9.8	1.1	(4)
Pyrolysis Chars				
Redwood (790° to 1020°F)	30.0	67.7	2.3	(2)
Redwood (800° to 1725°F)	23.9	72.0	4.1	(2)
Oak (820° to 1135°F)	25.8	59.3	14.9	(2)
Oak (1060°F)	27.1	55.6	17.3	(2)

(1) Bituminous Coal Research, Inc. 1974. *Gas Generator Research and Development, Phase II. Process and Equipment Development.* OCR-20-F; PB-235530/3GI.

(2) Howlett, K. and Gamache, A. 1977. *Forest and Mill Residues as Potential Sources of Biomass.* Vol. VI. Final Report. McLean, VA; The Mitre Corporation/Metrek Division; ERDA Contract No. E (49-18) 2081; MTR 7347.

(3) Boley, C. C. and Landers, W. S. 1969, *Entrainment Drying and Carbonization of Wood Waste.* Washington, D.C.: Bureau of Mines; Report of Investigations 7282.

(4) Klass, D. L. and Ghosh, S. 1973. "Fuel from Organic Wastes." *Chemical Technology*, p. 689.

Source: Reed 1981

Table 3-4. Approximate Moisture Contents of Typical Biomass Fuels

Biomass Fuel	Moisture Content (wt % Wet Basis)	(wt % Dry Basis)	Reference
Woody biomass, green	40-60	67-150	(1)
Woody biomass, dried	15	17	(1)
Straws	15	17	(1)
Stalks, cobs, hulls	15	17	(1)
Bagasse	70	230	(1)
Municipal refuse	35	55	(2)
Peat	90	900	(1)
Air dry feedlot waste	11	12	(2)

(1) Miles 1982

(2) Reed 1981

Table 3-5. Effect of Moisture Content on Heat Recovery and Combustion Efficiency[a]

Moisture (wt %) Dry Basis	Wet Basis	Recoverable Heat[b] (Btu/lb)	Combustion Efficiency (%)
0.00	0.00	7097	82.5
4.76	4.54	7036	81.8
9.09	8.33	6975	81.1
13.04	11.54	6912	80.4
16.67	14.29	6853	79.7
20.00	16.67	6791	78.9
23.08	18.75	6730	78.3
28.57	22.22	6604	76.8
33.33	25.00	6482	75.4
42.86	30.00	6178	71.8
50.00	33.33	5868	68.2
60.00	37.50	5252	61.1
66.67	40.00	4639	53.9
71.43	41.67	4019	46.7

[a]From Bliss, C. and Black, D. O. 1977. *Silvicultural Biomass Farms, Vol. 5, Conversion Processes and Costs.* McLean, VA: Mitre Corporation; ERDA Contract No. EX-76-C-01-2081.

[b]Theoretical values based on a maximum heating value of 8600 Btu/lb, an initial wood temperature of 62°F, a flue gas temperature of 450°F, an initial air temperature of 62°F and 50% excess air.

Source: Reed 1981

$$HHV = [34.1\ C + 132.2\ H + 6.8\ S - 1.53\ A - 12.0\ (O+N)]\ kJ/g \quad (3\text{-}5)$$

$$HHV = [146.6\ C + 568.8\ H + 29.4\ S - 6.6\ A - 51.5\ (O+N)] \times 10^2\ Btu/lb \quad (3\text{-}6)$$

where C, H, S, A, O, and N are the wt % of carbon, hydrogen, sulfur, ash, oxygen, and nitrogen in the fuel. The calculated value agrees with the measured value with an absolute error of 2.1% for a large number of biomass materials (Reed 1981).

3.2.2 Physical Tests

One of the most important physical characteristics of biomass fuel is the **bulk density**. The bulk density is the weight of biomass packed loosely in a container divided by the volume occupied. Clearly, it is not an exact number, depending on the exact packing of the particles.

The fuel shape and feeding characteristics determine whether it will be feasible to simply use gravity feeding techniques, or whether assistance, such as stirring and shaking, will be required. The **angle of repose** for a particular fuel type is generally measured by filling a large tube with the fuel, and then lifting the tube and allowing the fuel to form a pile. The angle of repose is the angle from the horizontal to the sides of the pile. The basic feed characteristic is more easily judged from the **dugout angle of repose**, the steepest angle (measured from the horizontal) formed by the sides of a pile of fuel when material is removed from the bottom of the pile. Angles approaching or exceeding 90° are a good indication of the tendency of the fuel to bridge or tunnel in the gasifier.

3.3 Other Fuel Parameters

The tests and analyses just mentioned are in widespread use because they were developed for use in other industries. However, many more tests need to be developed specifically for gasification processes. This section addresses the effects of other fuel parameters on biomass gasification, illustrating the need for more specific testing procedures. The basic fuel parameters important in gasifier design are

- particle size and shape
- particle size distribution
- char durability and fixed-carbon content
- ash fusion temperature
- ash content
- moisture content
- heating value.

3.3.1 Particle Size and Shape

The size and shape of the fuel particles are important for determining the difficulty of moving and delivering the fuel, as well as the behavior of the fuel once it is in the gasifier. Good fuel hopper design calls for a cone angle that is double the dugout angle of repose. With an angle of repose over 45°, the fuel may not flow even in a straight cylinder and will require either an inverted cone or some agitation (Perry 1973). **Smooth** hopper walls are always desirable.

Gasifiers frequently suffer from bridging and channeling of the fuel. The size and size distribution of the fuel

Table 3-6. Ultimate Analysis Data for Selected Solid Fuels and Biomass Materials (Dry Basis, Weight Percent)

Material	C	H	N	S	O	Ash	Higher Heating Value (kJ/g)	Higher Heating Value (Btu/lb)	Reference
Pittsburgh seam coal	75.5	5.0	1.2	3.1	4.9	10.3	31.67	13,650	(1)
West Kentucky No. 11 coal	74.4	5.1	1.5	3.8	7.9	7.3	31.23	13,460	(2)
Utah coal	77.9	6.0	1.5	0.6	9.9	4.1	32.87	14,170	(1)
Wyoming Elkol coal	71.5	5.3	1.2	0.9	16.9	4.2	29.49	12,710	(2)
Lignite	64.0	4.2	0.9	1.3	19.2	10.4	24.85	10,712	(2)
Charcoal	80.3	3.1	0.2	0.0	11.3	3.4	31.02	13,370	(1)
Douglas fir	52.3	6.3	0.1	0.0	40.5	0.8	21.0	9,050	(1)
Doublas fir bark	56.2	5.9	0.0	0.0	36.7	1.2	22.0	9,500	(1)
Pine bark	52.3	5.8	0.2	0.0	38.8	2.9	20.4	8,780	(1)
Western hemlock	50.4	5.8	0.1	0.1	41.4	2.2	20.0	8,620	(1)
Redwood	53.5	5.9	0.1	0.0	40.3	0.2	21.0	9,040	(1)
Beech	51.6	6.3	0.0	0.0	41.5	0.6	20.3	8,760	(1)
Hickory	49.7	6.5	0.0	0.0	43.1	0.7	20.1	8,670	(1)
Maple	50.6	6.0	0.3	0.0	41.7	1.4	19.9	8,580	(1)
Poplar	51.6	6.3	0.0	0.0	41.5	0.6	20.7	8,920	(1)
Rice hulls	38.5	5.7	0.5	0.0	39.8	15.5	15.3	6,610	(1)
Rice straw	39.2	5.1	0.6	0.1	35.8	19.2	15.8	6,540	(1)
Sawdust pellets	47.2	6.5	0.0	0.0	45.4	1.0	20.45	8,814	(3)
Paper	43.4	5.8	0.3	0.2	44.3	6.0	17.57	7,572	(4)
Redwood wastewood	53.4	6.0	0.1	0.1	39.9	0.6	21.26	9,163	(5)
Alabama oak woodwaste	49.5	5.7	0.2	0.0	41.3	3.3	19.18	8,266	(5)
Animal waste	42.7	5.5	2.4	0.3	31.3	17.8	17.1	7,380	(1)
Municipal solid waste	47.6	6.0	1.2	0.3	32.9	12.0	19.83	8,546	(6)

(1) Tillman, D. A. 1978. *Wood as an Energy Resource*. New York: Academic Press.

(2) Bituminous Coal Research, Inc. 1974. *Gas Generator Research and Development, Phase II. Process and Equipment Development*. OCR-20-F; PB-235530/3GI.

(3) Wen, C. Y., Bailie, R. C., Lin, C. Y., and O'Brien, W. S. 1974. "Production of Low Btu Gas Involving Coal Pyrolysis and Gasification." *Advances in Chemistry Series*. Vol. 131. Washington, D.C.: American Chemical Society.

(4) Bowerman, F. R. 1969. Introductory Chapter to *Principles and Practices of Incineration*. Corey, R. C., editor. New York: John Wiley and Sons.

(5) Boley, C. C. and Landers, W. S. 1969. *Entrainment Drying and Carbonization of Wood Waste*. Washington, D.C.: U.S. Bureau of Mines; Report of Investigations 7282.

(6) Sanner, W. S., Ortuglio, C., Walters, J. G., and Wolfson, D. E. 1970. *Conversion of Municipal and Industrial Refuse into Useful Materials by Pyrolysis*. U.S. Bureau of Mines; Aug; RI 7428.

Source: Reed 1981

determine the thickness of the gasification zone, the pressure drop through the bed, and the minimum and maximum hearth load for satisfactory operation. A uniform particle size helps overcome some problems. Improving the grate design, as well as added agitation or stirring, can go a long way to give trouble-free gasifier operation and to broaden the range of fuel shapes suitable for gasification.

At the same time, it is important to realize that excessive agitation results in excess carbon carryover, which in turn reduces the efficiency of the gasifier. In addition, carbon carryover reduces the oxygen/fuel ratio, since the carbon requires more oxygen than the biomass for gasification. This in turn reduces the oxygen available for flaming pyrolysis and increases the rate of tar formation.

3.3.2 Charcoal and Char Properties

Carbon is the name applied to a chemical element that occurs in dozens of physical forms, both pure (such as diamond and graphite) and impure (such as coke, charcoal, and soot). **Charcoal** refers to the 10% to 30% solid carbon product from biomass pyrolysis. Its composition can vary from 50% carbon to more than 80% carbon, depending on the temperature and conditions of pyrolysis (see Table 3-7). Also, since it contains most of the original ash from the biomass, charcoal typically contains from 2% to 10% mineral matter (Emrich 1985).

Charcoal manufacture dates to prehistoric times and is a well-established industry today with standards for its various uses. Charcoal is simpler to gasify, and it is easier to clean up the gas for engine use than biomass

Table 3-7. Ultimate Analysis Data for Selected Pyrolysis Chars (Dry Basis, Weight Percent)

Material	C	H	N	S	O	Ash	Higher Heating Value (kJ/g)	Higher Heating Value (Btu/lb)	Reference
Fir bark char	49.9	4.0	0.1	0.1	24.5	21.4	19.2	8,260	(1)
Rice hull char	36.0	2.6	0.4	0.1	11.7	49.2	14.2	6,100	(1)
Grass straw char	51.0	3.7	0.5	0.8	19.7	24.3	19.3	8,300	(1)
Animal waste char[a]	34.5	2.2	1.9	0.9	7.9	48.8	12.6	5,450	(1)
Municipal solid waste char (high temperature)	54.9	0.8	1.1	0.2	1.8	41.2	18.6	8,020	(2)
Redwood charcoal (790° to 1020°F)	75.6	3.3	0.2	0.2	18.4	2.3	28.8	12,400	(3)
Redwood charcoal (860° to 1725°F)	78.8	3.5	0.2	0.2	13.2	4.1	30.4	13,100	(3)
Oak charcoal (820° to 1185°F)	67.7	2.4	0.4	0.2	14.4	14.9	24.7	10,660	(3)
Oak charcoal (1060°F)	64.6	2.1	0.4	0.1	15.5	17.3	23.0	9,910	(3)

[a]Contains 3.7% chlorine lumped with oxygen

(1) Pober, K. W. and Bauer, H. F. 1977. "The Nature of Pyrolytic Oil from Municipal Solid Waste." *Fuels from Waste*. Anderson, L. L. and Tillman, D. A., Editors. New York: Academic Press, pp. 73-86.

(2) Sanner, W. S., Ortuglio, C., Walters, J. G., and Wolfson, D. E. 1970. *Conversion of Municipal and Industrial Refuse into Useful Materials by Pyrolysis*. U.S. Bureau of Mines; Aug; RI 7428.

(3) Boley, C. C. and Landers, W. S. 1969. *Entrainment Drying and Carbonization of Wood Waste*. Washington, D.C.: U.S. Bureau of Mines; Report of Investigations 7282.

Source: Reed 1981

gas is because of charcoal's low volatile content. At the beginning of World War II, most gasifiers used charcoal. However, charcoal manufacture wastes approximately 50% of the energy of biomass and usually requires hardwood biomass as a starting material. By the end of World War II, most gasifiers used wood instead of charcoal (Gengas 1950). Today, a large number of gasifiers built in the Philippines use charcoal, and charcoal is used in some other countries as well (Foley 1983; Kjellstrom 1983). It seems wise and probable that any long-term development of biomass gasification will ultimately use biomass again, rather than charcoal.

As charcoal is converted to gas in a gasifier, the ash content rises. We use the term **char-ash** to describe the end product from char gasification; although the char-ash is still black, it may contain up to 50% ash. The incoming oxygen/air/steam in updraft gasifiers contacts the char-ash at the grate and burns out the carbon, leaving a white ash. The principal problem in updraft gasifiers is to avoid ash slagging (melting), since it will plug the grate. In downdraft gasifiers, the char-ash reacts with CO_2 and H_2O, and is not contacted by oxygen so the carbon is normally not completely consumed in a downdraft gasifier. The result is black char-ash with 70% to 80% carbon. This carbon gives a good resistance to slagging. However, fuels with a high ash content can cause slagging in the area of the tuyeres, if they are used.

Thus in combustion and updraft gasifiers the fuel passes through the stages

Biomass → Charcoal → Char-Ash → Ash → Slag

and in downdraft gasifiers this process stops at char-ash.

Charcoal durability depends on the resistance of the charcoal to powdering (duffing) during transport or char gasification. Ideally, the charcoal should maintain its size until the carbon reaches the end of the reduction zone. In practice, a wide range of char particles are produced in the reduction zone, and these can cause a plugging problem if they are not removed. Stirring and augering out char and ash are effective techniques for preventing this plugging problem (Rogers 1985; Kaupp 1984b). Figure 3-3 shows the char ash content as a function of particle size and the relation between carbon conversion and char size for a stratified-bed gasifier. The fuel starts as biomass (1-in. birch dowels) on the far right of Fig. 3-3. Ash is 0.5% and carbon conversion is zero, of course. After flaming pyrolysis half of the carbon has been converted yet the resulting charcoal is only slightly smaller than its original size (25% - 35% shrinkage). The char then undergoes gasification reactions with hot pyrolysis combustion products, which consume the carbon on both the surface and in the interior of the particle. As interior carbon is consumed the char shrinks, causing fractures, and the particle loses mechanical strength, causing crumbling. The small fragments are swept away by gas velocity. Returning to Fig. 3-3, we see a plateau after pyrolysis and that the char ash remains between 2% and 3% all the way down to under 1000 μm (1 mm) particle size, indicating that this size particle has not engaged in much char gasification. Below 500 μm (0.5 mm) we see a second plateau, indicating the end of char gasification, and

Table 3-8. Sulfur Content of Biomass Fuels

Biomass Fuels	% Sulfur, Dry Weight Basis	Reference
Alfalfa seed straw	0.3	(1)
Almond shells	<0.02	(1)
Barley straw	0.14	(1)
Coffee hulls	0.2	(2)
Corn cobs	0.001-0.007	(1,3)
Corn fodder	0.15	(1,2)
Corn stalks	0.05	(1)
Oat straw	0.23	(2)
Cotton gin trash	0.26-0.31	(1)
Flax straw, pelleted	<0.01	(1)
Furfural residue	0.4	(4)
Olive pits	0.02	(1)
Peach pits	0.04	(1)
Peanut husks	0.1	(4)
Peat (Finnish)	0.05-0.2	(5)
Peat, general	1.5-2.0	(6)
Rice hulls	0.16	(1)
Rice straw	0.10	(1)
Walnut shells	0.03-0.09	(1)
Wheat straw	0.17	(2)
Wood, chipped	0.08	(1)
Wood, general	0.02	(1,7)
Wood, pine bark	0.1	(4)
Wood, green fir	0.06	(4)
Wood, kiln dried	1.0	(4)
Wood, air dried	0.08	(4)

(1) *Gasification Project Ultimate Chemical Analysis Log,* Agricultural Engineering Dept., University of California, Davis, 1979.

(2) Partridge, J. R., "Manitoba Crops as an Energy Source," Sixth Annual Conference, Biomass Energy Institute, Winnipeg, Manitoba, Canada, Oct. 13, 1977.

(3) Payne, F. A., *et al.,* "Gasification-Combustion of Corncobs and Analysis of Exhaust," American Society of Agricultural Engineers Summer Meeting, San Antonio, TX, Paper #80-3025, 1980.

(4) Bailie, R. C., "Current Developments and Problems in Biomass Gasification," Sixth Annual Meeting, Biomass Energy Institute, Winnipeg, Manitoba, Canada, Oct. 13, 1977.

(5) Ekman, E. and Asplund, D., *A Review of Research of Peat Gasification in Finland.* Technical Research Centre of Finland, Fuel and Lubricant Research Laboratory, Espoo, Finland.

(6) Rambush, N. E., *Modern Gas Producers,* New York: Van Nostrand, 1923.

(7) Jenkins, B., *Downdraft Gasification Characteristics of Major California Biomass-Derived Fuels,* Ph.D. Thesis, Department of Agricultural Engineering, University of California, Davis, 1980.

Source: Kaupp 1984a

that there is very little additional activity. It is clear that larger particles carry more unreacted carbon with them than do smaller particles. Therefore, the conversion efficiency will be maximized if removal of large char is kept to a minimum. The balance between conversion efficiency and ash removal will be fuel-specific.

The final weight of the char-ash residue is usually 2% to 10% of the biomass weight, depending on the char-ash removal rate and the char durability. However, the char-ash residue has a very low density and so may occupy up to 20% of the volume of the original biomass, depending on the completeness of char gasification. Therefore, it is important to provide for adequate removal of this bulky material.

Because charcoal often has a high value, gasifiers are sometimes operated to produce up to 10% charcoal by augering out the charcoal at the end of the flaming combustion zone (Pyrenco). This reduces the requirement for oxygen (air) and increases gas quality to more than 6.8 MJ/Nm^3, but also increases tar content. However, no current commercially successful small-scale charcoal production in gasifiers is known to the authors.

Charcoal is manufactured all over the world, and standards determine the quality and suitability for various uses (Emrich 1985). Recent tests at the Colorado School of Mines have tested char pellet strength at various stages of gasification (Hubis 1983).

3.3.3 Biomass Ash Content and Effects

Fuels with a high ash content require much greater attention to grate design, gas disengagement, and positive char-ash removal. The slagging behavior of various crop residues and wood is shown in Table 3-10.

Table 3-9. Nitrogen Content of Biomass Fuels

Biomass Fuels	% Nitrogen Dry Weight Basis	Reference
Barley straw	0.59	(1)
Corn cobs	0.16-0.56	(1)
Corn fodder	0.94	(1)
Cotton gin trash	1.34-2.09	(1)
Corn stalks	1.28	(1)
Flax straw, pelleted	1.1	(1)
Oat straw	0.66	(2)
Olive pits	0.36	(1)
Peach pits	1.74	(1)
Peat	0.5-3.0	(3)
Prune pits	0.32	(1)
Rice hulls, pelleted	0.57	(1)
Safflower straw	0.62	(1)
Walnut shells	0.260-0.4	(1)
Wood, general	0.009-2.0	(1,4)
Coal Fuels		
Anthracite	<1.5	(4)
German and English bituminous coal	0.5-1.9	(4)
American coal	0.5-2	(4)
Brown coal and lignites	0.5-2	(4)

(1) *Gasification Project Ultimate Chemical Analysis Log,* Agricultural Engineering Department, University of California, Davis, 1979.

(2) Partridge, J. R., "Manitoba Crops as an Energy Source," Sixth Annual Conference, Biomass Energy Institute, Winnipeg, Manitoba, Canada, Oct. 13, 1977.

(3) Ekman, E. and Asplund, D., *A Review of Research of Peat Gasification in Finland,* Technical Research Centre of Finland, Fuel and Lubricant Research Laboratory, Espoo, Finland.

(4) Rambush, N. E., *Modern Gas Producers,* New York: Van Nostrand, 1923.

Source: Kaupp 1984a

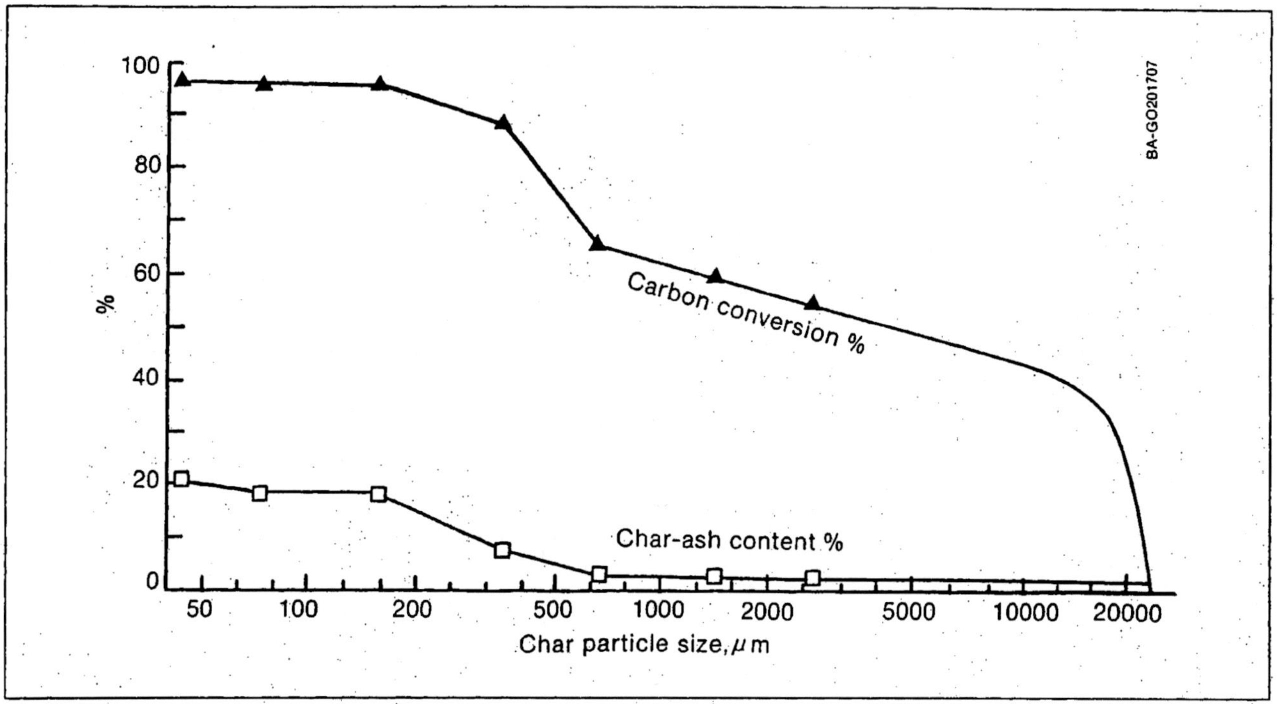

Fig. 3-3. Char-ash content and carbon content versus char particle size for a stratified bed gasifier (Source: Das 1985)

Table 3-10. Slagging Behavior of Crop Residues and Wood

Slagging Fuels	% Ash	Degree of Slagging	Non-Slagging Fuels	% Ash
Barley straw mix	10.3	Severe	Cubed alfalfa seed straw	6.0
Bean straw	10.2	Severe	Almond shell	4.8
Corn stalks	6.4	Moderate	Corn cobs	1.5
Cotton gin trash	17.6	Severe	Olive pits	3.2
Cubed cotton stalks	17.2	Severe	Peach pits	0.9
RDF pellets	10.4	Severe	Prune pits	0.5
Pelleted rice hulls	14.9	Severe	Walnut shell (cracked)	1.1
Safflower straw	6.0	Minor	Douglas Fir wood blocks	0.2
1/4" pelleted walnut shell mix	5.8	Moderate	Municipal tree prunings	3.0
Wheat straw and corn stalks	7.4	Severe	Hogged wood manufacturing residue	0.3
			Whole log wood chips	0.1

Source: Kaupp 1984a

3.3.4 Biomass Moisture Content and Effects

The fuel moisture content greatly affects both the operation of the gasifier and the quality of the product gas. These issues are addressed in the following sections.

3.3.5 Biomass Heating Value

It can be seen in Table 3-6 that there is a wide range of heating values for various biomass forms. A larger collection of heating values has recently been published showing a variation of 5-25 kJ/g (2000-10,000 Btu/lb) for various biomass forms (Domalski 1986). However, most of this variation is due to the variability of MAF content; and if reduced to a MAF basis, the variation is much less.

3.4 Beneficiation of Biomass Fuels

Chunky fuels (such as mill ends, chips, and corn cobs), which have at least one dimension larger than a few millimeters, can be used in fixed-bed gasifiers without further size reduction, though they may require separation from fines and dirt. Bulky fuels, such as logs, branches, and straw, require chipping or chopping and possibly densification before use in most gasifiers.

3.4.1 Densifying Biomass Fuels

Biomass fuels usually have bulk densities from one-half to one-tenth that of coal as shown in Table 3-11, presenting a drawback for shipping, storage, and gasification. Biomass fuels also come in a wide range of sizes, many of which are not suitable for fixed-bed gasification (such as sawdust, sander dust, shredder fines, straw, and husks).

However, biomass residues can be used in fixed-bed gasifiers if they are first densified to suitably sized pellets or cubes using commercially available equipment such as that shown in Fig. 3-4 (Reed 1978b). They make excellent gasifier fuels and allow the fuel to be stored at much higher densities. Densification typically consumes only 1% to 2% of the energy contained in the biomass; for some residues, drying may also require additional energy, but drying simultaneously increases the fuel value of the biomass.

Some biomass forms, with high ash or dirt contents, are difficult to densify because they cause excessive wear of the die. Also, densification is an additional expense, so its justification will depend on a comparison of the

Table 3-11. Bulk Density of Various Fuels

Fuel	Grading	Bulk Density kg/m^3	Reference
Sawdust	loose	177	(1)
Sawdust	briquets 100 mm long 75 mm diameter	555	(1)
Peat	dust	350-440	(2)
	briquets 45 x 65 x 60 mm	350-620	(2)
	hand cut	180-400	(2)
Charcoal (10% moisture)	beech	210-230	(3)
	birch	180-2003	
	softwood blocks	150-170	(3)
	softwood slabs	130-150	(3)
	mixed 60% hard/40% soft	170-190	(3)
Wood	hardwood	330	(3)
	softwood	250	(3)
	mixed 50/50	290	(3)
Straw	loose	80	—
	bales	320	—
Alfalfa seed straw	cube 30 x 30 x 50 mm, 7% moisture	298	(4)
Barley straw	cube 30 x 30 x 50 mm, 7% moisture	300	(4)
Bean straw	cube 30 x 30 x 50 mm, 7% moisture	440	(4)
Corn cobs	11% moisture	304	(4)
Corn stalks	cube 30 x 30 x 50 mm	391	(4)
Cotton gin trash	23% moisture	343	(4)
Peach pits	11% moisture	474	(4)
Olive pits	10% moisture	567	(4)
Prune pits	8% moisture	514	(4)
Rice hulls	cube 30 x 30 x 50 mm	679	(4)
Safflower straw	cube 30 x 30 x 50 mm	203	(4)
Walnut shells	cracked	336	(4)
	8 mm pollotc	559	(4)
Wood, blocks	17% moisture	256	(4)
chips	10% moisture	167	(4)
Coal	anthracite	830-900	(1)
	bituminous	770-930	(1)
Coke	hard	380-530	(1)
	soft	360-470	(1)
Brown coal	air dry lumps	650-670	(1)

(1) Rambush, N. E., *Modern Gas Producers*, New York: Van Nostrand, 1923.

(2) Ekman, E. and Asplund, D., *A Review of Research of Peat Gasification in Finland,* Technical Research Centre of Finland, Fuel and Lubricant Research Laboratory, Espoo, Finland.

(3) *Generator Gas, The Swedish Experience From 1939-1945,* Solar Energy Research Institute, Golden, CO, SERI/SP 33-140, 1979.

(4) Jenkins, B. M., *Downdraft Gasification Characteristics of Major California Residue-Derived Fuels,* Ph.D. Thesis, Department of Engineering, University of California, Davis, 1980.

Source: Kaupp 1984a

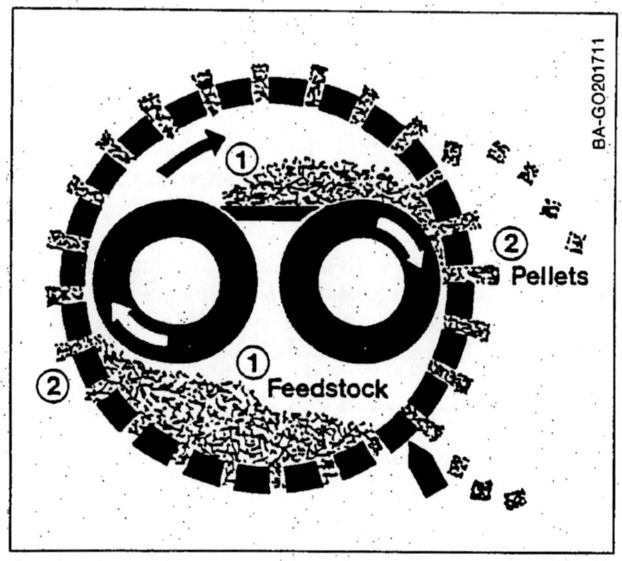

Fig. 3-4. Pelleting process (Source: Reed 1978b)

final fuel cost versus other alternatives (such as different fuels or other types of gasifiers).

3.4.2 Drying Biomass Fuels

The moisture content of the biomass fuel affects the quality of the gas that will be produced. Water requires about 2300 kJ/kg (1000 Btu/lb) to vaporize and 1500 kJ/kg to raise to 700°C during pyrolysis/gasification. Therefore, this energy must be subtracted from the heat budget of the gasifer. Although it is physically possible to gasify moderately high-moisture fuels in some gasifiers, fuel moisture reduces the quality of the gas as shown in Fig. 3-5. It also reduces the throughput of the gasifier and increases tar production. On the other hand, charcoal gasification is just the opposite; inadequate moisture input reduces the quality of char gas. Figure 3-5(b) combines char gasification and wood gasification data to illustrate the impact of total water inputs on gas quality. Total water input includes fuel moisture, chemically bound water, and air blast humidity (i.e., all mass inputs in the ratio H_2O). We see in Fig. 3-5(b) that starting with dry gasification, gas heating value increases with increased moisture input up to a peak between 30% and 40% total moisture input. The gas heating value then declines with additional moisture input.

Biomass can be considered as a source of water and charcoal using the generic formula for biomass

$$CH_{1.4}O_{0.6} = CH_{0.2}\,(0.6\ H_2O) \qquad (3\text{-}7)$$

so the chemical moisture M_C in bone dry biomass is

$$M_C = \frac{\text{Chemical Moisture}}{\text{Fuel Weight}}$$

$$= \frac{0.6[2(1) + 16]\,(100\%)}{12 + 0.2(1) + 0.6[2(1) + 16]}$$

$$= 47\% \qquad (3\text{-}8)$$

and the total moisture input M_T is

$$M_T = \frac{\text{Fuel Moisture + Chemical Moisture}}{\text{Wet Fuel Weight}}$$

$$= M_C + \frac{(100 - M_C)}{100} M_F$$

$$= 47 + 0.53\ M_F \qquad (3\text{-}9)$$

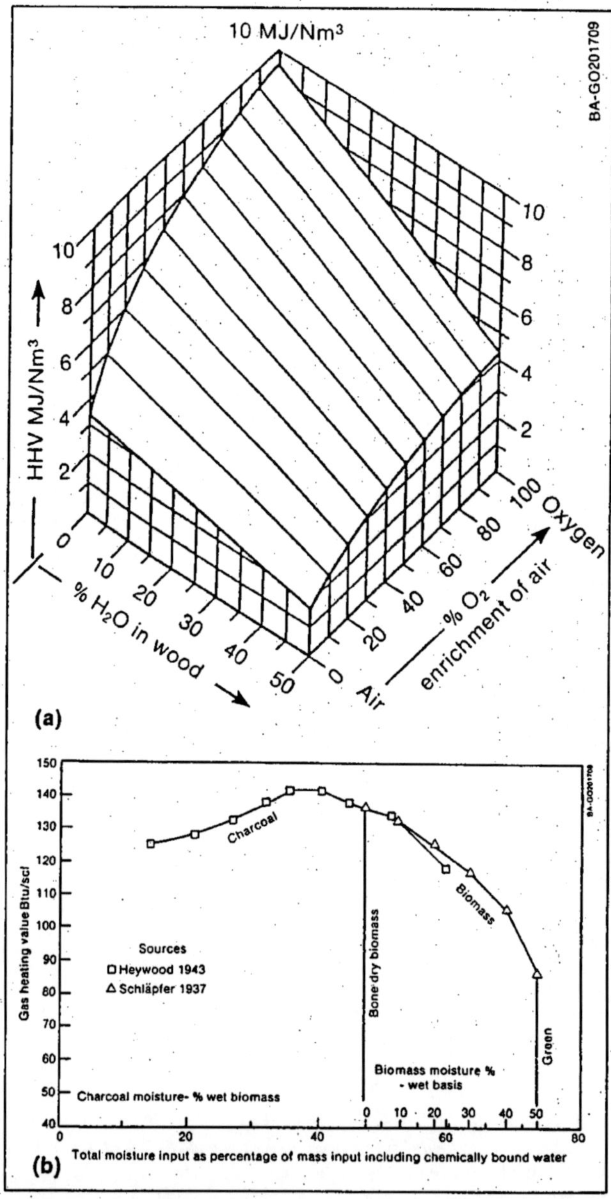

Fig. 3-5. (a) Effect of fuel moisture and oxygen on gas heating value (Source: Overend 1982, Fig. 5B)
(b) Effect of total moisture input on gas heating value

where M_F is the fuel moisture %. We see then in Fig. 3-5(b) that bone dry biomass corresponds to 47% total moisture input. The chemical moisture in bone dry biomass provides more moisture than is needed for peak heating value, and all fuel moisture reduces gas heating value.

Biomass can contain more than 50% moisture (wet basis) when it is cut; it is generally desirable to dry biomass containing more than 25% moisture (wet basis) before gasification. Drying often can be accomplished using waste heat or solar energy. If the temperature of the drying air is too high, the outer surfaces of the chunk will become dry and begin to pyrolyze before the heat can reach the center. For efficient drying, hot air, which if cooled to 60°-80°C would be moisture saturated, is preferred. The moisture slows feedstock drying (as well as slowing surface pyrolysis). Thus more air is required, improving the drying process (Thompson 1981). During operation of a gasifier and engine combination, 1-in. wood chips can be dried from 50% to 5% moisture content, with drying capacity to spare, using a 20-minute residence time with the hot engine exhaust, tempered with 90% recycle of dryer gases.

Commercial dryers are available in many forms and sizes, and it is beyond the scope of this handbook to recommend such equipment for commercial-scale operations. A simple batch dryer for drying small quantities in shown in Fig. 3-6 and a commercial dryer is shown in Fig. 3-7.

3.5 Biomass Fuel Emissions

The sulfur content of biomass fuels is usually very low compared with fossil fuels, as can be seen from Tables 3-6 and 3-8. Since sulfur oxides are corrosive, they make a major contribution to engine wear. The absence of sulfur in biomass fuels could allow a longer life for an engine operating on producer gas rather than on petroleum fuels, provided that the producer gas is free of other contaminants.

The nitrogen content of biomass fuels depends on the species of biomass used, as well as the harvest time, as shown in Table 3-9. Wood, dried stalks, hulls, and cobs have a very low nitrogen content, while leaves, seeds, and bark have a higher nitrogen content. Depending on the temperature of gasification and combustion, this may significantly lower the nitrogen oxide emissions

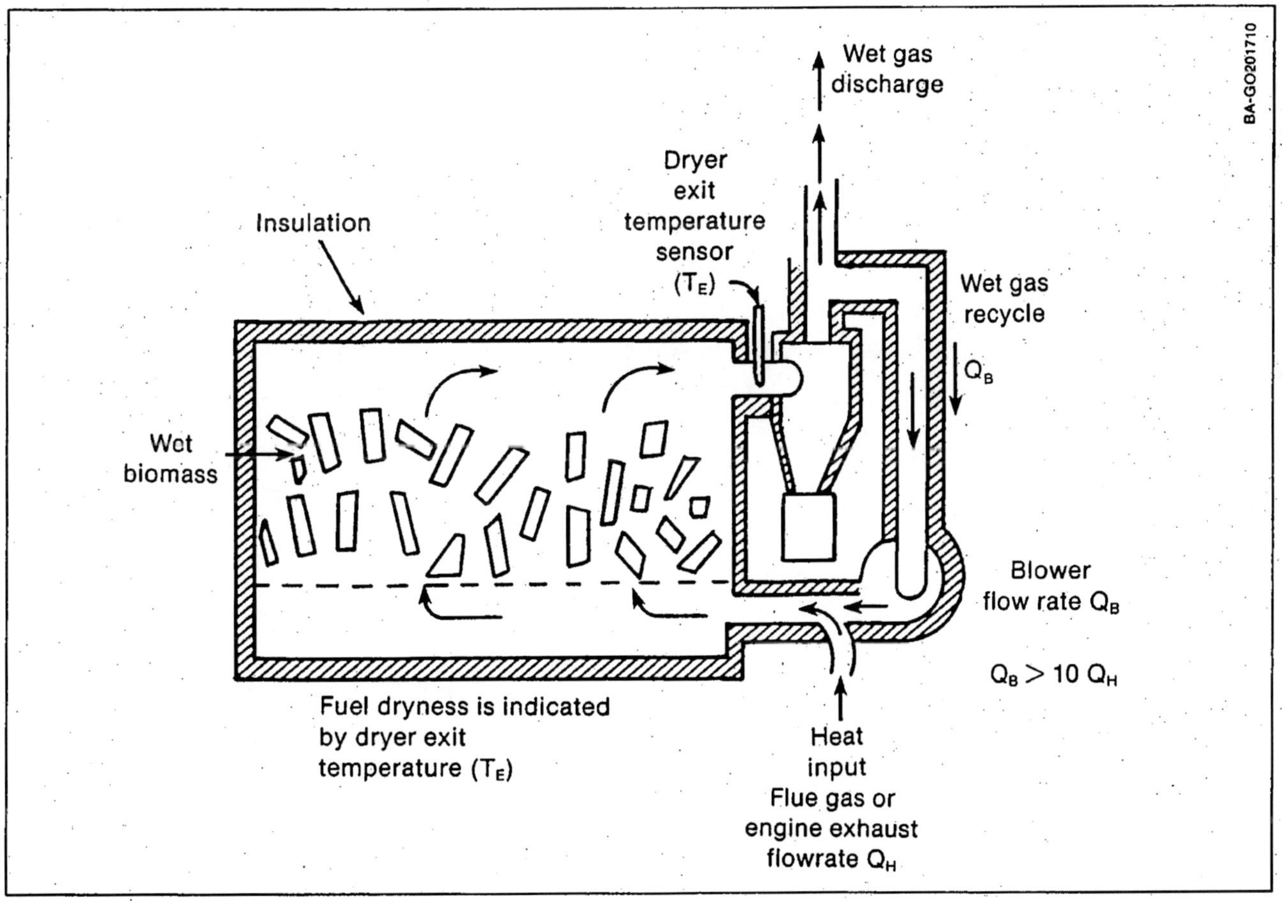

Fig. 3-6. Small batch dryer (Source: Das 1985)

from gasifier systems relative to those from fossil fuel systems. However, the final emissions depend specifically on the properties of the gasifier and the subsequent combustion of the gas, so that it is difficult to make a general statement about producer gas emissions.

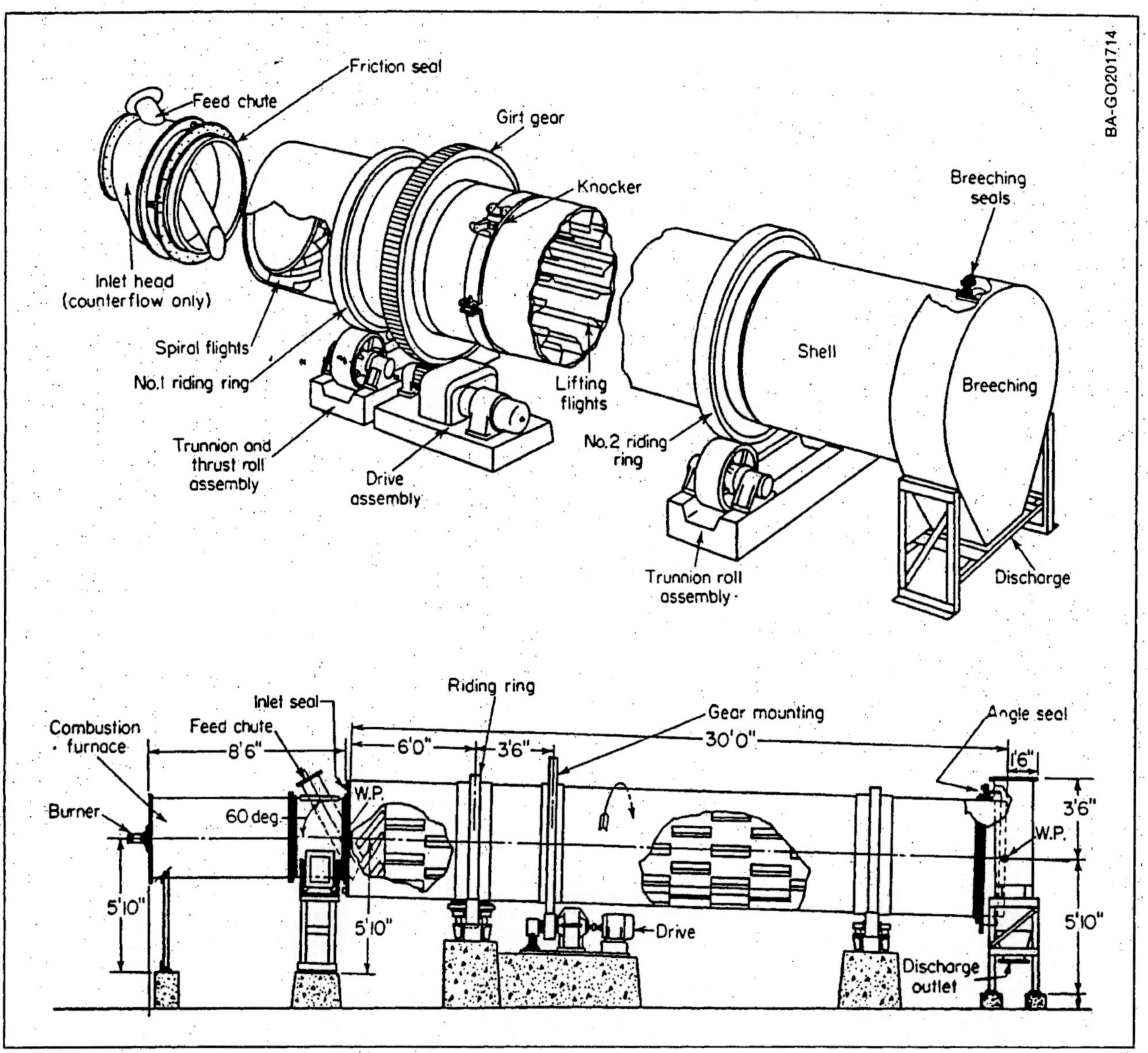

Fig. 3-7. Direct-heat rotary dryer (Source: Perry 1973, Figs. 20-35, 20-36)

Chapter 4
Principles of Gasification

4.1 Introduction

Gasifiers are relatively simple devices. The mechanics of their operation, such as feeding and gas cleanup, also are simple. The successful operation of gasifiers, however, is not so simple. No neat rules exist because the thermodynamics of gasifier operation are not well understood. Yet, nontrivial thermodynamic principles dictate the temperature, air supply, and other operating variables of the reactors that we build. It is a tribute to the persistence of experimentalists that so much progress has been made in the face of so little understanding. Nevertheless, it has been the experience in related fields (such as oil, gas, and coal combustion) that once the mechanisms at work are understood, the engineer is able to develop cleaner, more efficient processes. Fortunately, much of the knowledge acquired in these fields can be applied to enhance our understanding of gasification processes.

In this chapter, we present a summary of the underlying processes that occur during biomass gasification. We will attempt to keep the explanation simple because each fundamental process is basically simple. Chapter 5 gives a more extensive description of the operation of specific gasifiers. Details are available from the literature for those interested in a more thorough explanation (Reed 1982; Kaupp 1984a; Reed 1985b).

4.2 Biomass Thermal Conversion Processes

4.2.1 Introduction

Thermal conversion processes for biomass involve some or all of the following processes:

Pyrolysis: Biomass + Heat → Charcoal, oil, gas

Gasification: Biomass + Limited oxygen → Fuel gas

Combustion: Biomass + Stoichiometric* oxygen → Hot combustion products

Thermal processes typically have high throughputs and can, in principle, operate on any biomass form. (Biological processes only operate on some of the components of biomass, usually the cellulose.)

Cellulose is a linear polymer of anhydroglucose units; **hemicellulose** is a mixture of polymers of 5- and 6-carbon anhydrosugars, and **lignin** is an irregular polymer of phenyl propane units. In biomass, these three polymers form an interpenetrating system, or block copolymer, that varies in composition across the cell wall. Nevertheless, in large samples, there is a relatively constant atomic ratio of $CH_{1.4}O_{0.6}$. (The ratios will vary slightly with species. Coal is typically about $CH_{0.9}O_{0.1}$ but varies more widely in composition.) The relationship between solid, liquid, and gaseous fuels is easily seen in Fig. 4-1(a) where the relative atomic concentrations of carbon, hydrogen, and oxygen are plotted for a variety of fuels. Here it is seen that the solid fuels, biomass, coal and charcoal, lie in the lower left segment of the diagram; liquid and gaseous hydrocarbon fuels lie in the upper left section; CO and H_2 are joined by the bisector of the triangle; and the combustion products of fuels, CO_2 and H_2O, lie on a vertical line on the right.

Thermal conversion processes for biomass are indicated by the arrows of Fig. 4-1(b). Here it is seen that the conversion processes move the chemical composition of biomass to liquid or solid fuel regions, either by biological or thermal means. In some cases (such as oxygen/air gasification), the processes are spontaneous; in other cases (such as steam gasification) considerable energy must be expended to cause the change.

4.2.2 Biomass Pyrolysis

Pyrolysis is the breaking down (lysis) of a material by heat (pyro). It is the first step in the combustion or gasification of biomass. When biomass is heated in the absence of air to about 350°C (pyrolysis), it forms charcoal (chemical symbol: C), gases (CO, CO_2, H_2, H_2O, CH_4), and tar vapors (with an approximate atomic makeup of $CH_{1.2}O_{0.5}$). The tar vapors are gases at the temperature of pyrolysis but condense to form a smoke composed of fine tar droplets as they cool.

All the processes involved in pyrolysis, gasification, and combustion can be seen in the flaming match of Fig. 4-2. The flame provides heat for pyrolysis, and the resulting gases and vapors burn in the luminous flame in a process called flaming combustion. After the flame passes a given point, the char may or may not continue to burn (some matches are chemically treated to prevent the charcoal from smouldering). When the match is extinguished, the remaining wood continues to undergo residual pyrolysis, generating a visible smoke composed of the condensed tar droplets.

*"stoichiometric," that quantity required for a complete chemical reaction

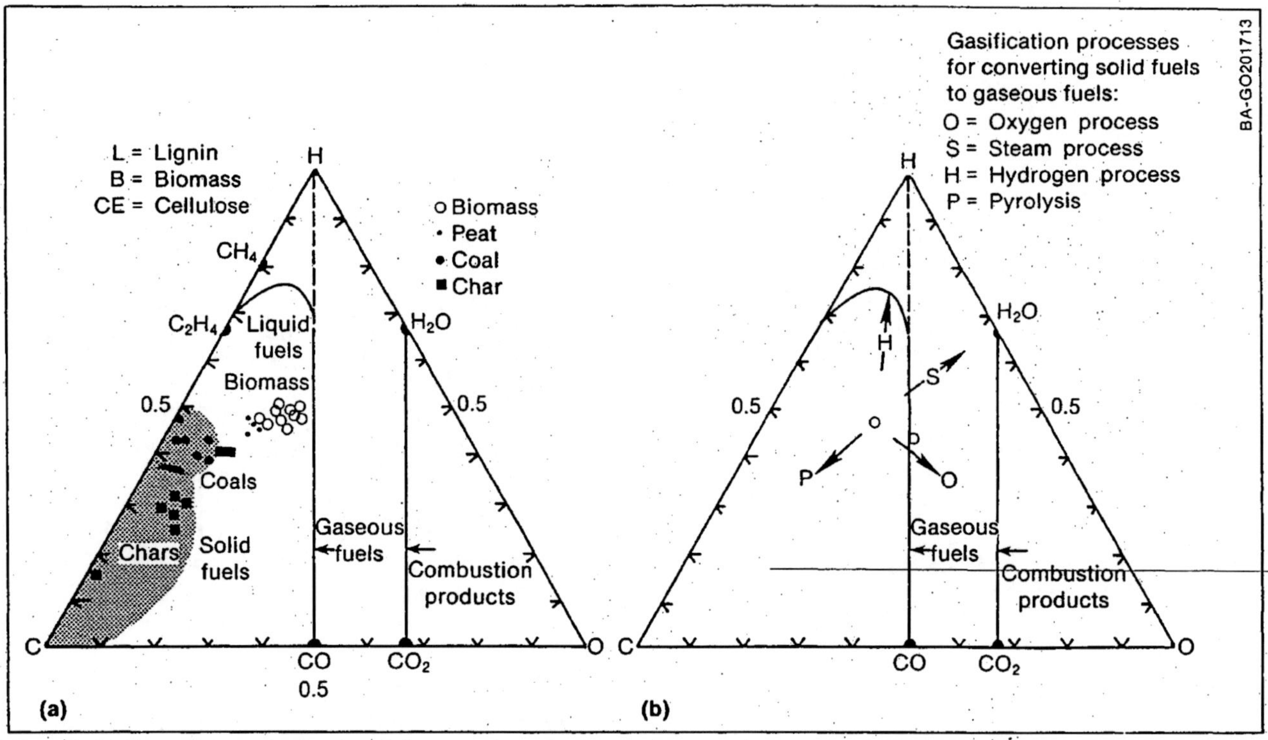

Fig. 4-1. (a) Phase diagram showing the relative proportions of carbon, hydrogen, and oxygen in solid, liquid, and gaseous fuels (b) Chemical changes during biomass conversion processes (Source: Reed 1981)

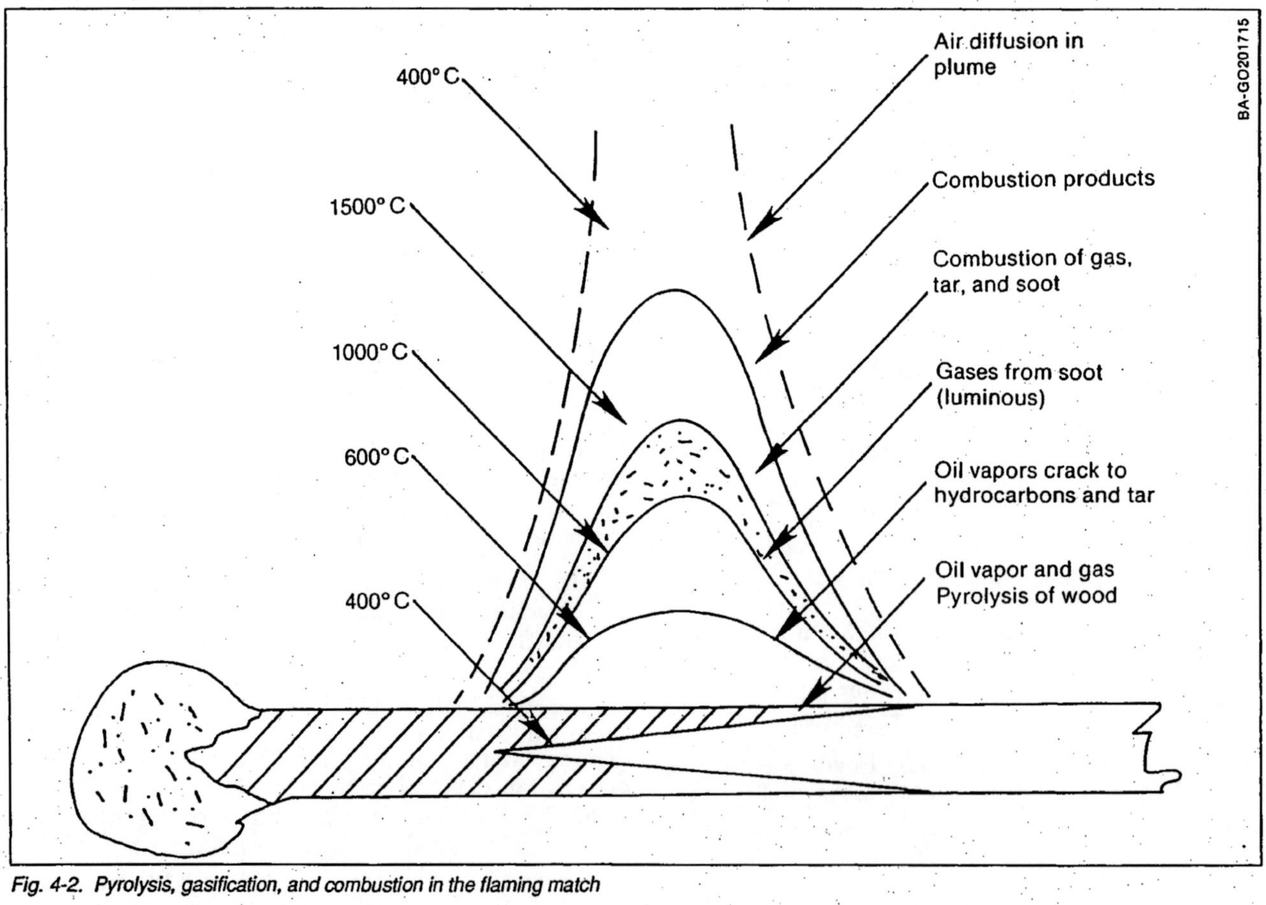

Fig. 4-2. Pyrolysis, gasification, and combustion in the flaming match

A more quantitative picture of pyrolysis is obtained through thermogravimetric analysis (TGA). In this technique, a small piece of biomass is suspended on a balance pan in a furnace, and the temperature is increased with time at a known rate. An example of the residual weight change experienced by a small sample of flax shives heated at a rate of 40°C/min is shown in Fig. 4-3. One sees that moisture is released first, at 100°C, followed by the volatile materials at 250°-450°C; these temperatures are important in understanding pyrolysis, gasification, and combustion. According to the figure, a fraction of char and ash remains in the end. If air is allowed to enter the system after pyrolysis, the carbon (char) will burn, leaving the ash as the final product. Each form of biomass produces slightly different quantities of char, volatile material, and ash. Knowledge of these quantities, as well as the temperature dependencies of the reaction and associated weight losses, are useful in understanding gasifier operation and design.

The results shown in Fig. 4-3 are qualitatively similar to those obtained in a proximate analysis of most biomass but are not identical because heating rates are higher and samples are smaller in TGA (see Chapter 3 and Table 3-3). The curve of Fig. 4-3 represents pure pyrolysis in an inert gas (such as nitrogen or argon). If pyrolysis occurs in air, the curve drops more steeply within the region from 250°-400°C because the char and products are oxidized also. As the char burns, it eventually reaches the ash line between 400° and 500°C.

In Fig. 4-3, more than 80% of the total dry mass of the sample is volatilized below 500°C, leaving an additional 10% to 20% of the original mass of carbon for conversion to gas. It is now recognized that the volatile matter is composed of monomers (as well as other fragments) of the cellulose, hemicellulose, and lignin polymer that make up biomass (Evans 1984). It is also recognized that up to 65% of the biomass dry weight can be converted to this water-soluble "wood oil," which potentially may form the basis of new processes for wood liquefaction (Roy 1983; Scott 1983; Diebold 1984). Unfortunately, these oils are corrosive and highly oxygenated, so that further processing will be required to make a high-grade liquid fuel (Diebold 1986). However, they have been burned successfully in industrial boilers and turbines with only minor modifications required for the burners (Bowen 1978; Jasas 1982).

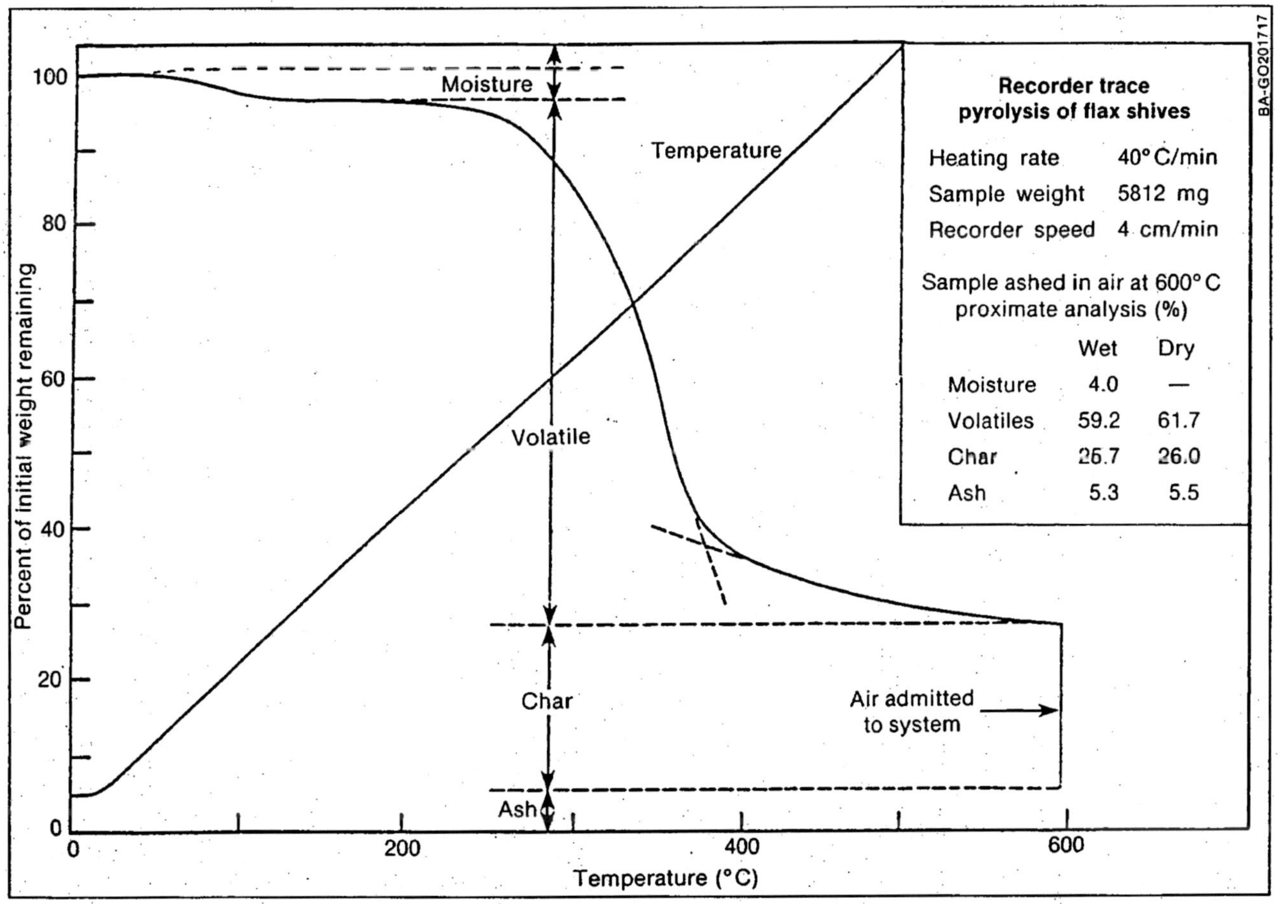

Fig. 4-3. Thermogravimetric analysis of a typical biomass sample heated in the absence of air (Source: Reed 1981, Fig. 5-2)

4.2.3 Combustion of Biomass

Biomass **combustion** is more complex than either pyrolysis or gasification since the biomass must first pyrolyze, then be partially combusted (gasified) before it is fully combusted.

However, the overall global reaction of biomass combustion can be represented by

$$CH_{1.4}O_{0.6} + 1.05\ O_2 + (3.95\ N_2) \rightarrow CO_2 + 0.7\ H_2O + (3.95\ N_2) \quad (4\text{-}1)$$

where $CH_{1.4}O_{0.6}$ is an average formula for typical biomass. (Actual composition for specific biomass is shown in Tables 3-3, 3-4, 3-6, and 3-7). The nitrogen is shown in parentheses because it is an inert portion of air and does not take part in the reaction. For oxygen combustion of biomass it would be omitted.

This combustion produces 20.9 kJ/g (8990 Btu/lb) when the temperature of the combustion products is low enough for all the liquid to be water, and this is the value that would be measured in a **bomb calorimeter** and reported as the high heat of combustion or HHV as shown in Tables 3-6 and 4-1. In most practical combustion devices, the water escapes to the atmosphere as a gas, and the heat of vaporization of the water is not recovered. In this case, the low heating value, LHV, 20.4 kJ/g (8770 Btu/lb), would be the maximum heat that could be generated. The difference between LHV and HHV is small for dry wood but increases rapidly with moisture content of the wood. (In the United States the HHV is normally used for rating the efficiency of stoves; in Europe the LHV is used. As a result, European wood stoves are typically quoted as 10% more efficient than comparable U.S. wood stoves.)

Table 4-1. Thermal Properties of Typical Biomass

Typical dry biomass formula: (moisture- and ash-free [MAF] basis) $CH_{1.4}O_{0.6}$	C	H	O
Composition (weight %)	52.2	4.3	41.7
Composition (mole %)	33.3	46.7	20.0
High Heating Value[a]	20.9 kJ/g	(8990 Btu/lb)	
Low Heating Value	20.4 kJ/g	(8770 Btu/lb)	

[a]The high heating value (HHV) is the value that is usually measured in the laboratory and would be obtained during combustion if liquid water was allowed to condense out as a liquid. The low heating value (LHV) is obtained when water is produced as a vapor. The high heating value of typical biomass fuels will be decreased in proportion to the water and ash content, according to the relation:

$$LHV(Net) = HHV(MAF)/(1 + M + A)$$

where M is the fraction of moisture (wet basis), A is the fraction of ash, and MAF designates the moisture- and ash-free basis. The air/biomass ratio required for total combustion is 6.27 kg/kg (lb/lb).

The LHV can be related to the HHV and an analysis of the combustion products as:

$$HHV = LHV + F_m h_w$$

where F_m is the weight fraction of moisture produced in the combustion gases, and h_w is the heat of vaporization of water, 2283 J/g (980 Btu/lb).

Source: Modified from data in Reed 1981.

4.2.4 Chemistry of Biomass Gasification

The change in composition produced by air or oxygen gasification is shown in Fig. 4-1(b). Ideally one would like to add the smallest amount of oxygen possible to carry the solid composition to the composition O in Fig. 4-1(b), a mixture of CO and H_2, according to the formula

$$CH_{1.4}O_{0.6} + 0.2\ O_2 \rightarrow CO + 0.7\ H_2 \quad (4\text{-}2)$$

Unfortunately, there is more energy contained in the CO and H_2 than is contained in the biomass, so that this reaction would require the transfer of energy from some external source, which would greatly complicate the process.

In practice, some excess oxygen must then be added for gasification (carrying the reaction to point O in Fig. 4-1(b)), producing some CO_2 and H_2O according to

$$CH_{1.4}O_{0.6} + 0.4\ O_2 \rightarrow 0.7\ CO + 0.3\ CO_2 + 0.6\ H_2 + 0.1\ H_2O \quad (4\text{-}3)$$

Typically a few percent of methane are formed as well. Typical properties of producer gas from biomass are shown in Table 4-2.

Table 4-2. Typical Properties of Producer Gas from Biomass[a]

Compound	Symbol	Gas (vol.%)	Dry Gas (vol.%)
Carbon monoxide	CO	21.0	22.1
Carbon dioxide	CO_2	9.7	10.2
Hydrogen	H_2	14.5	15.2
Water (v)	H_2O	4.8	—
Methane	CH_4	1.6	1.7
Nitrogen	N_2	48.4	50.8
Gas High Heating Value:			
Generator gas (wet basis)[b]	5506 kJ/Nm^3 (135.4 Btu/scf)		
Generator gas (dry basis)[b]	5800 kJ/Nm^3 (142.5 Btu/scf)		
Air Ratio Required for Gasification:	2.38 kg wood/kg air (lb/lb)		
Air Ratio Required for Gas Combustion:	1.15 kg wood/kg air (lb/lb)		

[a]These values are based on ash- and moisture-free biomass with the composition given in Table 4-1. The wet-gas composition is the most important property of the gas for mass and energy balances, but the dry-gas composition is usually reported because of the difficulty in measuring moisture. The heating value of the gas is usually calculated from the gas composition, using a value of 13,400 kJ/Nm^3 (330 Btu/scf) for H_2 and CO, and 41,900 kJ/Nm^3 (1030 Btu/scf) for methane.

[b]These are typical values for downdraft air gasifiers, but they can vary between 4880 and 7320 kJ/Nm^3 (120-180 Btu/scf), depending on variables such as gasifier heat loss, biomass moisture content, and char removal at the grate.

Source: Modified from data in Reed 1981.

The ratio CO/CO_2 (or H_2/H_2O) is a measure of the producer gas quality. Approximately 30% of the biomass is burned to provide the energy for gasification of the rest. The exact amount of excess oxygen required depends on the efficiency of the process. It can be improved in practice with insulation, by drying, or by preheating the reactants. A fascinating question in gasification is how the reacting products "know" how much oxygen to use (see below).

4.2.5 Thermodynamics of Gasification

Thermodynamics is the bookkeeping of energy. Although thermodynamics cannot always predict what will happen for a particular process, it can rule out many things that cannot happen. It was mentioned above that Eq. (4-2) is thermodynamically impossible in the absence of added heat and that Eq. (4-3) actually governs the reaction. How is this determined?

At the high temperature where gasification takes place (typically 700°-1000°C), there are only a few stable combinations of the principal elements of biomass—carbon, hydrogen, and oxygen. These are C, CO, CO_2, CH_4, H_2, and H_2O. The relative concentration of these species that will be reached at equilibrium can be predicted from the pressure, the amount of each element, and the equilibrium constant determined from the thermodynamic properties and temperature, subject to an energy balance. It is then possible to determine the species that would form at equilibrium as a function of the amount of oxygen added to the system. The results of calculations of this type are shown in Figs. 4-4 and 3-5.

The **adiabatic reaction temperature** of biomass with air or oxygen, determined in this manner, is shown in Fig. 4-4(a). This is the temperature that would be reached if biomass came to equilibrium with the specified amount of air or oxygen. (There is no guarantee that equilibrium will be reached in any given gasifier, but downdraft gasifiers approach equilibrium quite closely - see below.)

The oxygen used in a process determines the products and temperature of the reaction. The oxygen consumed is typically plotted as the **equivalence ratio,** ϕ - the oxygen used relative to that required for complete combustion. (Complete oxidation of biomass with oxygen requires a weight ratio of 1.476 [mass of oxygen/mass of biomass]; with air, a ratio of 6.36.) A very low or zero oxygen use is indicative of pyrolysis, shown at the left of the figure; a ϕ of about 0.25 is typical of the gasification region at the middle; and combustion is indicated by a $\phi \geq 1$ at the right.

The composition of the gas produced is shown in Fig. 4-4(b). The amount of energy remaining in the char and converted from solid to gas is shown in Fig. 4-4(c). The low heating value of the gas is shown in Fig. 4-4(d). From these figures it is seen that at an equivalence ratio ϕ of about 0.25 all of the char is converted to gas, and the fraction of energy in the wood converted to gas reaches a maximum. With less oxygen, some of the char is not converted; with more oxygen, some of the gas is burned and the temperature rises very rapidly as shown in Fig. 4-4(a). Thus, it is desirable to operate as close to an equivalence ratio of 0.25 as possible.

How is it possible to operate exactly at this ratio of 0.25? In a fixed bed gasifier, operation at lower values of ϕ would cause charcoal to be produced (as shown for low ϕ in Fig. 4-4(c)), and it would build up in the reactor unless it is augered or shaken out. Operation at values of ϕ above 0.25 consumes charcoal and the temperature goes up rapidly. Hence, maintaining the bed at a constant level automatically ensures the correct oxygen input.

4.3 Indirect and Direct Gasification Processes

4.3.1 Indirect (Pyrolitic) Gasification

It is now recognized that wood-oil vapor is unstable at temperatures above 600°C and cracks rapidly at 700° to 800°C to form hydrocarbon gases (such as methane, ethane, and ethylene), H_2, CO, and CO_2. In addition, one obtains a 1% to 5% yield of a tar composed of polynuclear aromatics and phenols similar to those found in coal tar (Antal 1979; Diebold 1984; Diebold 1985).

Pyrolytic gasification is accomplished when a portion of the fuel or char is burned in an external vessel with air, and the resulting heat is used to supply the energy necessary to pyrolyze the biomass. The principal advantage of this process is that a medium-energy gas is produced without using oxygen. The higher energy content may be required for long-distance pipeline delivery. The disadvantage is that a significant fraction of tar may be produced, and indirect heat or mass transfer is required, which complicates the apparatus and the process. Pyrolytic gasification will not be discussed further because it is only practical in large installations and is not as well-developed as direct gasification with oxygen or air.

4.3.2 Direct Gasification

Pyrolysis and gasification processes are endothermic, so heat must be supplied in order for the processes to occur. In fact, the heat required to accomplish pyrolysis and raise the products to 600°C is about 1.6-2.2 kJ/g (700-800 Btu/lb), representing 6% to 10% of the heat of combustion of the dry biomass (Reed 1984). This heat is supplied directly by partially combusting the volatile tars in downdraft gasifiers; in updraft gasifiers, it comes from the sensible heat of the gases resulting from charcoal gasification. This combustion then dilutes the product gas with CO_2 and H_2O, the products of combustion with oxygen. If the combustion is accomplished with air, the gas is also diluted with about 50% nitrogen from the air.

The principal advantages of direct gasification are that the one-stage process is very simple, the direct heat transfer from the gases to the biomass is very efficient, and the process is largely self-regulating. If air is used, the resulting gas is diluted with atmospheric nitrogen to a producer gas value of 5800-7700 kJ/Nm^3 (150-200 Btu/scf). When oxygen is used for gasification, a medium-energy gas containing 11,500 kJ/Nm^3 (300 Btu/scf) is obtained (Reed 1982). Medium-energy gas can be distributed economically for short distances

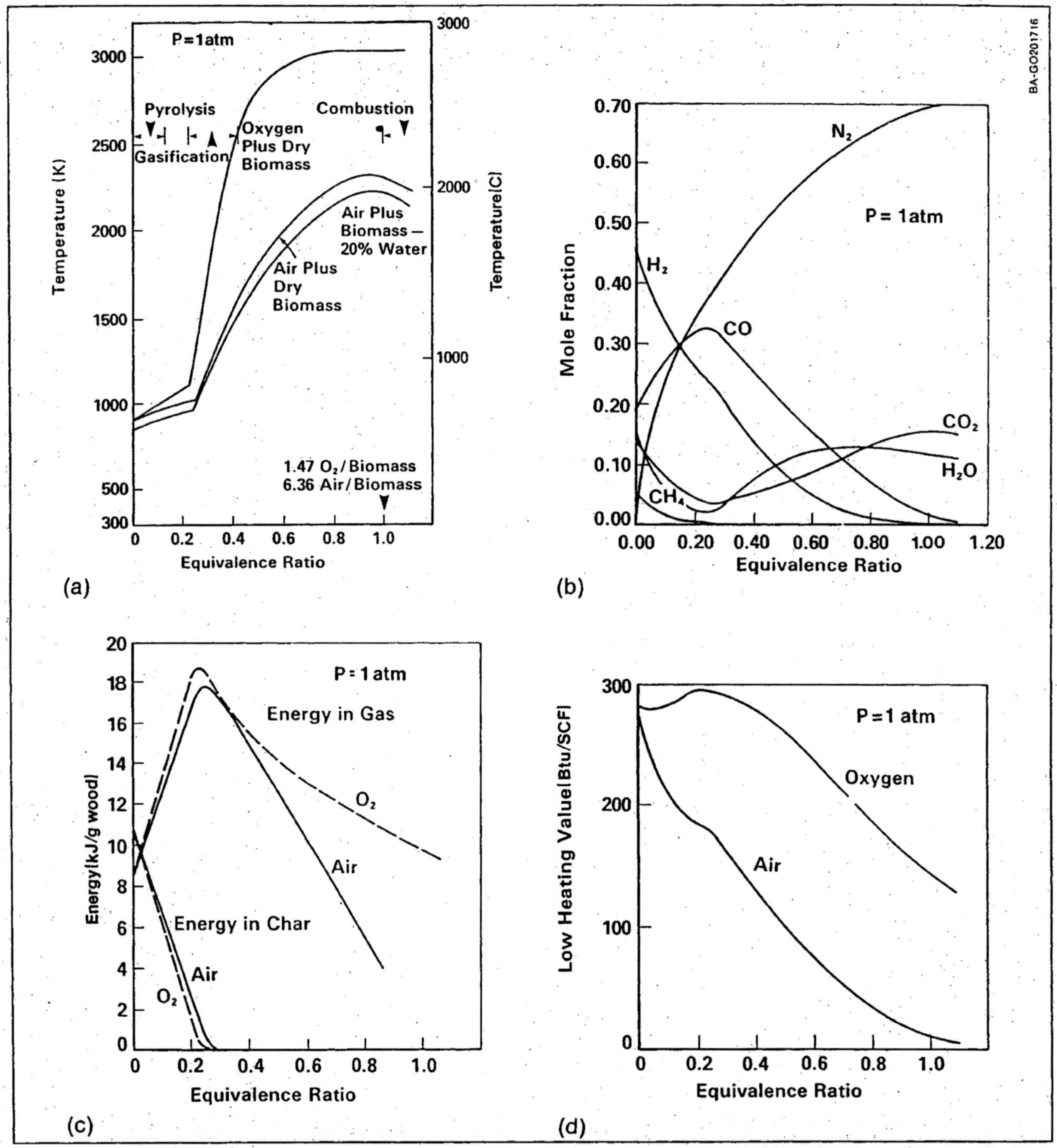

Fig. 4-4. (a) Abiabatic reaction temperature for biomass of atomic composition $CH_{1.4}O_{0.6}$ reacting with oxygen and air, plotted against the equivalence ratio, ϕ, the ratio of oxygen to that required for complete combustion (b) Equilibrium gas composition for reaction with air (c) Energy in solid and gas (d) Energy per volume of gas (Source: Reed 1981, Figs. S-2 - S-5)

(up to one mile) in pipelines. It is also called synthesis gas, since it can be used as a feedstock for the chemical synthesis of methanol, ammonia, methane, and gasoline. The oxygen must be either purchased or produced on-site, making it economically prudent only in larger installations. It has been reported that pipeline distribution of low-energy gas is also economically practical for distances up to one mile if the air used for gasification is compressed, rather than compressing the larger volume of producer gas (McGowan 1984).

There are many types of direct gasifiers, each with its special virtues and defects. They will be discussed in Chapter 5.

4.4 Principles of Operation of Direct Gasifiers

4.4.1 Introduction

Since volatile organic molecules make up approximately 80% of the products from biomass pyrolysis (Diebold 1985b), the principal task in biomass (but not coal) gasification is to convert this condensible volatile matter to permanent gases. A secondary task is to convert the resulting charcoal also to gas.

The most important types of fixed-bed gasifiers for this task are the **updraft** and **downdraft** gasifiers of Fig. 4-5. These gasifiers will be discussed in greater detail in Chapter 5, but a brief introduction here will facilitate understanding of the fundamental principles involved.

The terms "updraft gasifier" and "downdraft gasifier" may seem like trivial mechanical descriptions of gas flow patterns. In practice, however, updraft biomass gasifiers can tolerate high moisture feeds and thus have some advantages for producing gas for combustion in a burner. However, **updraft** gasifiers produce 5% to 20% volatile tar-oils and so are unsuitable for operation of engines. Downdraft gasifiers produce typically less than 1% tar-oils and so are used widely for engine operation. The reasons for this difference are given below.

4.4.2 Operation of the Updraft Gasifier

The updraft gasifier is shown schematically in Fig. 4-5(a). Biomass enters through an air seal (lock hopper) at the top and travels downward into a rising stream of hot gas. In the pyrolysis section, the hot gas pyrolyzes the biomass to tar-oil, charcoal, and some gases. In the reduction zone the charcoal thus formed reacts with rising CO_2 and H_2O to make CO and H_2. Finally, below the reduction zone incoming air burns the charcoal to produce CO_2 and heat (Desrosiers 1982; Reed 1985b). Note that the combustion to CO_2 is exothermic, and the heat produced in the gas here is absorbed in the endothermic reduction and pyrolysis reactions above.

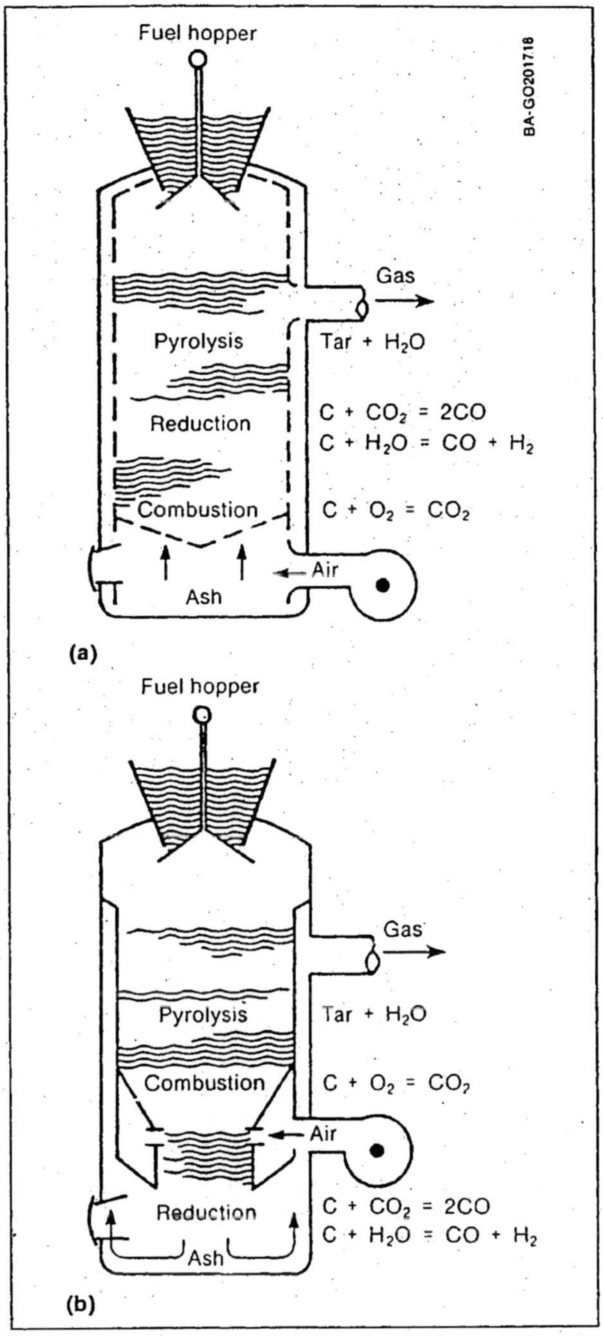

Fig. 4-5. Schematic diagram of (a) updraft and (b) downdraft gasifier showing reactions occurring in each zone (Source: Reed 1981, Figs. 8-6, 8-7)

Depending upon the pyrolysis conditions in a gasifier, one can generate a wide range of vapors (wood oil and wood tar) in the hot gas. If the pyrolysis products are to be burned immediately for heat in a boiler or for drying (close-coupled operation), then the presence of condensible vapors in the gas is of little importance. In fact, the condensible tars represent a high-energy fuel and greatly enhance the energy obtained from each unit volume of biomass.

If the volatile materials are condensed, they produce tars and oils known commonly as creosote. These materials collect in the chimneys of airtight wood stoves, the piping of gasifiers, and the valves of engines. Most of the companies advertising and selling updraft gasifiers at a 1979 conference no longer produce them (Reed 1979).

If the gas is to be conveyed over a distance in a pipeline, burned in any form of engine, or used as a chemical feedstock, the condensing tars will plug pipes sometimes in only a few minutes. In these cases, it is necessary to use a mode of gasification that succeeds in converting the tars to gas. This can be accomplished either by cracking (secondary pyrolysis) or by partial oxidation in flaming pyrolysis.

4.4.3 Operation of the Downdraft Gasifier

Downdraft gasifiers have been very successful for operating engines because of the low tar content. Most of the work reported in this book was performed on downdraft systems, and they will be the principal gasifier discussed in the balance of this book.

In the **downdraft** gasifier of Fig. 4-5(b), air contacts the pyrolyzing biomass before it contacts the char and supports a flame similar to the flame that is generated by the match in Fig. 4-2. As in the case of the match, the heat from the burning volatiles maintains the pyrolysis. When this phenomenon occurs within a gasifier, the limited air supply in the gasifier is rapidly consumed, so that the flame gets richer as pyrolysis proceeds. At the end of the pyrolysis zone, the gases consist mostly of about equal parts of CO_2, H_2O, CO, and H_2. We call this flame in a limited air supply "flaming pyrolysis," thus distinguishing it from open wood flames with unlimited access to air (Reed 1983a). Flaming pyrolysis produces most of the combustible gases generated during downdraft gasification and simultaneously consumes 99% of the tars. It is the principal mechanism for gas generation in downdraft gasifiers.

If the formula for biomass oil is taken as approximately $CH_{1.2}O_{0.5}$, then partial combustion of these vapors can be represented approximately by the reaction:

$$CH_{1.2}O_{0.5} + 0.6\ O_2 \rightarrow 0.5\ CO + 0.5\ CO_2 + 0.4\ H_2 + 0.2\ H_2O \qquad (4\text{-}4)$$

(The exact O_2-to-vapor-ratio will depend on the exact vapor composition and gasifier conditions.) Downdraft gasifiers usually produce vapors that are less than 1% condensible oil/tar, the reason behind the almost exclusive use of downdraft gasifiers as an energy source for operating engines.

Although flaming pyrolysis is a new concept in explaining biomass gasification, **partial oxidation** of small and large hydrocarbon molecules to CO and H_2 is a standard industrial process. Texaco has used an oxygen gasifier to oxidize hydrocarbons to CO and H_2, as in the following reaction for a typical oil:

$$C_{10}H_{22} + 5\ O_2 \rightarrow 10\ CO + 11\ H_2 \qquad (4\text{-}5)$$

The resulting gas, called synthesis gas, can be used to manufacture methanol, hydrogen, or ammonia. There is some interest in using the Texaco system to gasify biomass (Stevenson 1982).

4.4.4 Factors Controlling Stability of Gasifier Operation

Gasifer operating temperature is a function of the amount of oxygen fed to the gasifier (Fig. 4-4(a)). The temperature response, however, changes abruptly at an equivalence ratio (ER) of approximately 0.25. This change point, or knee, occurs for temperatures of 600° to 800°C (900-1100 K), depending on oxygen source. Gasifier pyrolysis produces oils and tars that are stable for periods of 1 second or more at temperatures below 600°C. Since updraft gasifiers operate below an ER of 0.25 (temperatures less than 600°C), considerable quantities of tars are emitted with the product gas.

In the gasifier of Fig. 4-5(b), air is injected at the interface between the incoming biomass and the char. If too much char is produced, the air consumes the excess char rather than biomass; if the char is consumed too fast, more biomass is consumed. Thus, the Imbert gasifier is self regulating. At SERI we have built the oxygen gasifier shown in Fig. 5-12. We operate this with a fixed flow of oxygen and add biomass faster or slower to maintain a fixed bed level. In the Buck Rogers gasifier of Fig. 5-11, a fraction of air is introduced through the rotating nozzles and maintains the zone at that level (Walawender 1985).

Some gasifiers operate at lower values of ϕ on purpose by augering charcoal out of the char zone in order to produce charcoal—a valuable byproduct—and to yield the higher gas heating value shown at low ϕ in Fig. 4-4(d). Such operation is not a true gasification but might be called "gas/charification." In entrained or fluidized bed operation, the ratio of biomass to oxygen can be varied independently. In this case ϕ must be set, typically by fixing oxidant flow and varying fuel flow to maintain a constant temperature.

4.5 Charcoal Gasification

The manufacture of charcoal for use as a synthetic fuel dates back at least 10,000 years and is closely associated with the development of our civilization. Today, charcoal is used as the prime source of heat for cooking in less developed countries and also is used for the reduction of many ores in smelting processes.

The charcoal yield from a biomass feedstock is highly dependent on the rate of heating and the size of the biomass particles. Industrial charcoal manufacture uses very slow heating rates to achieve charcoal yields of more than 30% of the initial dry weight of the biomass. The intermediate heating rates used in proximate analysis usually produce charcoal yields of 15% to 20%. The very rapid heating rates encountered when small biomass particles are gasified and combusted realize charcoal yields of less than 15% of the initial dry weight of the biomass; larger size feedstocks produce 15% to 25% charcoal.

During updraft or downdraft gasification, 10% to 20% of the biomass will remain as charcoal after pyrolysis is complete. In an updraft gasifier, air entering at the grate initially burns this char to liberate heat and CO_2 according to the reaction:

$$C + O_2 \rightarrow CO_2 + \text{heat} \qquad (4\text{-}6)$$

Almost immediately, or even simultaneously, the CO_2 and any H_2O present in the gasifier react with the char to produce the fuel gases CO and H_2 according to the following reactions:

$$C + CO_2 \rightarrow 2\,CO \qquad (4\text{-}7)$$

$$C + H_2O \rightarrow CO + H_2 \qquad (4\text{-}8)$$

The first reaction is called the Boudouard reaction, and the second is called the water-gas reaction. They have been studied extensively for the last 100 years in connection with coal and biomass gasification, since the principal product of coal pyrolysis is coke (carbon). The rate of the reaction has been studied by measuring the rate of disappearance of carbon, coal, or charcoal while passing H_2O or CO_2 over the solid (Nandi 1985; Edrich 1985).

Both of these reactions require heat (i.e., they are **endothermic** reactions) and therefore cool the gas about 25°C for every 1% of CO_2 that reacts. These reactions occur very rapidly at temperatures over 900°C, and their cooling effect helps to keep the gas temperature from rising above this temperature. Below 800°C, the reactions become sluggish and very little product forms. We have modeled the reactions of downdraft char gasification using known kinetic values and find that the temperatures measured in char gasification correspond to those observed in the gasifier (Reed 1983a; Reed 1984). We refer to the process observed in an actual bed of char as **adiabatic** (no heat input) **char gasification.**

The CO and H_2 formed in the hot char zone can react below 900°C to form methane according to the reaction:

$$CO + 3\,H_2 \rightarrow CH_4 + H_2O \qquad (4\text{-}9)$$

This reaction proceeds slowly unless there is a catalyst present; however, it is quite exothermic and can supply heat if suitably catalyzed.

Concurrent with the emergence of biomass as an important energy source, it was natural that coal gasification interpretations would be carried over to explain biomass gasification. Even today, most articles on biomass gasification use only Eqs. (4-7) and (4-8) to explain biomass gasification and ignore Eq. (4-4), even though Eq. (4-4) applies to the 80% biomass volatiles. Biomass pyrolysis produces only 10% to 20% charcoal, and the charcoal is very reactive. Therefore, this cannot be the primary explanation for the conversion of biomass to gas.

4.6 Summary

In summary, the task of a gasifier is threefold:

- to pyrolyze biomass to produce volatile matter, gas, and carbon
- to convert the volatile matter to the permanent gases, CO, H_2, and CH_4
- to convert the carbon to CO and H_2.

These tasks are accomplished by partial oxidation or pyrolysis in various types of gasifiers.

Chapter 5
Gasifier Designs

5.1 Introduction

Many different designs of gasifiers have been built and are described in the extensive literature on this subject (see especially Gengas 1950; Skov 1974; Foley 1983; Kjellstrom 1983, 1985; Kaupp 1984a; NAS 1983). Much of this material has been collected by A. Kaupp of the University of California at Davis. (Copies of these papers are also at SERI and the German Appropriate Technology Exchange [GATE] in Eschborn, West Germany.) Anyone interested in design modification and improvement would be well-advised to become acquainted with this material before repeating tried and tested techniques. However, many of the documented design variations are minor.

We believe that future improvements to gasifiers will be based on a better understanding of the basic processes, combined with improved measurements of gasifier behavior and better regulation of fuel properties. Work is under way at various private and public centers to increase our understanding of the gasification process. Consequently, gasifier design is in a state of flux. This makes it difficult to organize a "handbook of gasifier design" without having it out of date before the ink is dry.

To avoid this problem, we will first describe the construction and operation of a number of historical gasifiers described in the literature to aid in understanding various tradeoffs still under development. The reader must remember that the choice of gasifier is dictated both by the fuels that will be used and the use to which the gas will be put. We will then describe some gasifiers currently under development.

5.2 Basic Gasifier Types

Fixed bed (sometimes called **moving bed**) gasifiers use a bed of solid fuel particles through which air and gas pass either up or down. They are the simplest type of gasifiers and are the only ones suitable for small-scale application.

The **downdraft gasifier** (Figs. 4-5(b), 5-1, and 5-2) was developed to convert high volatile fuels (wood, biomass) to low tar gas and therefore has proven to be the most successful design for power generation. We concern ourselves primarily with several forms of downdraft gasifiers in this chapter.

The **updraft gasifier** (Figs. 4-5(a), 5-3, and 5-4) is widely used for coal gasification and nonvolatile fuels such as charcoal. However, the high rate of tar production

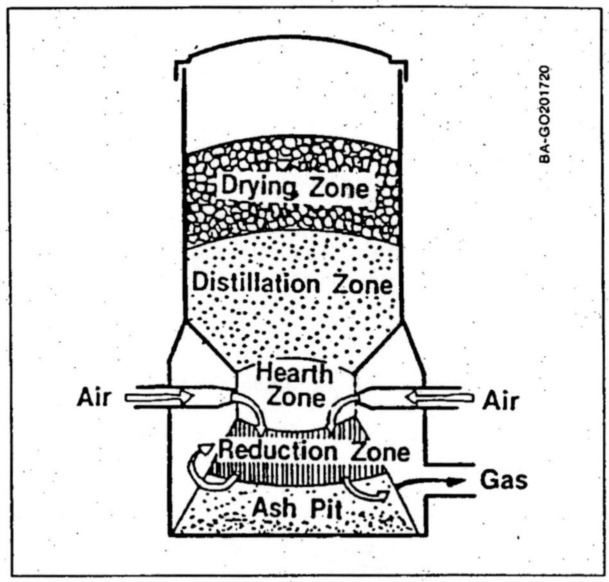

Fig. 5-1. Diagram of downdraft gasification (Source: Skov 1974, Fig. 14. © 1974. Used with permission of Biomass Energy Foundation, Inc.)

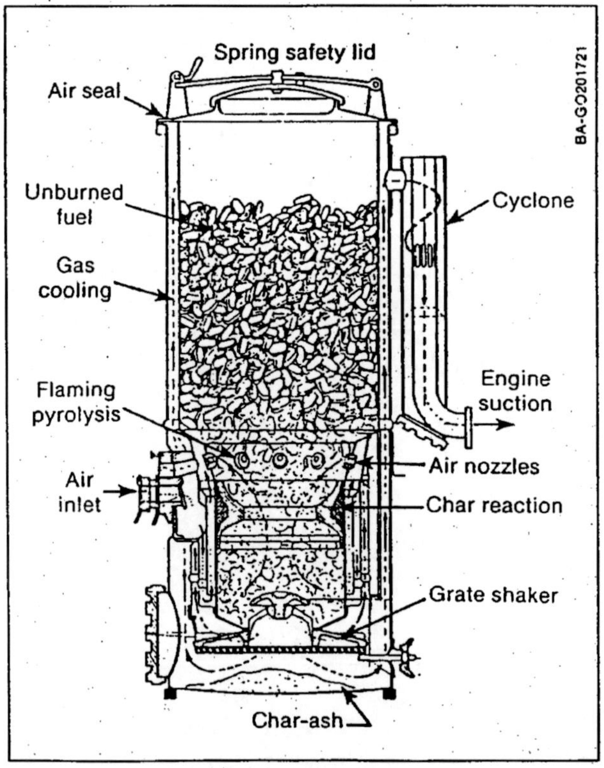

Fig. 5-2. Imbert (nozzle and constricted hearth) gasifier (Source: Gengas 1950, Fig. 75)

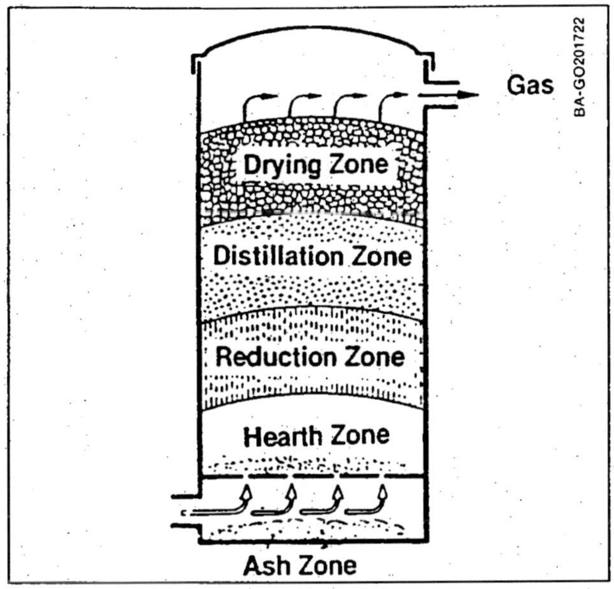

Fig. 5-3. Diagram of updraft gasification (Source: Skov 1974 Fig. 9. © 1974. Used with permission of Biomass Energy Foundation, Inc.)

(5%-20%) (Desrosiers 1982) makes them impractical for high volatile fuels where a clean gas is required.

Fluidized beds are favored by many designers for gasifiers producing more than 40 GJ(th)/h* [40 MBtu(th)/h] and for gasifiers using smaller particle feedstock sizes. In a fluidized bed, air rises through a grate at high enough velocity to levitate the particles above the grate, thus forming a "fluidized bed." Above the bed itself the vessel increases in diameter, lowering the gas velocity and causing particles to recirculate within the bed itself. The recirculation results in high heat and mass transfer between particle and gas stream.

Suspended particle gasifiers move a suspension of biomass particles through a hot furnace, causing pyrolysis, combustion, and reduction to give producer gas. Neither fluidized bed nor suspended particle gasifiers have been developed for small-scale engine use.

We have already mentioned that gasifier designs will differ for different feedstocks, and special gasifiers have been developed to handle specific forms of biomass feedstocks, such as municipal solid wastes (MSW) and rice hulls.

The manner in which ash is removed determines whether the gasifier is classified as either a dry ash (ash is removed as a powder) or slagging (ash is removed as a molten slag) gasifier. Slagging updraft gasifiers for biomass and coal have been operated at only a very large scale.

*The units J(th) and Btu(th) refer to the thermal or chemical energy produced. This can be converted to electricity with an efficiency of 10% to 40%, so the electrical energy content (J or Btu) will be proportionally lower.

5.3 Charcoal Gasifiers

Updraft charcoal gasifiers were the first to be developed for vehicle operation. They are suitable only for low-tar fuels such as charcoal and coke. Figure 5-4 shows an updraft charcoal gasifier that was used in the early part of World War II. Air enters the updraft gasifier from below the grate and flows upward through the bed to produce a combustible gas (Kaupp 1984a). High temperatures at the air inlet can easily cause slagging or destruction of the grate, and often some steam or CO_2 is added to the inlet air to moderate the grate temperature. Charcoal updraft gasifiers are characterized by comparatively long starting times and poor response because of the large thermal mass of the hearth and fuel zone.

Charcoal manufacture is relatively simple and is carried on in most countries. However, it requires tight controls on manufacturing conditions to produce a charcoal low in volatile content that is suitable for use in charcoal gasifiers.

5.4 Charcoal versus Biomass Fuels

High-grade charcoal is an attractive fuel for gasifiers because producer gas from charcoal, which contains very little tar and condensate, is the simplest gas to clean. Charcoal gasifiers were restricted over much of Europe during the later years of World War II because charcoal

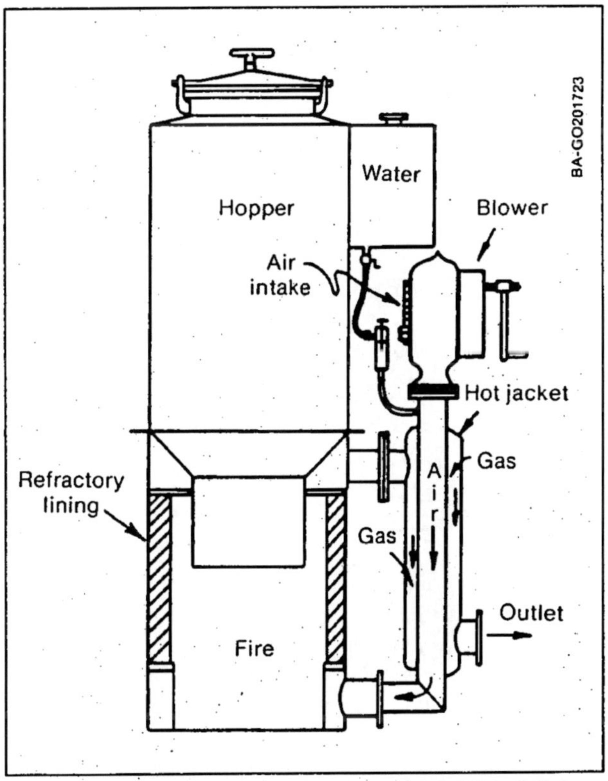

Fig. 5-4. Updraft coke and charcoal gasifier, early World War II (Source: Kaupp 1984a, Fig. 27)

manufacture wastes half of the energy in the wood (Gengas 1950). On the other hand, Australia worked almost exclusively with charcoal during this period because of that country's large forest acreage and small number of vehicles.

Nevertheless, the simplicity of charcoal gasification has attracted many investigators, and more than 2000 charcoal systems have been manufactured in the Philippines. A large number are not currently working (Kadyszewski 1986).

5.5 The Crossdraft Gasifier

The crossdraft gasifier shown in Fig. 5-5 is the simplest and lightest gasifier. Air enters at high velocity through a single nozzle, induces substantial circulation, and flows across the bed of fuel and char. This produces very high temperatures in a very small volume and results in production of a low-tar gas, permitting rapid adjustment to engine load changes. The fuel and ash serve as insulation for the walls of the gasifier, permitting mild-steel construction for all parts except the nozzles and grates, which may require refractory alloys or some cooling. Air-cooled or water-cooled nozzles are often required. The high temperatures reached require a low-ash fuel to prevent slagging (Kaupp 1984a).

The crossdraft gasifier is generally considered suitable only for low-tar fuels. Some success has been observed with unpyrolyzed biomass, but the nozzle-to-grate spacing is critical (Das 1986). Unscreened fuels that do not feed into the gasifier freely are prone to bridging and channeling, and the collapse of bridges fills the hearth zone with unpyrolyzed biomass, leading to momentarily high rates of tar production. The fuel size also is very important for proper operation. Crossdraft gasifiers have the fastest response time and the smallest thermal mass of any gas producers because there is a minimum inventory of hot charcoal. In one design, a downdraft gasifier could be operated in a crossdraft scheme during startup in order to minimize the startup time (Kaupp 1984a).

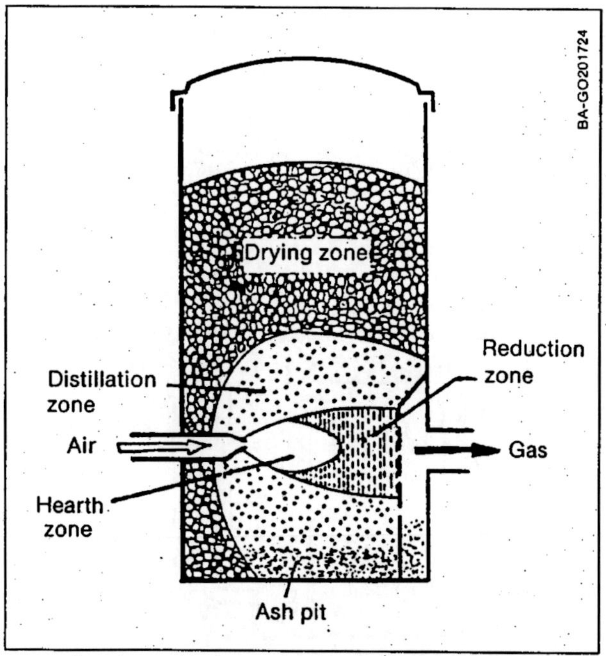

Fig. 5-5. Diagram of crossdraft gasification (Source: Skov 1974, Fig. 18. © 1974. Used with permission of Biomass Energy Foundation,

5.6 The Updraft Gasifier

The updraft gasifier has been the principal gasifier used for coal for 150 years, and there are dozens in operation around the world. In fact, World War II-type Lurgi gasifiers now produce a large share of the gasoline used in South Africa by oxygen gasification followed by Fischer-Tropsch catalytic conversion of the gas to gasoline.

The geometry of the updraft gasifier is shown in Figs. 4-5(a), 5-3, and 5-4. During operation, biomass is fed into the top while air and steam are fed through a grate, which often is covered with ash. The grate is at the base of the gasifier, and the air and steam react there with charcoal from the biomass to produce very hot CO_2 and H_2O. In turn, the CO_2 and H_2O react endothermically with the char to form CO and H_2 according to Eqs. (4-6) through (4-8). The temperatures at the grate must be limited by adding either steam or recycled exhaust gas to prevent damage to the grate and slagging from the high temperatures generated when carbon reacts with the air.

The ascending, hot, reducing gases pyrolyze the incoming biomass and cool down in the process. Usually, 5% to 20% of the tars and oils are produced at temperatures too low for significant cracking and are carried out in the gas stream (Desrosiers 1982). The remaining heat dries the incoming wet biomass, so that almost none of the energy is lost as sensible heat in the gas.

The updraft gasifier throughput is limited to about 10 GJ/h-m^2 (10^6 Btu/h-ft^2) either by bed stability or by incipient fluidization, slagging, and overheating. Large updraft gasifiers are sometimes operated in the slagging mode, in which all the ash is melted on a hearth. This is particularly useful for high-ash fuels such as MSW; both the Purox and Andco Torax processes operate in the slagging mode (Masuda 1980; Davidson 1978). Slagging updraft gasifiers have both a slow response time and a long startup period because of the large thermal mass involved.

5.7 The Imbert Downdraft Gasifier

5.7.1 Introduction

The nozzle (tuyere) and constricted hearth downdraft gasifier shown in Figs. 4-5(b), 5-4, and 5-5 is sometimes called the "Imbert" gasifier (after its entreprenurial inventor, Jacques Imbert) although it was produced by dozens of companies under other names during World War II. Approximately one million of these gasifiers were mass produced during World War II, at a cost of about $1000 U.S. (1983) each. It is important to realize that the cost of producing such a unit today would depend primarily on the degree to which it could be mass produced since none of the components are inherently expensive.

Air gasifiers can be operated either by forcing air through the fuel (pressurized) or by drawing the air through the fuel (suction). In practice, gasifiers that fuel engines generally use the suction of the engine to move air through the gasifier and cleanup train, and these are called "suction gasifiers." We will describe only suction gasifiers here; however, only minor modifications are required to build pressurized gasifiers. (See Chapter 8, which deals with the topics of blowers, fans, ejectors, and compressors).

A large number of descriptive articles on gasifiers appeared during World War II, but no detailed drawings have been located from that period. Fortunately, formulas for determining critical dimensions are given in a number of the older references (Gengas 1950; Schläpfer 1937).

Renewed interest in biomass gasification has manifested itself in the fact that a number of individuals and groups have built modern versions of the Imbert gasifier. Plans and manuals for constructing some of these designs are available from several groups (Mother 1982; Skov 1974; Nunnikhoven 1984; Rissler 1984). Some of these gasifiers have been attached to cars and trucks that have succeeded in traversing the country on several occasions. In particular, *Mother Earth News* and its subsidiary, *Experimental Vehicle News*, have performed extensive tests on gasifiers and have published informative articles and plans with photographs of fabrication steps. The plans are sufficiently detailed so that a skilled welder can fabricate a gasifier for a relatively small expense.

In 1978, a number of tests were performed under a SERI contract on a 75-hp "Hesselman" (Imbert-type) downdraft gasifier. This gasifier was built in Sweden at the end of World War II and was imported to this country by Professor Bailie of the University of West Virginia. Professor Bailie used the gasifier in tests during which the gasifier operated on wood, wood pellets, and oxygen (Bailie 1979). Subsequently, the gasifier was sent to SERI in Colorado for further testing with a 15-kW Onan electric generator. More recently, the gasifier has been used to gasify peat by Professor Goldhammer of Lowell University. The gasifier is now being used by Syngas Systems, Inc., to generate producer gas to test gas cleanup systems for use with its 750-kW power generator. Although much of the testing was qualitative in nature, the authors have had considerable experience in running this interesting technological antique.

5.7.2 Description of the Downdraft (Imbert) Gasifier

Referring to Figs. 5-1 and 5-2, the upper cylindrical part of the inner chamber is simply a magazine for the wood chips or other biomass fuel. During operation, this chamber is filled every few hours as required. The spring-loaded cover is opened to charge the gasifier, and then it is closed during gasifier operation. The spring permits the cover to pop open to relieve pressure in the case of a gas explosion, thus functioning as a safety valve.

About one-third of the way up from the bottom, there is a set of radially directed air nozzles that permit air to be drawn into the chips as they move down to be gasified. Typically, there are an odd number of nozzles so that the hot gases from one nozzle do not impinge on the opposite nozzle. The nozzles are attached to a distribution manifold that in turn is attached to the outer surface of the inner can. This manifold is connected through the outer can to a large air-entry port. One air nozzle is in line with this port, allowing the operator to ignite the charcoal bed through this nozzle.

During operation, the incoming air burns and pyrolyzes some of the wood, most of the tars and oils, and some of the charcoal that fills the gasifier below the nozzles. Most of the mass of biomass is converted to gas within this flaming combustion zone since biomass contains more than 80% volatile matter (Reed 1983a).

The gasifier is in many ways self-adjusting. If there is insufficient charcoal at the air nozzles, more wood is burned and pyrolyzed to make more charcoal. If too much char forms during high-load conditions, then the char level rises above the nozzles so that incoming air burns the char to reduce the char level. Thus, the reaction zone is maintained at the nozzles.

Below the air nozzle zone lies the gas-reduction zone, usually consisting of a classical Imbert hearth (Fig. 5-2) or in later years, of the "V" hearth (Fig. 5-6). Most recently, the flat-plate hearth constriction (Fig. 5-7) has been introduced. The latter two hearth designs accumulate a layer of retained ash to form a high-quality, self-repairing insulation.

Improved insulation in the hearth results in lower tar production and a higher efficiency over a wider range of operating conditions.

After the combustion/pyrolysis of wood and hot char at the nozzle level (see below), the resulting hot combustion gases (CO_2 and H_2O) pass into this hot char where they are partially reduced to the fuel gases CO and H_2 according to Eqs. (4-7) and (4-8). This procedure

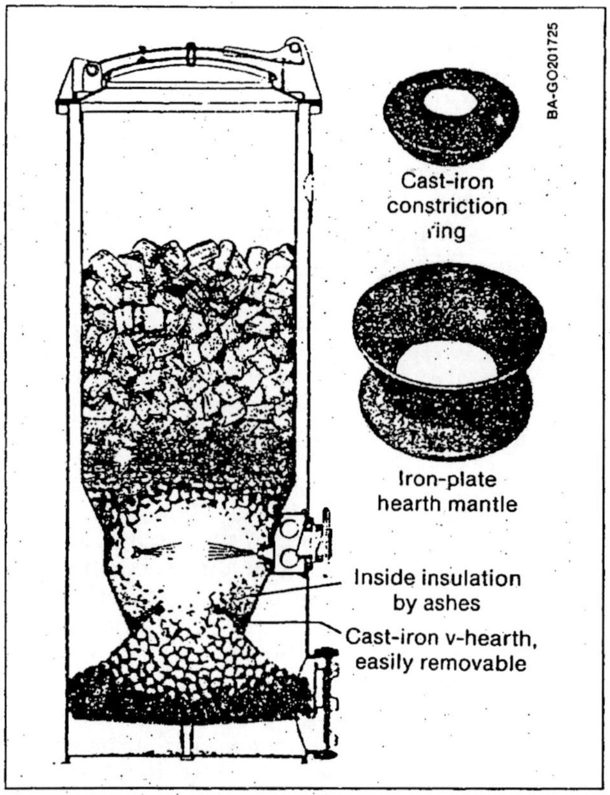

Fig. 5-6. V-hearth Imbert gasifier (Source: Gengas 1950, Fig. 74)

results in a marked cooling of the gas, as sensible-gas heat is converted into chemical energy. This removes most of the charcoal and improves the quality of the gas. Eventually, the charcoal is "dissolved" by these gases and disintegrates to smaller chunks and a fine powder that either is swept out with the gases to the cyclone separator or falls through the grate. Tars that have escaped combustion at the nozzle may crack further in the hot char although tar cracking is now thought to occur only above about 850°C (Kaupp 1984b; Diebold 1985).

The spaces between the nozzles (shown in Fig. 5-8) allow some unpyrolyzed biomass to pass through. The hearth constriction then causes all gases to pass through the hot zone at the constriction, thus giving maximum mixing and minimum heat loss. The highest temperatures are reached in this section so the hearth constriction should be replaceable. If tarry gas is produced from this type of gasifier, common practice is to reduce the hearth constriction area until a low-tar gas is produced. However, one should remember that hearth dimensions also play a role in the gas production rate (see below).

The fine char-ash dust can eventually clog the charcoal bed and will reduce the gas flow unless the dust is removed. The charcoal is supported by a movable grate that can be shaken at intervals. Ash builds up below the grate and can be removed during cleaning operations. Usually, wood contains less than 1% ash. However, as the charcoal is consumed, it eventually collapses to form a powdered char-ash that may represent 2% to 10% of the total biomass, in turn containing 10% to 50% ash. Ash contents depend on the char content of the wood and the degree of agitation. The greater the degree of char reduction, the smaller the resulting particles and the higher the ash, as shown in Fig. 3-3. The downdraft gasifier startup and response time is intermediate between the fast crossdraft gasifier and the slow updraft gasifier.

The Imbert gasifier requires a low-moisture (<20% moisture) and uniformly blocky fuel in order to allow easy gravity feeding through the constricted hearth. Twigs, sticks, and bark shreds must be completely removed. The reduction in area at the hearth and the protruding nozzles present hazards at which the passage of the fuel can be restricted, thus causing bridging and channeling followed by high tar output, as unpyrolyzed biomass falls into the reaction zone. The

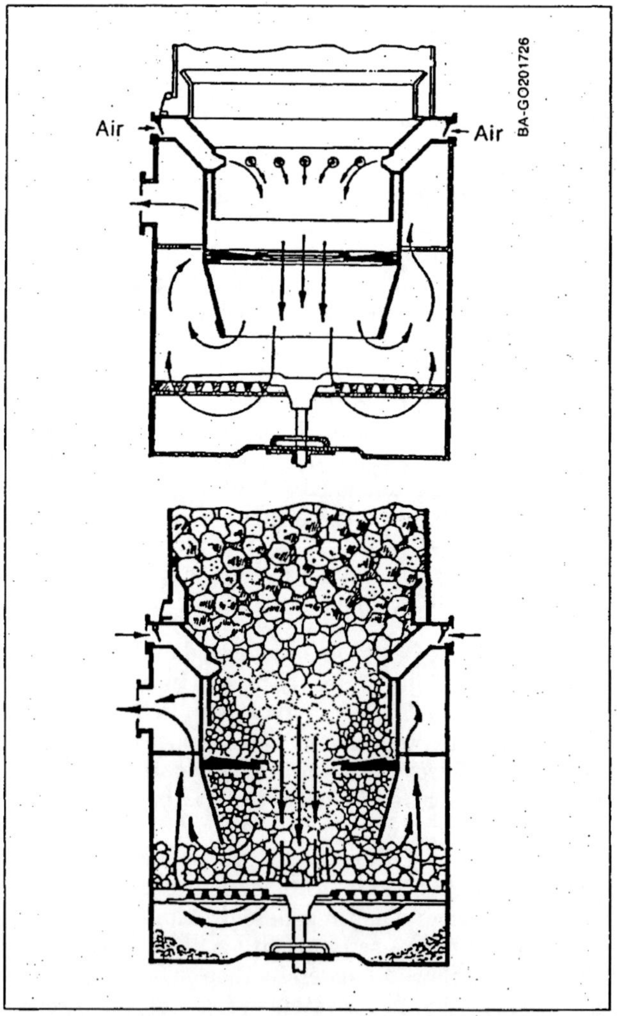

Fig. 5-7. Flat-plate hearth constriction (Source: Gengas 1950, Fig. 76)

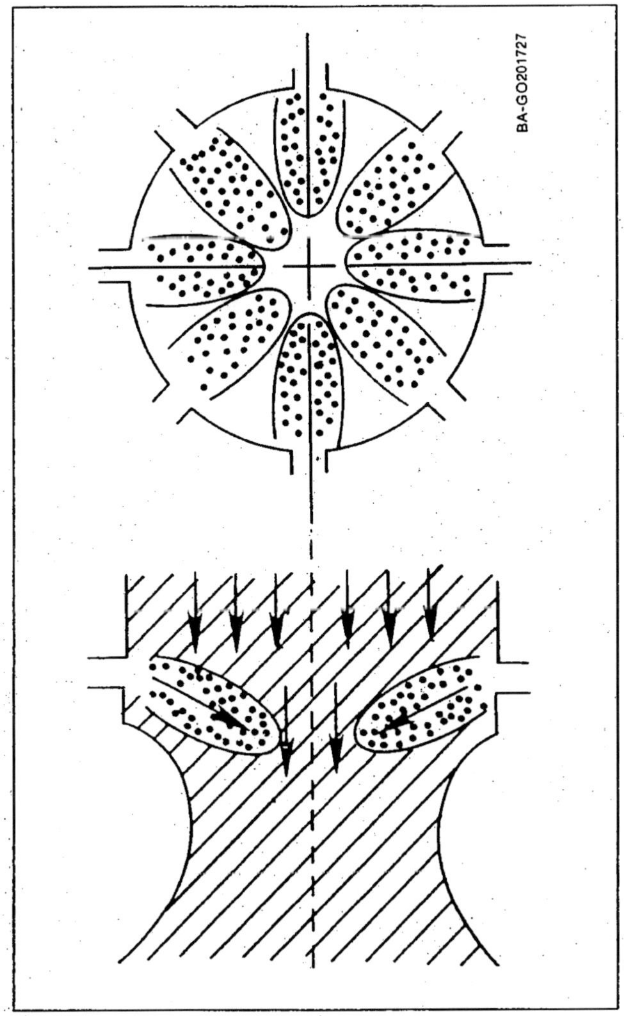

Fig. 5-8. High temperature zone of a downdraft gas producer with wall tuyeres (Source: Kaupp 1984a, Fig. 55)

vehicle units of the World War II era had ample vibration to jar the carefully sized wood blocks through. In fact, an entire industry emerged for preparing car wood at that time (Gengas 1950).

5.7.3 Superficial Velocity, Hearth Load, and Gasifier Sizing

An important factor used in choosing dimensions of any gasifier is the "superficial velocity, V_s," of the gas calculated where it passes through the narrowest part of the gasification zone. Although the units of V_s are length/time (e.g., m/s), one should think of the superficial velocity as gas production expressed in terms of gas volume/cross-sectional area-time (m^3/m^2-s), a specific gas production rate. It is called a **superficial** velocity since **actual** velocities will be three to six times higher due to the presence of the charcoal and the high temperatures existing at the throat. A closely related term is the maximum hearth load, B_h, expressed in gas volume/hearth area-h, expressed in practical units. This term enables one to compare the performance of a wide variety of gasifiers on a common basis. The maximum specific hearth loads for a number of gasifiers are shown in Table 5-1. The table was calculated from data available on gasifiers that have been thoroughly tested and lists the maximum superficial velocity and heating load reported. Note that in European literature, hearth load is reported in gas volume units; in the United States, it is reported in energy units.

In *Generator Gas* (Gengas 1950) a maximum hearth load (B_{hmax}) value for an Imbert-style gasifier is about 0.9 $Nm^3/h\text{-}cm^2$. In other words, 0.9 m^3 of gas is produced for each square centimeter of cross-sectional area at the constriction. This corresponds to a superficial gas velocity V_s of 2.5 m/s (8.2 ft/s) calculated at NTP* from the throat diameter and ignoring the presence of fuel. This corresponds to a specific gas production rate of 9000 m^3 of gas per square meter of cross-sectional area per hour (29,500 scf/ft^2-h). If the gas has a (typical) energy content of 6.1 MJ/Nm^3 (150 Btu/scf), this results in a specific energy rate of 54.8 GJ/m^2-h (4.4 MBtu/ft^2-h). The diameter of the pyrolysis zone at the air nozzles is typically about twice that at the throat, and Table 5-1 shows the hearth load on this basis also. This puts the hearth load for the Imbert type gasifier on a comparable basis to the stratified downdraft gasifier. Knowledge of maximum hearth load permits one to calculate the size of hearth needed for various engine or burner sizes. Dimensions for a variety of Imbert-type gasifiers are shown in Tables 5-2 and 5-3.

The maximum hearth load is limited by many factors, such as the mechanical integrity of the char bed structure within the gasifier, degree of agitation, and the time available for conversion. High velocities can disturb the char and fuel bed, causing instability. If char fragments become dislodged and airborne, they may plug the bed or form channels. Therefore, a little agitation can effectively increase the maximum specific hearth load.

The heating value of producer gas varies with flow rate, as shown in Fig. 7-20. Notice that the maximum efficiency for rice hulls occurs at twice the flow rate that produces the maximum heating value from rice hulls. This occurs because the combination of lower temperatures and low flow rate favors methane and tar production. Although the change in efficiency is small, the benefit of reducing tar production is substantial.

Closely related to hearth area (A_h) is the cross-sectional area of the air nozzles (tuyeres) (A_m). Early workers

---

*NTP refers to the European practice of correcting gas volume measurements to a "normal temperature and pressure" of 0°C and 1 atmosphere. In the United States it is conventional to correct measured volumes to STP, "standard temperature and pressure," 77°F (or 25°C) and 1 atmosphere.

Table 5-1. Maximum Reported Superficial Velocity and Hearth Load of Various Gasifiers[a]

Gasifier	Type[b]	D[c] Diameter m	ft	V_s[d] Superficial Vel m/s	ft/s	B_h[e] Hearth Load m³/cm²-h	MBtu/ft²-h	Reference
Imbert[c]	I-A	0.15	0.5	2.50	8.2	0.90	4.76	(Gengas 1950)
	I-A	0.30	1.0	0.63	2.1	0.23	1.19	(Gengas 1950)
Biomass Corp.[c]	I-A	0.30	1.0	0.95	3.1	0.34	1.81	(Graham 1983)
	I-A	0.61	2.0	0.24	0.8	0.09	0.45	(Graham 1983)
SERI Air/ox	S-A	0.15	0.5	0.28	0.9	0.10	0.53	(Reed 1982)
	S-O	0.15	0.5	0.24	0.8	0.09	0.90	(Reed 1982)
Buck Rogers	S-A	0.61	2.0	0.13	0.4	0.05	0.25	(Walawender 1985)
	S-A	0.61	2.0	0.23	0.7	0.08	0.43	(Chern 1985)
Syn-Gas, Inc.	S-A	0.76	2.5	1.71	5.6	0.62	3.26	(Graboski 1985)
	S-O	0.76	2.5	1.07	3.5	0.39	4.07	(Graboski 1985)

[a]Data in this table are based on reports on well-tested gasifiers, rather than manufacturers' claims, etc.

[b]I is the Imbert (nozzle and constricted throat) gasifier of WW II. S is the Stratified Downdraft Gasifier. A signifies operation on air, O on oxygen.

[c]For Imbert type gasifiers (I), the diameter is measured at the throat (upper value) or at the air entry level (lower value). For Stratified Downdraft Gasifiers (S), the diameter is constant at all levels.

[d]The superficial velocity is calculated as the volume of gas (taken at room temperature) passing through the area without regard to the presence of fuel. It has units of vol/area-time = length/time = velocity.

[e]The hearth load, B_h, is a practical measure of gasifier gas volume throughput (SI units) or energy throughput (English units).

observed an optimum relationship between the hearth and nozzle areas. For instance, maximum power was obtained from 130-mm hearths that had five 12-mm nozzles. Any variation of either the nozzle or hearth ring from these dimensions caused a power reduction. Table 5-2 shows successful nozzle sizes for wood-fueled Imbert gas producers and the wider variation for nozzles used in successful Imbert and SGB gasifiers. (SGB units were used for 2-cycle pulsating flow engines.)

A larger hearth diameter requires either a higher nozzle velocity or some other means to penetrate the deeper fuel bed. This leads to a higher pressure drop for larger hearths, placing an upper size limit on nozzle-fed downdraft gasifiers when gas flow is provided by engine vacuum. If the cross-sectional area of the nozzles is too small, there will be an excessive pressure drop in forming the air jets; if the cross-sectional area is too large, the air jets will have too low a velocity and the air will not penetrate the bed. The velocity of the air blast is shown in Table 5-2.

5.7.4 Turndown Ratio

Another important concept in sizing gasifiers is the "turndown ratio," the ratio of the highest practical gas generation rate to the lowest practical rate. The turndown ratio of World War II gasifiers varied between 3 for Imbert-style gasifiers with uninsulated V-hearth gasifiers and 18 for highly insulated V-hearth gasifiers. Vehicle operation requires turndown ratios of at least 8:1, making the need for insulation and proper sizing in high-turndown applications apparent. Although engineers often oversize equipment, this can be fatal in gasifier design. Heat losses tend to be independent of throughput and at low loads become disproportionately high. A low specific hearth load may also cause tar formation problems. A high turndown ratio is less important for electric generators and irrigation pumps that constantly operate at full capacity.

In summary, the Imbert gasifier design has survived the test of time and mass production. It is relatively inexpensive, uses simple materials of construction, is easy to fabricate, and can be operated by motorists with a minimum of training. It supplies low-tar gas from highly volatile fuels with a high turndown ratio.

5.7.5 Disadvantages of the Imbert Design

Although the Imbert gasifier has been the prototype downdraft gasifier, it has a number of disadvantages. The hearth constriction seriously limits the range of biomass fuel shapes that can be successfully gasified without expensive cubing or pelletizing pretreatment. (The stratified-bed gasifiers currently under development at SERI and other facilities and discussed in Section 5.8 are free of constrictions and promise to broaden the range of fuels that can be gasified.) The Imbert gasifier requires a high-grade, usually hardwood, fuel, generally at least 2 cm along the smallest dimension with no more than 20% moisture. During World War II, stringent specifications were maintained on fuel production, which was carried out at a number of licensed factories.

The Imbert design cannot be scaled-up to larger sizes because the air enters at the sides and is incapable of penetrating a large-diameter fuel bed unless the fuel size is increased proportionally. The tar level, while

Table 5-2. Imbert Nozzle and Hearth Diameters

d_r/d_h	d_h mm	d_r mm	$d_{r'}$ mm	h mm	H mm	R mm	A No.	d_m mm	$\frac{A_m \times 100}{A_h}$	$\frac{d_r}{d_h}$	$\frac{h}{d_h}$	Range of Gas Output max. Nm^3/h	Range of Gas Output min. Nm^3/h	Maximum Wood Consumption kg/h	Air Blast Velocity Vm m/s
268/60	60	268	150	80	256	100	5	7.5	7.8	4.5	1.33	30	4	14	22.4
268/80	80	268	176	95	256	100	5	9.0	6.4	3.3	1.19	44	5	21	23.0
268/100	100	268	202	100	256	100	5	10.5	5.5	2.7	1.00	63	8	30	24.2
268/120	120	268	216	110	256	100	5	12.0	5.0	2.2	0.92	90	12	42	26.0
300/100	100	300	208	100	275	115	5	10.5	5.5	3.0	1.00	77	10	36	29.4
300/115	115	300	228	105	275	115	5	11.5	5.0	2.6	0.92	95	12	45	30.3
300/130	130	300	248	110	275	115	5	12.5	4.6	2.3	0.85	115	15	55	31.5
300/150	150	300	258	120	275	115	5	14.0	4.4	2.0	0.80	140	18	67	30.0
400/130	130	400	258	110	370	155	7	10.5	4.6	3.1	0.85	120	17	57	32.6
400/150	135	400	258	120	370	155	7	12.0	4.5	2.7	0.80	150	21	71	32.6
400/175	175	400	308	130	370	155	7	13.5	4.2	2.3	0.74	190	26	90	31.4
400/200	200	400	318	145	370	153	7	16.0	3.9	2.0	0.73	230	33	110	31.2

Variables not given in figure are defined as follows:

d_m = inner diameter of the tuyere.

A_m = sum of cross sectional areas of the air jet openings in the tuyeres.

A_h = cross sectional area of the throat.

A = number of tuyeres.

Source: Kaupp 1984a, Table 5; Fig. 75.

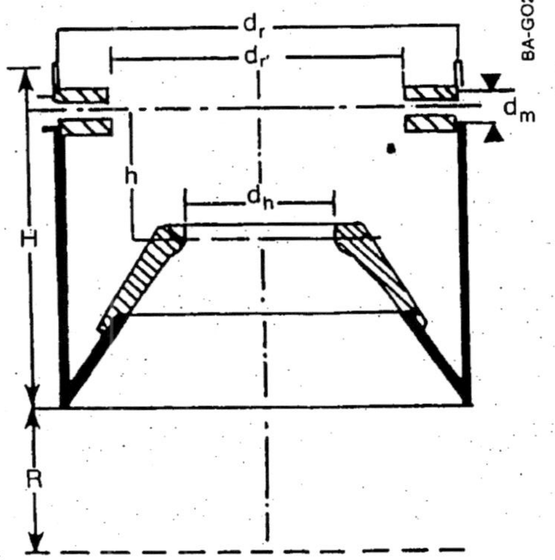

Table 5-3. Sizing Data

Engine			Maximum Generator Gas Needed at 2300 rpm, L/s	Power with Gasoline Operation, hp
Number of Cylinders	Cylinder Dimensions, mm	Cylinder Volume, L		
4	110 x 136	5.17	50	80
6	110 x 136	7.75	75	130
8	110 x 136	10.34	100	180

Note: At a heavy load, 170 mm cross section should be used instead of 150 mm cross section.

Source: Adapted from Gengas 1950, Table 32.

low (usually 5000 ppm), is still high enough to require extensive scrubbing and disposal procedures.

Groeneveld has studied the recycle of gases at the nozzle and developed improved understanding of the tar combustion and improved mixing methods shown in Fig. 5-9 to permit scaleup (Groeneveld 1980a,b). Unfortunately, there is no overall theory of operation for Imbert gasifiers that would permit sizing the gasifier for fuels other than hardwood blocks. The geometry and flow of fuel and air are quite complex, making any attempts to model the gasifier very difficult tasks indeed. (More information is provided in later discussions.)

Some efforts to scale the Imbert gasifier to larger sizes have realized a disastrous increase in tar production (Goss 1979; Graham 1983). However, researchers have met with more success when the fuel size has been increased with the gasifier size. Billets that were 8 cm in diameter and 15 cm long have operated well in large Imbert-style gasifiers used for heating applications (Makray 1984).

5.8 The Stratified Downdraft Gasifier

5.8.1 Introduction

A new type of gasifier, which we have named the "stratified downdraft gasifier," (also called "open-top" or "topless" gasifier) has been developed during the last few years through cooperative efforts among researchers at SERI (Reed 1982; 1983a,b; 1984), the University of California in Davis (Kaupp 1984a), the Open University in London (Reines 1983), the Buck Rogers Co. (Walawender 1985; Chern 1985) in Kansas, and in Florida (LaFontaine 1984). It is also related to the Chinese rice hull gasifier (Kaupp 1984b; Cruz 1984). The stratified downdraft gasifier overcomes many of the difficulties of the Imbert gasifier and may ultimately be the basis for improved gasifier designs. However, it has not been widely commercialized at this point; the reader must balance the proven reliability of the gasifiers discussed above against the promises of the stratified downdraft gasifier.

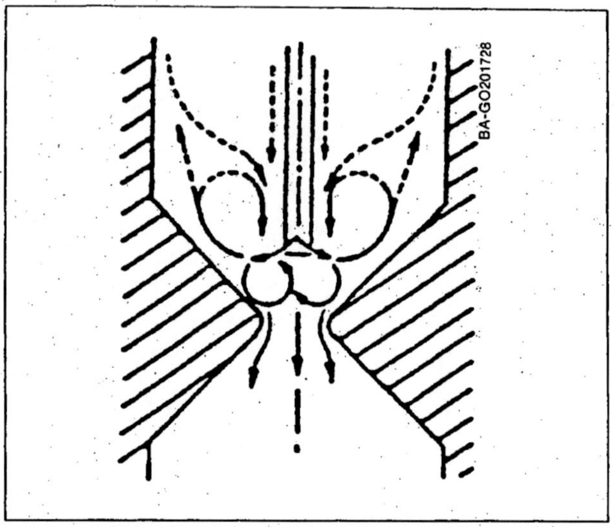

Fig. 5-9. Downdraft center nozzle gas producer (Source: Groeneveld 1980a)

5.8.2 Description of the Stratified Downdraft Gasifier

The stratified downdraft gasifier consists of a cylindrical vessel with a hearth on the bottom as shown in Figs. 5-10 to 5-12. During operation of the stratified downdraft gasifier, air and biomass pass uniformly downward through four zones, hence the name "stratified." The open top ensures uniform access of air or oxygen to the flaming pyrolysis zone, as opposed to the Imbert gasifier. The uppermost layer is composed of **unreacted biomass** fuel through which air enters. In the second layer, biomass reacts with air in **flaming pyrolysis**. The third layer, which is made up of **char** from the second layer, reduces the pyrolysis gases. **Inert char**, which constitutes the fourth layer, normally is too cool to cause further reactions. However, since the fourth layer is available to absorb heat or oxygen if conditions change, it serves both as a buffer and as a charcoal storage zone. The temperatures and chemical compositions in each zone are shown schematically in Fig. 5-10.

The top zone of the stratified downdraft gasifier may be adjusted to any depth during air operation and serves the same function as the fuel magazine in the Imbert gasifier. Fuel is added through the open top of the gasifier and should be replenished before the advancing pyrolysis front consumes all of the available fuel.

During oxygen operation, the advancing pyrolysis front moves much faster and is stabilized at the top of the second zone so there is no first zone of fuel storage. Biomass must then be fed regularly onto the top of the flaming pyrolysis zone, and the second zone must be closed and insulated above, forming a burner section.

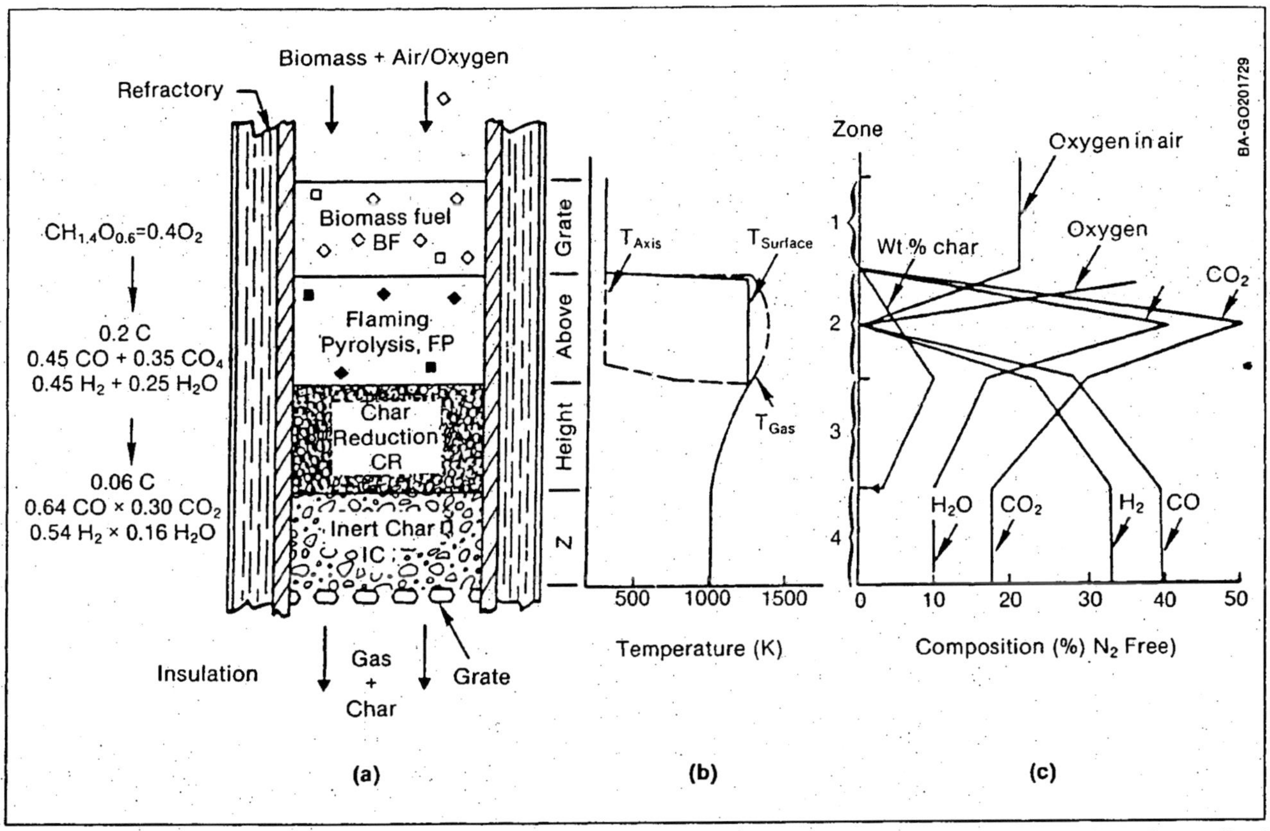

Fig. 5-10. Schematic of stratified downdraft gasifier showing (a) chemical reaction, (b) temperature profiles, and (c) composition Source: Reed 1984, p. 226)

Air reacts with pyrolyzing biomass in the second zone, and most of the volatile wood oil is burned to supply heat for this pyrolysis as explained in Chapter 4. We have called this process **"flaming pyrolysis,"** and distinguish it from **"flaming combustion,"** which occurs in the absence of solids with excess air or oxygen. At the bottom of the second zone, the biomass has been converted to charcoal, and all of the oxygen from the air has reacted. The final gas leaving the second zone contains CO and H_2, as well as the CO_2 and H_2O produced in the earlier stages of combustion, as shown in Fig. 5-10. The CO and H_2 mixture already is sufficiently concentrated to be a combustible gas at this point.

The hot gases produced in the flaming pyrolysis zone react with the charcoal in the third, or char gasification, zone to convert more of the CO_2 and H_2O to CO and H_2, through the Boudouard and water-gas reactions (Eqs. 4-7 and 4-8). We call this process **adiabatic char gasification** (adiabatic means no heat flows into or out of the section). During the reaction, sensible heat of the gas is converted into chemical energy of the fuel gas. This results in cooling the gas to about 800°C, a temperature at which no further reaction is possible.

Finally, there may be a zone of **unreacted charcoal** below the char gasification zone through which the gas must pass before it reaches the grate. This last zone has the disadvantage that char and ash from the char gasification zone also must pass through it to reach the grate. However, as we mentioned before, it provides a "buffer" or reservoir of charcoal that is available to accommodate changes in the power level, which otherwise might cause the grate to heat excessively.

The stratified downdraft design has a number of advantages over the Imbert gasifier. The open top permits fuel to be fed more easily and allows easy access for instruments to measure conditions within the bed. The uniform passage of air and fuel down the gasifier keeps local temperatures from becoming too high or too low while the average temperature is high. The cylindrical construction is easy to fabricate and permits continuous flow for otherwise troublesome fuels without causing bridging or channeling. Finally, the various strata are more accessible for measuring compositions and temperatures within the bed so that it is possible to compare modeling results with empirical observations.

The stratified downdraft gasifier is both conceptually and mathematically easier to comprehend. Quantitative descriptions and mathematical models of gas flows through the bed are thus facilitated.

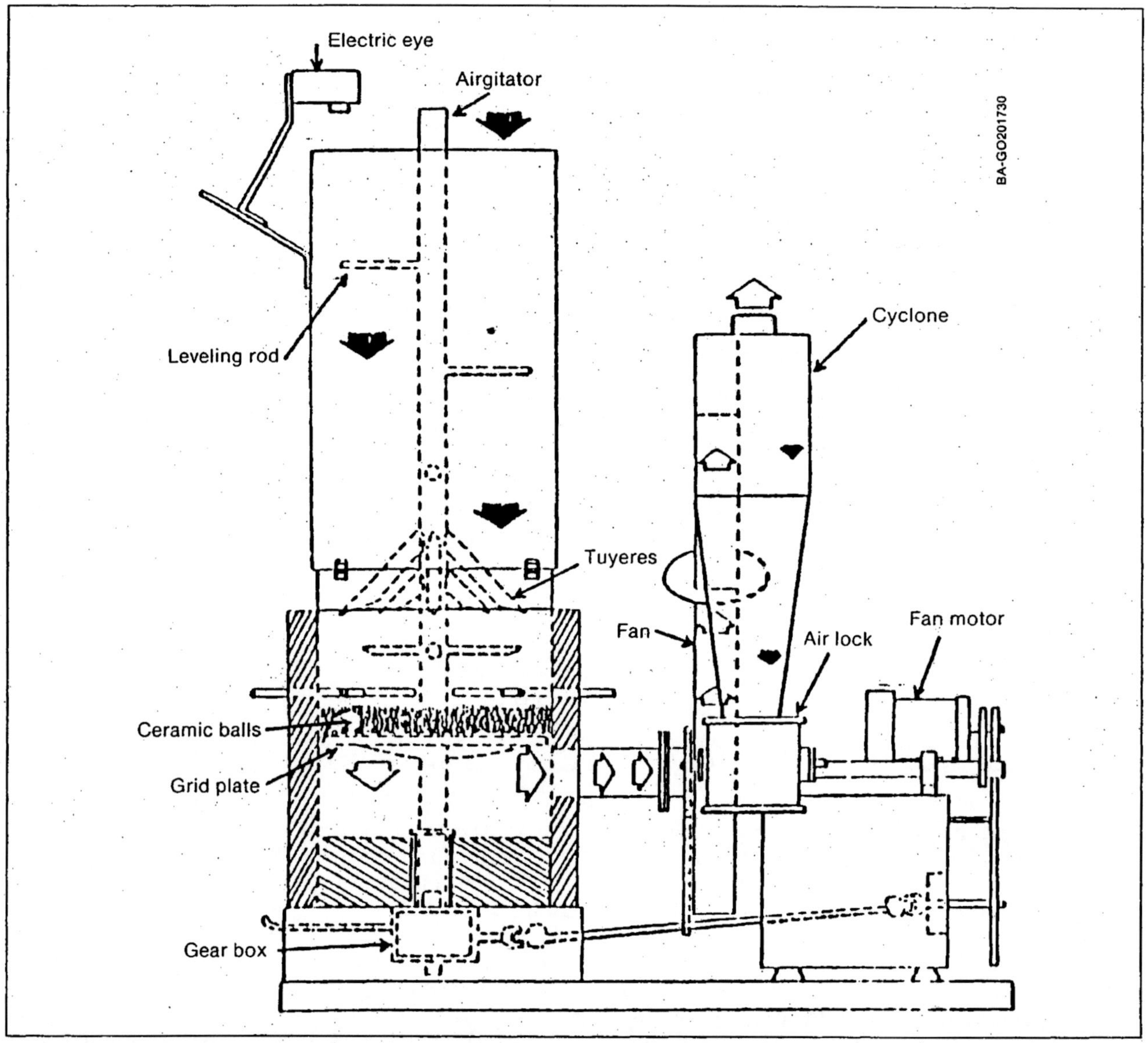

Fig. 5-11. Buck Rogers gasifier (Source: Walawender 1985, p. 913. © 1985. Used with permission of Elsevier Science Publishing Co., Inc.)

In principle, the gasifier can be scaled to large diameters because it operates as a plug-flow reactor, and the air and fuel are uniformly mixed. A 0.6 m (24 in.) internal diameter gasifier has been operated successfully by the Buck Rogers Co. of Kansas (Walawender 1985; Chern 1985). A 0.77 m (30 in.) internal diameter gasifier to produce 750 kW of power has been developed by Syngas Systems, Inc., (Graboski 1985) and is being operated extensively on air and oxygen (Graboski 1986).

We believe that these advantages, coupled with the design's simplicity, may ultimately allow the stratified downdraft gasifier to supplant the Imbert and other earlier gasifiers, and that a number of design variations will grow from the basic stratified downdraft gasifier described here. Nevertheless, several years of experience with these gasifiers (using both oxygen and air) have uncovered questions that must be understood and resolved in any commercial design.

5.8.3 Unanswered Questions About the Stratified Downdraft Gasifier

The foremost question about the stratified downdraft gasifier design concerns char and ash removal. As the charcoal reacts with the gases in the char gasification zone, it eventually reaches a very low density and breaks up into a dust containing all of the ash as well as a percentage of the original carbon. This dust may be carried away partially by the gas. However, sooner or later it will begin to plug the gasifier so it must be removed by shaking or stirring. (Imbert gasifiers have a provision for shaking the grate and, when used to power vehicles, are largely self-purging.) Yet, it is desirable to gasify as much of the char as possible before its packing increases the pressure drop. Minimal char-ash removal can be accomplished automatically by using pressure-sensing switches that activate the removal mechanism only when pressure begins to build. It is desirable to gasify more than 95% of the biomass, leaving only 5% char-ash.

The ability to remove variable amounts of char with a moving grate adds a second design issue to the stratified downdraft gasifier. Char consumes more than

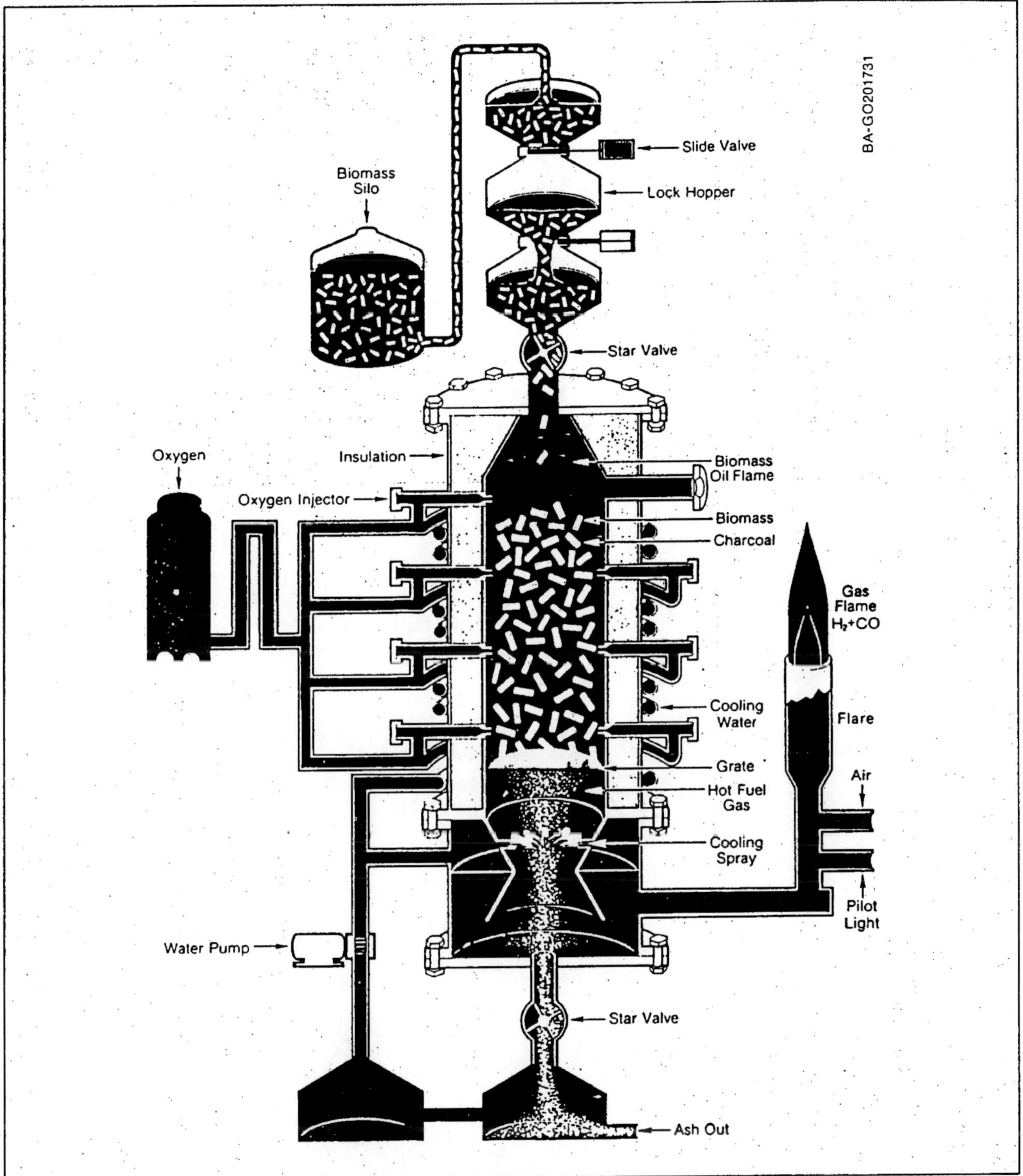

Fig. 5-12. SERI oxygen gasifier (Source: Reed 1985b, Fig. 3.4. © 1985. Used with permission of Plenum Publishing Corporation)

twice the mass of oxygen required for biomass pyrolysis, and hence high char conversion increases the overall oxygen/biomass ratio. If up to 10% of the biomass is removed as char-ash at the grate, then the oxygen/fuel ratio decreases; in turn, the temperatures of flaming combustion decrease, and the resulting gas has both a higher energy content and a higher tar content. This added control of the oxygen/biomass ratio has not been well-defined.

A third issue in the design of the stratified downdraft gasifier is the prevention of bridging and channeling. High-grade biomass fuels such as wood blocks or pellets will flow down through the gasifier under the influence of gravity. However, other fuels (such as stringy chips, sawdust, and rice hulls) can form a bridge, preventing continuous flow and giving rise to very high temperatures. Obviously, it is desirable to use these widely available biomass residues. Bridging can be prevented by stirring, shaking, or agitating the bed and, since the temperatures in the gasifier are relatively low, it is possible to use a stainless steel central stirring arm such as the one shown in the Buck Rogers gasifier of Fig. 5-11.

A fourth issue to be addressed in the design of stratified downdraft gasifiers is bed stabilization. When the gasifier operates at stable steady state, the flaming pyrolysis zone advances into the biomass at the same rate that the char is consumed, resulting in a stationary reaction zone at a fixed level in the gasifier. However, this zone can move up in very dry fuels, consuming the fuel reservoir and emerging at the top of the gasifier. Since no more fuel is available, the gasifier will operate in this "top-stabilized" mode, but there is a substantial heat loss through the open top. This results in lower conversion efficiency and could create a fire hazard. (The SERI and SGI gasifiers operate regularly in this top-stabilized mode with oxygen but have closed refractory tops and pressure-feeding apparatus.) Fuel must be added at a steady rate to prevent alternate pyrolysis and char gasifier operation, which can generate high tar levels. (We have observed higher tar levels from this pulsing pyrolysis process when using oxygen than when using air.)

Very wet fuels inhibit the flaming pyrolysis zone from advancing fast enough to keep up with the incoming fuel, and the zone subsequently moves toward the grate, consuming the dead char zone at the bottom. The zone may become "grate-stabilized" at this point, or it may continue to move to the grate and be extinguished.

Thus, control of the reaction zone position is very important in the stratified downdraft gasifier. A number of mechanisms seem to be effective in stabilizing this position and they are discussed in a recent paper (Reed 1985a).

In summary, both understanding and commercialization of the stratified downdraft gasifier have made remarkable progress in only a few years of work, but a great deal of effort still is in progress. It is not clear whether this design ultimately will displace conventional Imbert and other gasifiers.

5.8.4 Modeling the Stratified Downdraft Gasifier

A mathematical model has been developed at SERI to predict the behavior and dimensions of the stratified downdraft gasifier (Reed 1983a, 1984, 1985a). The model is based upon predicting the length of both the flaming pyrolysis and char reaction zones from the properties of the biomass fuels and the gasifier throughput. The zone lengths predicted for various sizes of biomass fuels are given in Table 5-4, and a diagram of the gasifier, dimensioned for dry wood chips, is shown in Fig. 5-13. The predicted pyrolysis and char bed temperatures and chemical composition measured in the char bed are shown in Fig. 5-14. The interested reader is referred to the papers for further details, which are beyond the scope of this handbook.

5.9 Tar-Cracking Gasifiers

5.9.1 Introduction

The cost of the gas cleanup system needed for engine use generally exceeds the cost of the gasifier. Furthermore, even if tar impurities are removed in filters and scrubbers (see Chapter 8), they still must undergo the difficult task of disposal. Therefore, one of the major areas for future gasifier development will be the design

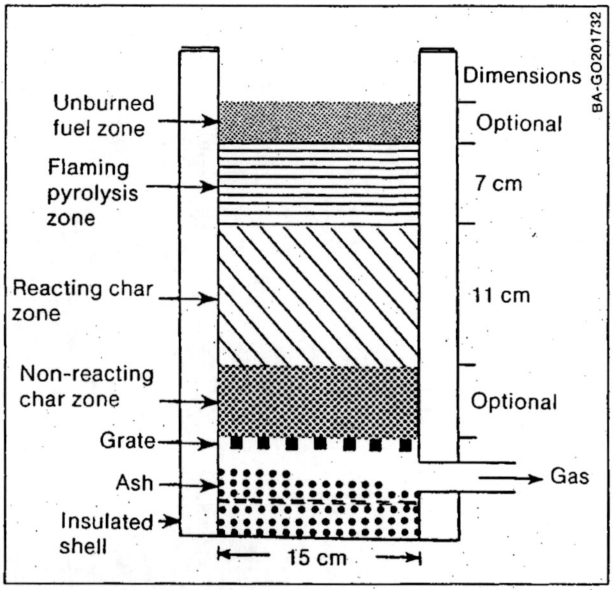

Fig. 5-13. Stratified downdraft gasifier design (Source: Reed 1984, Fig. 6)

Table 5-4. Prediction of Zone Length for Different Fuel Sizes

Parameters			Small Chips	Inch Chips	Sawdust Cubes	Peat Pellets
Gasifier Inputs						
Proximate Analysis: (Dry Basis)						
Volatile			.90	.803	.90	.65
Char			.10	.188	.10	.35
Ash			.01	.009	.01	.05
Water	F_w		.20	.027	.05	.25
Fuel Properties:						
Density	ρ	g/cm^3	.40	.40	1.00	1.10
Bulk Density		g/cm^3	.20	.15	.50	.45
Void Fraction	F_v		.50	.63	.50	.59
Length		cm	1.00	2.00	5.00	1.00
Width		cm	1.00	2.00	3.00	.30
Height		cm	.20	.50	3.00	.30
Average Equivalent Diameter		cm	4.41	1.56	4.41	.51
Volume	V	cm^3	.20	2.00	45.00	.07
Area	A	cm^2	2.80	12.00	78.00	1.38
Gasifier Conditions:						
Diameter	D_g	m	.15	.15	.15	.15
Heat Transfer	q	W/cm^2	2.00	2.00	2.00	2.00
Feed Rate	M	kg/h	10	10	10	10
Specific Feed Rate	m	kg/m^2h	566	566	566	566
Gasifier Parameters						
Pyrolysis Zone:						
Reaction Time	t_p	s	43	73	656	93
Fuel Velocity	v_f	cm/s	.079	.105	.031	.035
Pyrolysis Length	l_p	cm	3.38	7.64	20.33	3.26
Char Zone:						
Reaction Time	t_c	s	100	100	100	100
Char Zone	l_c	cm	7.90	10.50	3.10	3.50

Notes: Calculations based on the following equations and assumptions:
Time of pyrolysis: $t_p = (h_p + F_w h_w)\, \rho\, V/A\, q$
Fuel velocity: $v_f = m/D_g \rho\, (1-F_v)$
Flaming pyrolysis zone length: $l_p = v_f\, t_p$
Temperature of pyrolysis = 600°C
Heat of pyrolysis: h_p = 2081 J/g
Heat to vaporize water to 600°C: h_w = 3654 J/g
Char reaction time: t_c = 100 s
Char zone length: $l_c = (t_c)(v_f)$

Source: Reed 1984

of gasifiers that convert the maximum quantity of tar to gas during gasification.

5.9.2 Combustion of Tars

The tar levels from a number of gasifiers are shown in Table 5-5. (One cubic meter of producer gas weighs about 1 kg at NTP; therefore a tar level of 1 g/m^3 corresponds to a concentration of 1000 ppm or 0.1%; 1 mg/m^3 is 1 ppm by weight, and we shall use this equivalence in discussing tar levels.) It is important to note that updraft gasifiers generate 5% to 20% tar (50,000-200,000 ppm!) (Desrosiers 1982). The downdraft gasifiers of Table 5-5 produce tar in amounts at least an order of magnitude lower than the updraft gasifiers, and new developments are now reducing tars into the 100 to 1000 ppm level.

In his thesis, Groeneveld used cold flow models to investigate the flow of gases around a nozzle. He found that the incoming air stream entrains and burns tar-laden gas as shown in Fig. 5-9 if the gasifier is properly designed (Groeneveld 1980a,b). After publication of these results, a gasifier using this principle (a central air nozzle promotes recirculation and combustion of the volatiles produced in pyrolysis) was designed and marketed in the Netherlands.

The DeLaCotte tar-recycling gasifier (Fig. 5-15) was the first tar-burning gasifier. It has two solid-fuel chambers and a gas-combustion chamber on the side. Fuel is pyrolyzed in the upper part of the fuel chamber. Pyrolysis products are aspirated out the top to the side combustion chamber using the flow of combustion air in an ejector, where they burn completely at high

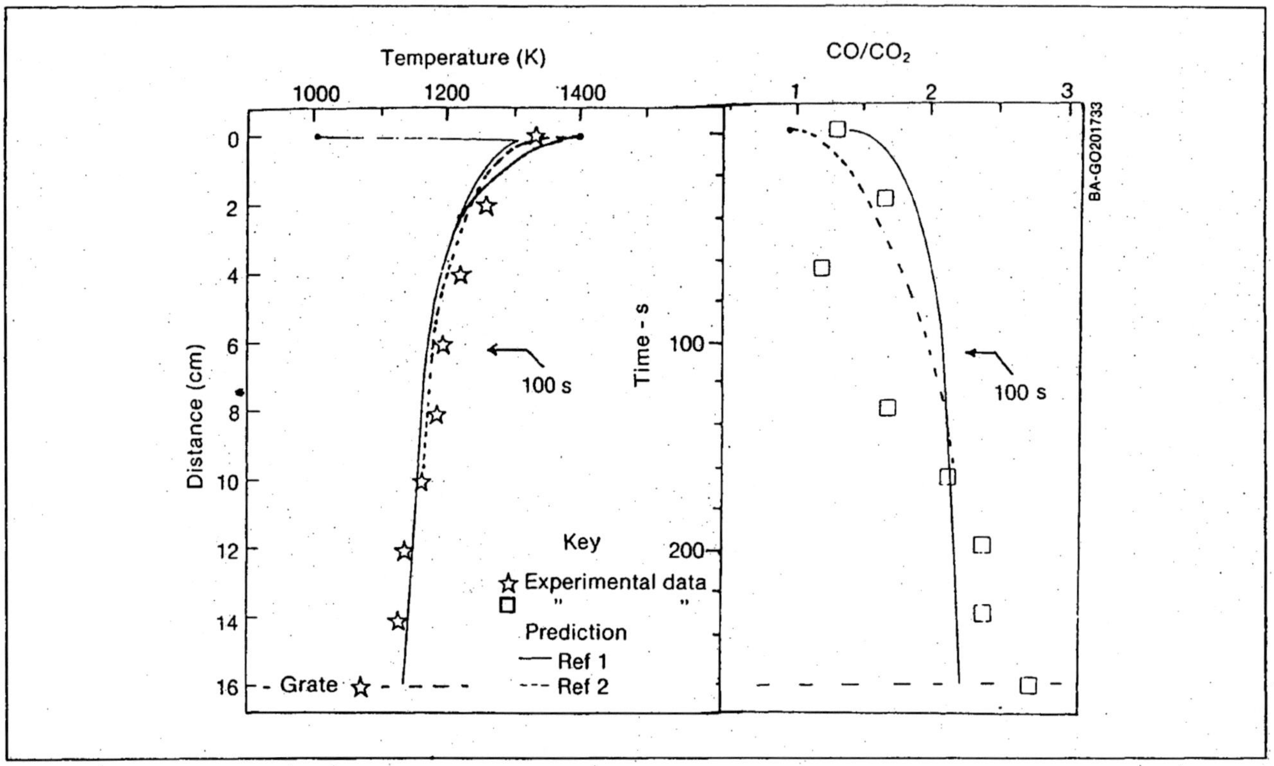

Fig. 5-14. Observed and calculated temperature and composition in abiabatic char conversion zone (Source Reed 1984, Fig. 5)

temperature in the absence of solids. (Combustion with air generates producer gas; combustion with oxygen could generate synthesis gas.) The hot combustion products (1000°-1100°C) are reinjected at the center of the gasifier. One-fourth of the gas rises through the upper chamber to assure pyrolysis of the biomass fuel. The remaining three-fourths travel down through the lower chamber containing the char produced from the biomass in the upper chamber. The char is gasified by reacting with the CO_2 and H_2O produced by combustion, as in other gasifiers. The high-temperature combustion chamber may permit more thorough destruction of the tars; in any case, this gasifier claims to produce very low tar levels.

Table 5-5. Tar Content in Product Gas from Downdraft Gasifiers

Gasifier	Capacity kg/h	Throat Diameter m	Specific Load kg/hm²	Tar Content g/Nm³
Kromag KS-12	15.0	0.12	1330	0.62
Kromag K-4	7.5	0.09	1180	1.90
Semmler	12.0	0.15	680	0.88
Danneberg	19.0	0.15	1075	0.70
Leobersdorfer	36.0	0.42	260	1.20
TH. Twente	20.0	0.20	640	0.50
Forintek Canada	50.0	?	?	3.00
Mini Gasifier	0.2	0.01	2550	3.00

Source: Susanto 1983

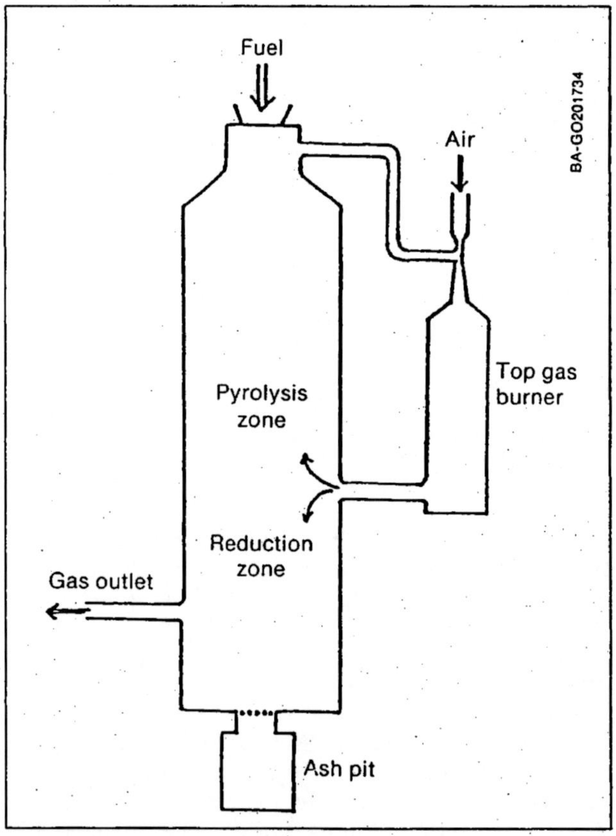

Fig. 5-15. DeLaCotte tar recycling gasifier (Source: Kaupp 1984a, Fig. 133)

Susanto and Beenackers have developed a gasifier that recycles tars internally in a similar manner, as shown in Fig. 5-16. In this case, the combustor is contained centrally in the lower (char) section of the gasifier and, therefore, has very little heat loss (Susanto 1983). Without recycle, this gasifier produced 1400 mg/Nm^3 (approximately 1400 ppm). With a gas/air recycle ratio of 0.85, the tars were reduced to the very low level of 48 ppm of tar as shown in Fig. 5-17.

The high degree of tar destruction in these two units is due to the high tar combustion temperature promoted by the positive circulation of tars upward away from the reduction zone and also to the more complete combustion that takes place in the absence of solids.

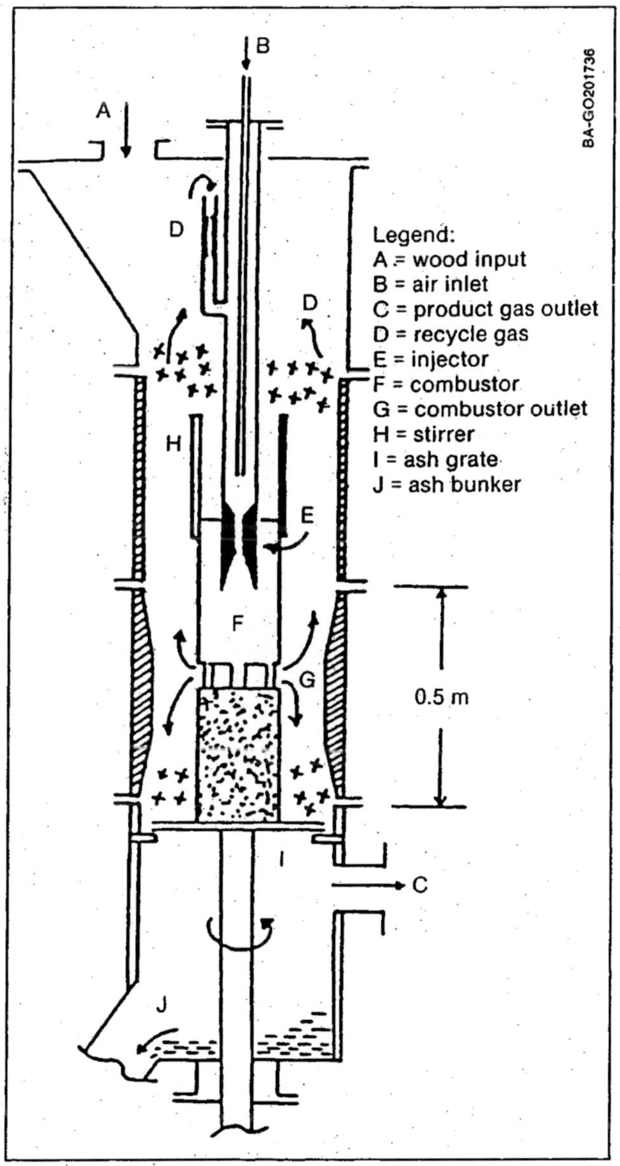

Fig. 5-16. Gasifier with internal tar recycle (Source: Susanto 1983, Fig. 3.2. © 1983. Used with permission of the Beijer Institute)

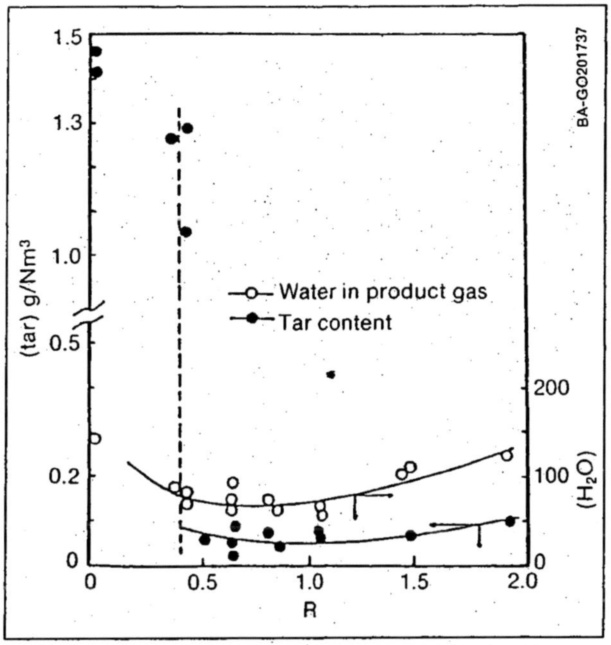

Fig. 5-17. Water and tar content in producer gas (dry basis) versus recycle ratio (Source: Susanto 1983, Fig. 2.7. © 1983. Used with permission of the Beijer Institute)

5.9.3 Thermal Tar Cracking

Temperatures above 800°C rapidly crack the primary pyrolysis oils to olefins and aromatic compounds. These compounds continue to react in the absence of oxygen to make polynuclear aromatic compounds (PNAs) and eventually soot. While high temperatures (above 800°C) can destroy tars rapidly, these same high temperatures also promote reaction with char, which in turn rapidly quenches the gas to 800°C. Therefore, the time available for tar cracking in a bed of hot charcoal is very short. For this reason, a bed of hot char may not be very effective in tar cracking as was originally believed (Reed 1982; Chittick 1983). The French Croisot Loire process allows a tarry gas from a fluidized bed to be burned further in a separate chamber at 1300°C (Bioenergy 1985), resulting in a final gas that has a very low tar content (Chrysostome 1985).

A laboratory transparent gasifier, a modification of the SERI stratified downdraft gasifier, shown in Fig. 5-18, has added a tar-cracking chamber in which small amounts of oxygen or air can be added to crack the final trace quantities of gas from the gasifier, in the absence of the quenching action of the charcoal. We have measured tar concentrations of 50-500 ppm at the exit from the cracking chamber. However, the difficulty of maintaining a large chamber at temperatures in excess of 900°C caused considerable loss in gas quality at this scale. Perhaps thermal cracking alone is practical in much larger gasifiers (Reed 1985c).

5.9.4 Catalytic Tar Cracking

Recent work in Europe has focused on producing synthesis gas for making methanol using oxygen and several schemes for eliminating tars and methane. In the Swedish MINO process, the tarry gas from an oxygen fluidized-bed is passed through a bed of hydrocarbon cracking catalyst at temperatures between 950° and 1040°C, resulting in a gas containing 10-100 ppm of tar (Ström 1985).

D'Eglise has studied the kinetics of cracking of pyrolysis oils generated at lower temperatures and found that more than 99.9% of these oils can be cracked by dolomite lime at temperatures as low as 500°C (Donnot 1985). However, these low temperature compounds are cracked and rearranged much more easily than the tars formed at high temperatures so the results may not be representative of the difficulty of gasifier tar cracking.

We used the laboratory sized transparent gasifier to generate typical gasifier tars for testing the kinetics of tar cracking with a number of catalysts in the apparatus shown in Fig. 5-19. The results for three catalysts are shown in Table 5-6. The variation of cracking rate with temperature is shown in Fig. 5-20 for a dolomite lime catalyst, a refinery silica-alumina cracking catalyst, and a silicalite molecular sieve type catalyst.

5.10 Summary

A large number of gasifiers have been developed over the last century using both experience and intuition. The most successful of these has been the Imbert downdraft gasifier, which produces relatively low levels of tar gas from uniformly high-grade fuels.

We are in a new period, during which the principles of combustion science are being applied to develop a better understanding of gasification. New gasifiers, such as the stratified downdraft gasifier and the tar-reburning gasifier, promise to expand the range of usable fuels and to produce an even cleaner product gas. Only time will tell whether this increased understanding will result in cleaner, more versatile gasifiers at an acceptable cost.

Table 5-6. Catalytic Cracking of Gasifier Tars by Several Catalysts

Run Conditions				Tar Concentration		
Temp. °C[a]	Flow Rate kg/h	Residence Time(t), s	Space Vel g/g-h	Before(C1) mg/Nm^3	After(C2) mg/Nm^3	Rate-k L/s[b]
Catalyst: Dolomite[c]						
600	0.73	0.34	0.33	9574	3597	2.84
750	0.73	0.29	0.29	3169	1294	3.05
820	0.73	0.28	0.27	18082	2674	6.95
960	0.73	0.24	0.24	8346	1113	8.26
750	0.20	1.07	1.04	3169	200	2.58
750	0.40	0.54	0.52	7537	2408	2.13
Catalyst: Si-Al Catalyst[d]						
432	0.51	1.54	2.02	18082	4654	0.88
432	0.51	1.54	2.02	22070	1313	1.83
552	0.56	1.20	1.57	5929	333	2.40
415	0.48	1.68	2.19	4863	695	1.16
343	0.49	1.84	2.40	4654	847	0.93
287	0.51	1.94	2.54	5605	780	1.02
Catalyst: Crystalline Silica S-155[e]						
416	0.59	0.91	1.68	8280	790	2.58
406	0.39	1.40	2.57	15237	2303	1.35
469	0.34	1.47	2.70	15189	3359	1.03
505	0.48	0.99	1.83	11725	3131	1.33
613	0.47	0.89	1.64	25305	3930	2.10
812	0.42	0.81	1.50	9184	1075	2.64

[a] Average temperature over a 20 cm length of furnace.

[b] Rate calculated from $k = -\ln(C1/C2)/t$

[c] 1.03 kg of dolomite lime contained in 20 cm length of 5 cm i.d. stainless pipe. Bulk density = 3346 kg/m^3; void volume = 0.25 cm^3/g. Dolomite limestone decomposes to dolomite lime in the range 600-1000°C. Particle size 5 mm.

[d] Si-Al cracking catalyst (Davison Chemicals, Gr 980-13) 510 g sample, bulk density = 7655 kg/m^3; void volume = 0.85 cm^3/g. Particle size 1mm d x 5 mm long cylinders.

[e] Crystalline silica catalyst S-155 (Union Carbide) 543 g charge; bulk density = 0.76 g/cm^3; void volume = 0.80 cm^3/g

Source: Reed 1986b

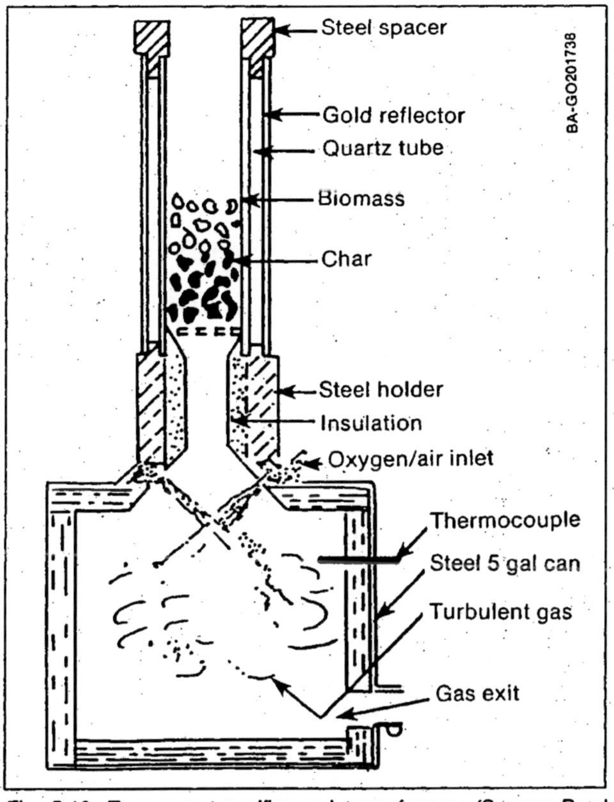

Fig. 5-18. Transparent gasifier and tar reformer (Source: Reed 1985d, Vol. I, p. 210)

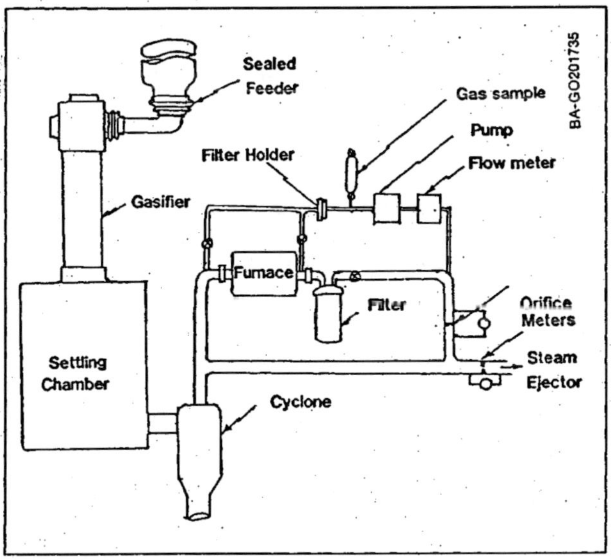

Fig. 5-19. Tar cracking apparatus (Source: Reed 1985c)

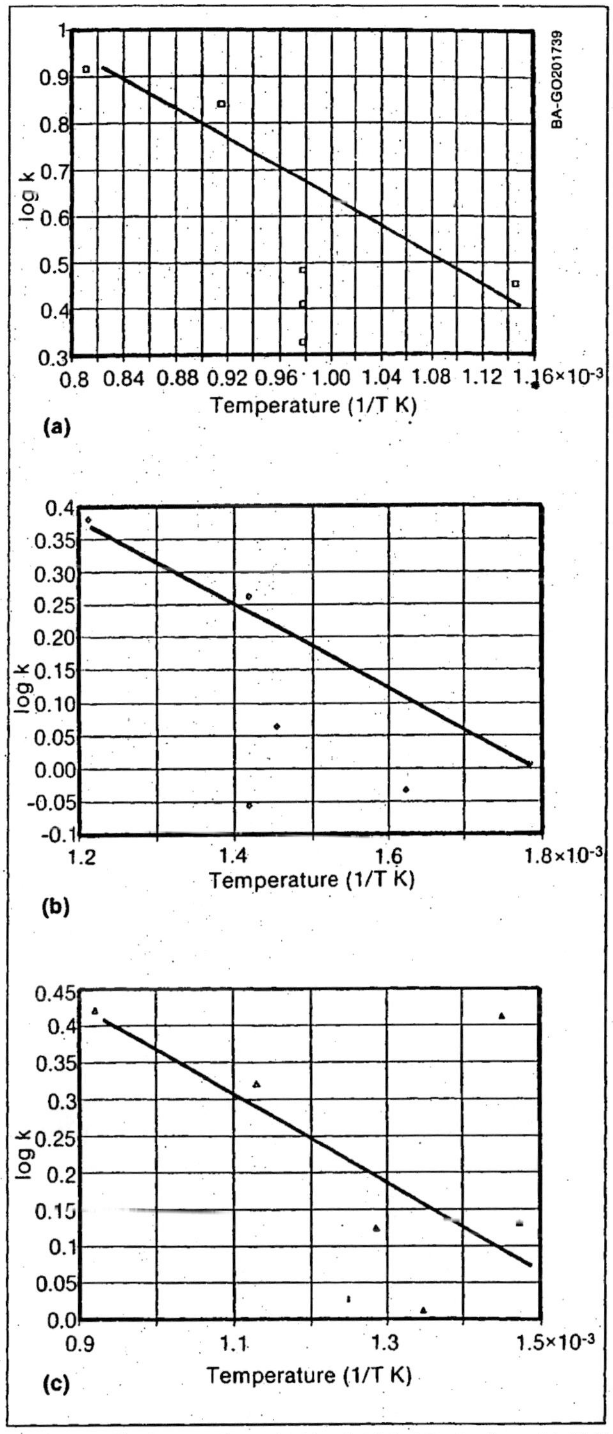

Fig. 5-20. Kinetic rate of tar cracking for (a) dolomite lime, (b) Si-Al cracking catalyst, and (c) silicalite cracking catalyst

Chapter 6
Gasifier Fabrication and Manufacture

6.1 Introduction

Gasifier construction is a relatively simple task and can be accomplished in any well-equipped shop using basic sheet metal and welding assembly techniques. Indeed, the task is so simple that it was possible for the countries of wartime Europe to construct almost one million gasifiers in just a few years in spite of wartime shortages (Egloff 1943; Gengas 1950). Nevertheless, a number of new materials and fabrication techniques have become available since World War II, and we shall call attention to these improvements in this discussion.

According to Kaupp (1984a), "the construction of a small gasifier, including the purification system, does not require sophisticated equipment or highly skilled mechanics. It can be built in workshops comparable to the auto repair shops found in most third world countries."

Fabrication refers to the construction of a single gasifier for use or for an experiment. **Manufacture** commences when one undertakes the construction of a number of identical units.

A general discussion of drilling, welding, and assembly procedures pertinent to gasifier fabrication and manufacture is beyond the scope of this manual. Instead, we shall comment upon specific techniques of fabrication, as well as the wide range of materials specifically applicable to gasifiers.

6.2 Materials of Construction

Gasifiers are usually constructed from commercially available materials such as steel pipe, sheet, and plate. When choosing materials, one should (where possible) select those that are readily available and use off-the-shelf equipment and materials that are available in bulk quantities. One should avoid exotic alloys, special shapes, and custom fabrication techniques that require large initial setup and tooling costs, except in cases where their use is justified.

Smaller, atmospheric-pressure gasifiers require a minimum metal thickness of 20-gauge, with double-thickness reinforcements extending a few centimeters (1 in.) around all fittings and fastenings (Freeth 1939). The maximum mild-steel service temperature is 480°C (900°F) (MASEC). Although the metal temperatures encountered in well-designed air gasifiers do not usually exceed the softening point of mild steel, certain stainless steels or inconel may give the extra temperature resistance necessary for critical areas such as the grate, hearth, or nozzles.

Some of the mild-steel components may suffer chemical corrosion in certain parts. Corrosion is likely to occur in areas where water condenses or collects since gasifier water often contains organic acids. Water collection is especially a problem in regions such as the upper magazine of Imbert gasifiers, as well as in some wet-scrubber systems. In these instances, the steel should be replaced by corrosion-resistant materials such as copper, brass, epoxy lined steel, or stainless steel as required. Stainless steel usually costs two to three times as much as mild steel and requires inert gas welding techniques. Copper and brass cost five times as much as mild steel but can be joined by brazing or hard-soldering using an acetylene torch. Aluminum is particularly vulnerable to corrosion in alkaline environments, and its use should be avoided there.

Some wartime gasifiers, particularly the stationary ones, contained massive but fragile firebrick insulation. We are fortunate today to have lightweight insulating materials based on spun alumino-silicate that are capable of withstanding temperatures up to 1500°C, far beyond the requirements of gasifiers. The alumino-silicate insulation also offers many times the durability and heat-flow resistance of firebrick, at a fraction of the weight (Perry 1973). It is relatively inexpensive and is available in a wide variety of forms. The 2- to 5-cm-thick felt blankets and vacuum preformed cylinders (or "risa sleeves") are particularly recommended for insulating the reaction zone. "Moldable ceramics" that come as a wet putty can be shaped to corners and edges, and thus are also very useful.

Finally, plastics can be used in certain applications. Some plastic pipe will perform acceptably up to the boiling point of water, is more flexible than metal pipe, and will not corrode. Plastic liners, such as epoxy paints, can sometimes provide the corrosion resistance needed in critical areas, provided temperatures are not greater than 120°C.

6.3 Methods of Construction

A gasifier is built much like a water heater, and the same methods of construction are used. The workshop should be equipped with tools for performing tasks such as shearing sheet metal, rolling cylinders and cones, drilling, riveting, grinding, painting, sawing, tube cutting, and pipe threading.

An oxyacetylene torch is valuable for cutting and welding tasks, but an arc welder is preferred for mild-steel welding. When aluminum, stainless steel, or inconel is used, it will be necessary to use either TIG (tungsten-inert gas) or MIG (metal-inert gas) welding techniques.

All seals must be made gas-tight; threaded and welded fittings are preferred at all points, and exhaust-pipe-type gaskets can be used if necessary. High-temperature, anti-sieze pipe dope should be used on all pipe joints. High-temperature applications will require ceramic fiber or asbestos gaskets. Silicone sealant is appropriate at temperatures below 300°C and rubber or Viton "O" rings and gaskets will perform excellently at room temperature. The system should be leak-tested before the initial startup, as well as after modifications. Leak-testing is accomplished by plugging the system and pressurizing it to 25 cm (10 in.) of water with a blower. A thick soap solution is applied to all fittings and joints, and they are checked for emerging soap bubbles. Leak-testing should also be performed as a standard test in the regular maintenance schedule.

6.4 Sizing and Laying out the Pipes

When designing a gasifier, it is important to keep the pressure drop in the system as small as possible. Because there are unavoidable pressure drops associated with the gasifier, the cyclone separator, and the cleanup system, it is very important to use adequately sized pipe. The pressure drop associated with standard runs of pipes is shown in Fig. 6-1 (Perry 1973). Engineering and plumbing handbooks also list the pressure drops associated with pipe fittings such as elbows and couplings.

On the other hand, gas velocities within the pipes should be adequate so that entrained solids will be conveyed to their proper point of removal, as shown in Table 6-1 rather than deposited inside the pipe.

When laying out pipe connections for a gasifier system, it is important to allow access to various parts that may require cleaning or adjustment. It is recommended that new systems be assembled with a large number of pipe unions to facilitate cleaning out the pipes, as well as future design modifications. In general, it is better to use a pipe "T" with a plug rather than an elbow, to allow for instrument mounting and other additions at a later time. After a unit has been thoroughly tested and is in production, such provisions can be omitted.

Table 6-1. Gas Velocity Requirements for Conveying Solids

The conveying velocities in pipes are dependent upon the nature of the contaminant. Recommended minimum gas velocities are:

Contaminant	Velocity
Smoke, fumes, very light dust	10 m/s
Dry medium density dust (sawdust, grain)	15 m/s
Heavy dust (metal turnings)	25 m/s

Source: Kaupp 1984a

6.5 Instruments and Controls

The gasifiers of the past were crude, inconvenient devices. Today's gasifiers are evolving toward safer, automated processes that make use of a wide range of present-day instruments and controls. An extended discussion of the system instrumentation and control requirements is found in Chapter 10.

6.5.1 Temperature

Thermocouples (such as chromel-alumel type K) should be used to measure various gasifier temperatures, especially below the grate, as a check for normal or abnormal operation. Temperatures at the grate should not exceed 800°C; higher temperatures indicate abnormal function. Consequently, the signal from the thermocouple can be used by a control system or an alarm system.

6.5.2 Pressure

Manometers are required to measure pressure drops across the bed, cyclone, filters, and other components. (Usually, these pressure drops amount to only a few centimeters of water pressure.) The manometers are available as tubes filled with colored liquid or, more conveniently, bellows manometers (such as a Dwyer Magnehelic gauge). Both types give a direct reading of the pressure drop and can be equipped with limit switches that will sound an alarm to warn when preset flow levels have been violated, and/or activate control valves to regulate those flows. Also, electrical transducers are available that convert pressure difference into an electrical signal suitable for readout or control processing.

6.5.3 Gas Mixture

Oxygen sensors have been developed by the automotive industry to measure the small changes in oxygen concentration required to control the air/fuel ratio for clean, efficient combustion. They are relatively inexpensive and, in principle, can be adapted to gasifier systems for similar functions (though this has not yet been done).

6.5.4 Automatic Controls

The fact that an operator will be required for both large- and small-scale gasifiers is a fixed-cost scale factor that naturally causes larger systems to be favored. Automatic unattended operation is therefore essential to the economic viability of small gasifier systems. Automatic fuel feed and char-ash removal equipment are already well-developed for stoker-fed boilers and could be adapted to automatic gasifier operation.

Finally, we can look forward to gasifier systems of the future that will use inexpensive microprocessors to integrate the signals from these sensors into the automatic and unattended operation of highly efficient gasifiers, making gasifiers as simple to use as a car or home furnace.

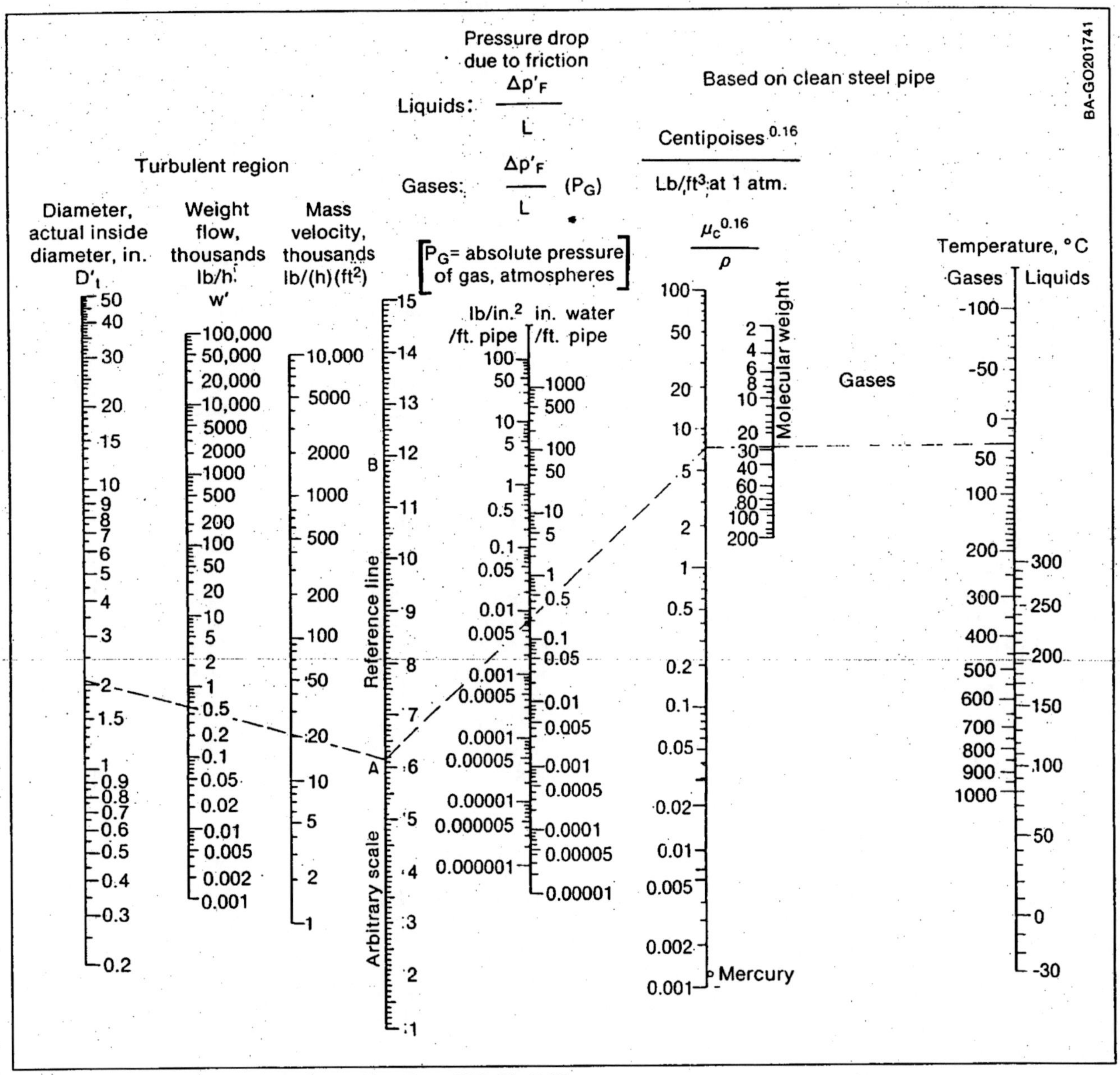

Fig. 6-1. Piping flow chart (Source: Adapted from Perry 1973, Fig. 5-27)

Chapter 7
Gas Testing

7.1 Introduction

It is relatively simple to build a gasifier and operate an engine for a short time. However, the commercial success of gasification ultimately depends on long-term, reliable operation of gasifier systems. Many gasifier systems have failed after less than 100 h of operation because of tar buildup in either the system or the engine. Destroying an engine is a costly method for determining whether a gas is sufficiently clean for engine operation. A quantitative knowledge of gas quality and cleanliness is necessary for the designer, developer, buyer, and user of gasifier equipment.

This chapter will describe simple and inexpensive tests of the physical and chemical properties of producer gas. Using these tests will allow one to determine whether the gas is suitable for its intended purpose.

7.2 Gas-Quality Measurements and Requirements

During gasifier system development, one may need to be able to measure:

- Gas composition: The volume percent of CO, CO_2, H_2, H_2O, CH_4, C_2 and higher hydrocarbons, and N_2 to calculate the gas energy content or to analyze gasifier operation.
- Gas energy content: Can be calculated from gas composition, or it can also be measured calorimetrically without the need to know composition. The gas must have an energy content greater than 4 MJ/Nm^3 (100 Btu/scf) for most applications. (See Appendix for definition of scf and Nm^3.)
- Quantity of tars: The quantity of condensible organics in raw gas is a measure of gasifier performance and determines whether the gas can be cleaned. Above 5000 mg/Nm^3 tars, the gas is difficult to clean up and is suitable only for close coupled direct combustion. Gas cleanup equipment should reduce the tar level to below 10 mg/Nm^3.
- Quantity and size of particulates: The nature and quantity of char-ash and soot entrained in the gas stream can help to design filters. Particles larger than 10 μm must be removed to a level below 10 mg/Nm^3 for engine applications.
- Water content of gas: The water content of the gas helps to calculate cooling requirements.

7.3 Description of Producer Gas and Its Contaminants

7.3.1 The Gas Analysis

A typical raw gas analysis from a recent SERI test of corncob gasification is given in Table 7-1. This analysis includes volume concentrations of each major chemical constituent, as well as the physical contaminants of the gas. The energy content of the gas can be calculated from the energy content of the components using the high or low heating values (HHV or LHV) for each gas, as shown in Tables 7-1 and 7-2. The LHV also can be determined graphically from Fig. 7-1.

7.3.2 Particulates

Some particulate levels reported for wood and charcoal gasifiers are listed in Figs. 7-2 and 7-3.

In order to remove particles with the appropriate equipment, it is necessary to know their nature and size distribution. Particle size distributions shown in Fig. 7-4, Table 7-3, and Fig. 7-5 were obtained by mechanical screen separation of the cyclone contents for SERI and Imbert tests. The results of both tests are plotted in Fig. 7-4 on log probability paper for ease of analysis, displaying the distinctive slope common to most fine powders produced by fragmentation.

The potential particle-size range of a wide variety of particles and their characteristics are shown in Fig. 7-6,

Table 7-1. Composition of Producer Gas from Corn Cobs after Cyclone Separation

Physical Composition		
Tar content	1300 mg/m^3	1300 ppm[a]
Particulate	330 mg/m^3	330 ppm
Ash content of particulate	30 mg/m^3	30 ppm
H_2O	7.1 wt %	71,000 ppm
Chemical Composition		
CO	19 Vol % x 322 Btu/scf = 61	
CO_2	14 Vol % x 0	
H_2	17 Vol % x 325 Btu/scf = 55	
CH_4	2 Vol % x 1031 Btu/scf = 20	
N_2	48 Vol % x 0	
	136	
Dry Gas (HHV)[b] 136 Btu/scf (60°F, 30 in. Hg Dry)		

[a] 1 Nm^3 of gas weighs about 1 kg, so that 1 mg/Nm^3 = 1 ppm

[b] The gas heating value may be calculated from the gas analysis using Table 7-2.

along with the equipment appropriate for separation of each size range.

It is important to distinguish between the various forms of particulates that result from biomass combustion and gasification. Starting with full-sized biomass fuel and 0.5% ash, we can use this ash as a tracer to follow conversion in the gasification process.

Although combustion and updraft gasifiers leave a white ash, downdraft gasifiers do not produce this white mineral ash because there is no oxygen present when the final charcoal breakup occurs. Freshly produced charcoal, just after it has finished flaming pyrolysis, is only slightly smaller than it started out and should not be able to pass through the grate. As gasification of the charcoal proceeds, carbon is

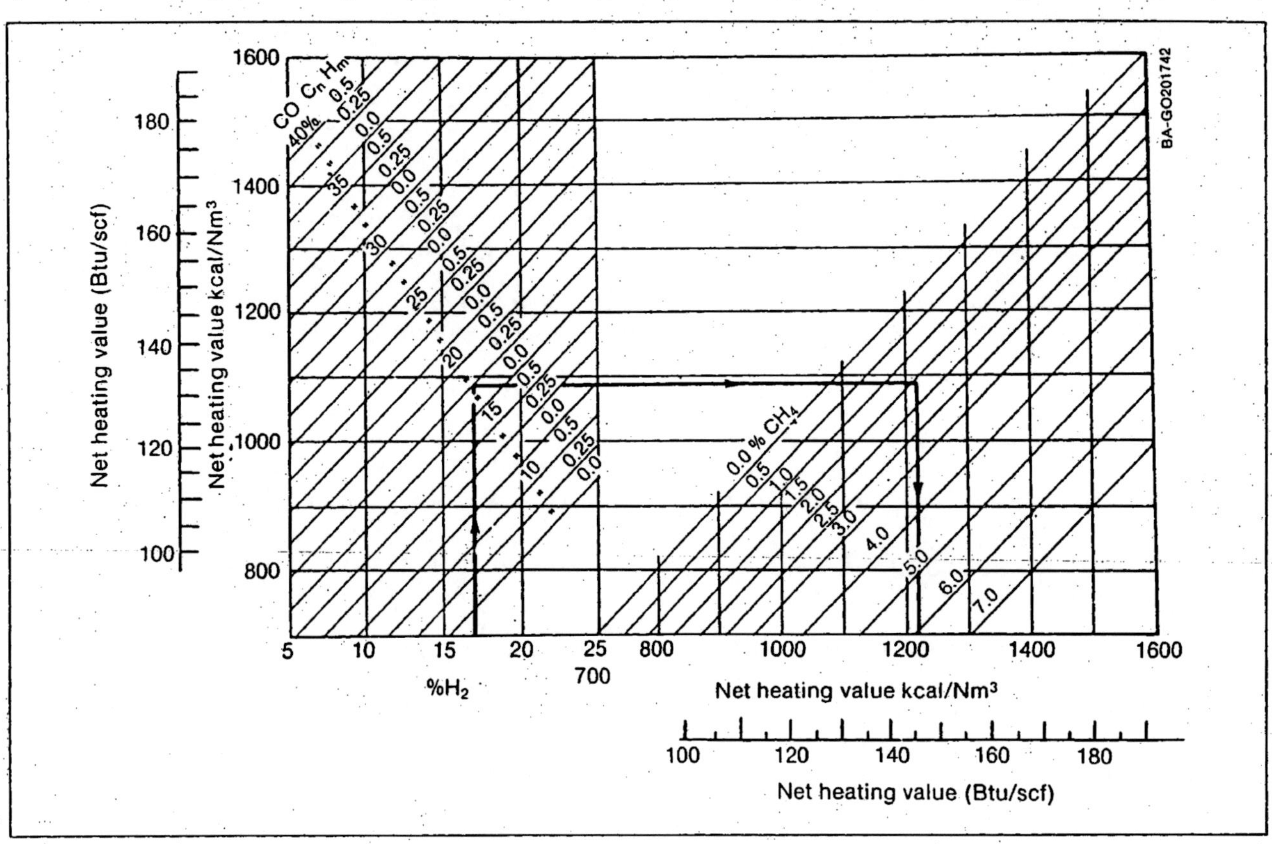

Fig. 7-1. Nomogram for lower heating value of producer gas (Source: Kaupp 1984a)

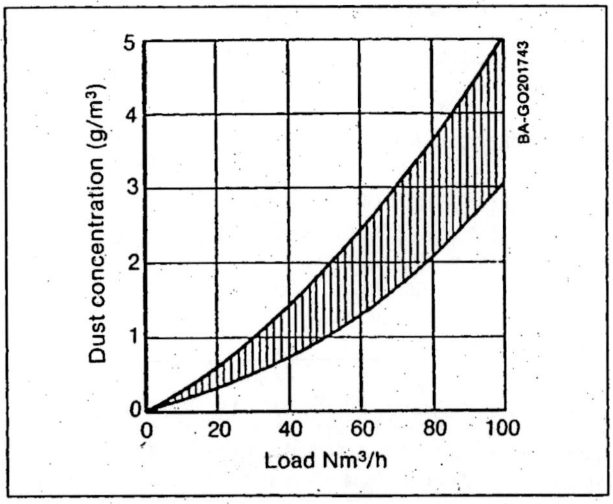

Fig. 7-2. Dust concentrations for wood gas generators of various makes as a function of load (Source: Gengas 1950, Fig. 87)

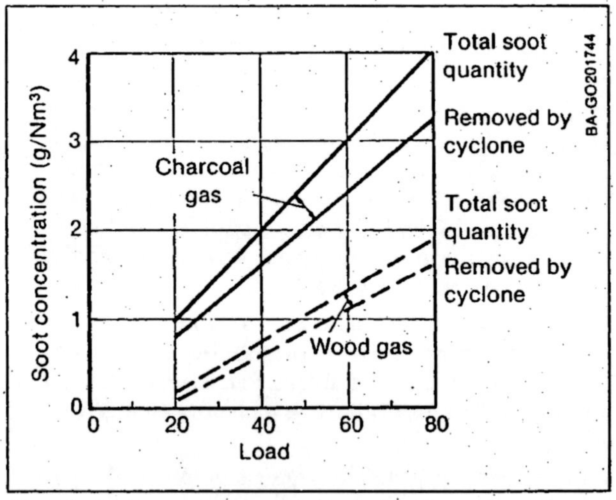

Fig. 7-3. Dust concentration related to the load in wood and charcoal gas generators (Source: Gengas 1950, Fig. 86)

removed from both the surface and interior of the char particle by motion of the hot gases. The char particle cracks and crumbles as carbon is converted to ash.

The term char-ash refers to the black dust that falls naturally through the grate in a downdraft gasifier when gasification is as complete as it will go. Char-ash is produced during the final breakdown of the charcoal mechanical structure as the charcoal reacts with pyrolysis gases. Char-ash from downdraft gasifiers still contains 50% to 80% carbon (Fig. 3-3), which is enough carbon to give char-ash a black color. Char-ash usually is collected below the grate or in the cyclone separators.

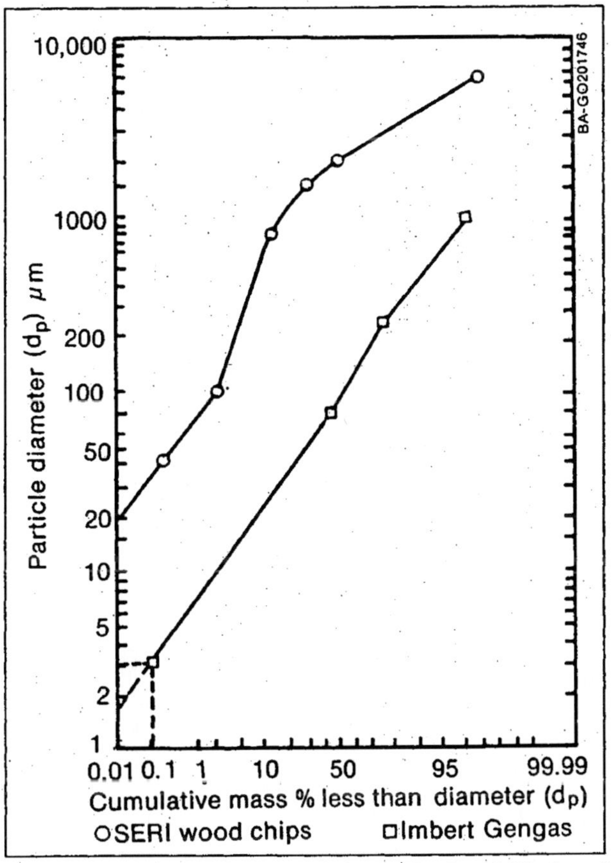

Fig. 7-4. Imbert and SERI downdraft particle size distribution of producer gas char dusts

If the fine char-ash from downdraft gasifiers is about 20% ash, then this represents a 95% mass conversion of the fuel since we started with 0.5% ash biomass. We can also see from Fig. 3-3 that it is desirable to keep the char larger than 500 μm (0.5mm) in the gasifier to boost efficiency. Also, particles under 500 μm have completed their task and should be removed as thoroughly as possible.

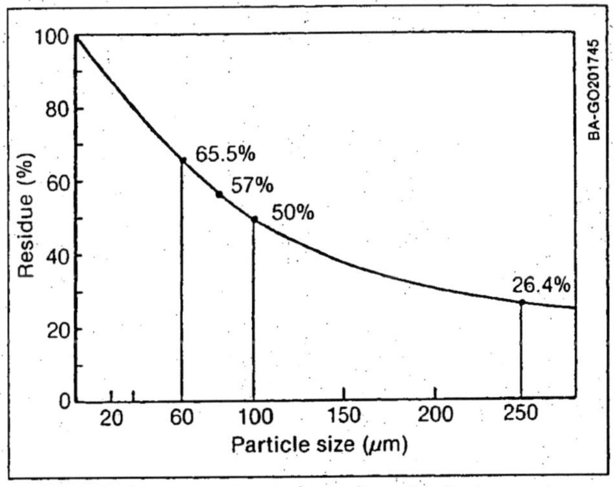

Fig. 7-5. Residue curve for the screening of Imbert generator gas (Source: Gengas 1950, Fig. 88)

Although the unconverted carbon found in char-ash represents an energy loss, it also has several benefits. The final ash from combustion is less than 1 μm in size and can be captured only in expensive bag house filters. The char-ash holds the ash in a 10 μm matrix which is captured by cyclones. The char-ash may have considerable value as a charcoal or as a soil conditioner.

Char-ash particles smaller than the cyclone separator's cut-point pass through the cyclone separator. Smaller particles normally are higher in ash content, as shown by Fig. 3-3. Higher ash content is more abrasive; however, solids smaller than the oil film thickness do not cause major engine wear. Ash that has been subjected to slagging is much harder and more abrasive than nonslagging ash, which crumbles easily.

All biomass contains some ash (typically a few percent), but some fuels, such as rice hulls or MSW, can

Table 7-2. High Heating Value and Low Heating Value of Gas Components[a]

Component	Symbol	HHV[b] (MJ/Nm^3)	HHV[c] (Btu/scf)	LHV[b] (MJ/Nm^3)	LHV[c] (Btu/scf)
Hydrogen	H_2	13.2	325	11.2	275
Carbon monoxide	CO	13.1	322	13.1	322
Methane	CH_4	41.2	1013	37.1	913

[a] 1 Btu/scf = 8.26 kcal/Nm^3 = 40.672 kJ/Nm^3

[b] Standard conditions, 0°C and 760 mm Hg Dry

[c] Standard conditions, 60°F and 30 in. Hg Dry

Source: Adapted from Perry 1973, Table 9-18.

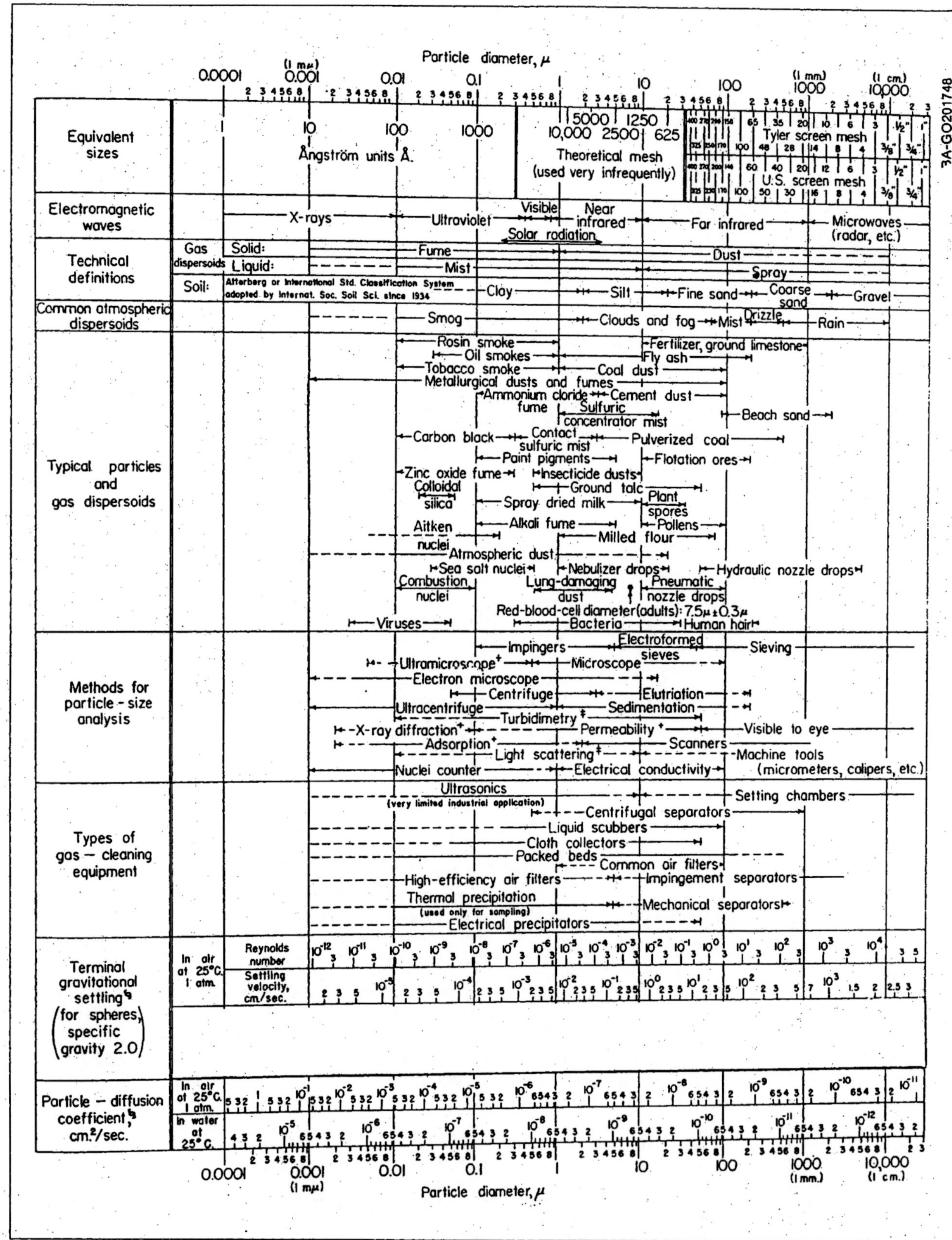

Fig. 7-6. Characteristics of particles and particle dispersoids (Source: Adapted from Perry 1973, Fig. 20-92)

Table 7-3. Analysis of Wood Gas Dust

	% Ash
Over 1000 μm (1 mm screen)	1.7
1000 - 250 μm	24.7
250 - 102 μm	23.7
102 - 75 μm	7.1
75 - 60 μm	8.3
Under 60 μm	30.3
Losses	4.2
	100.0
Water content	3.2
Ash content, dry sample	10.6
Loss due to burning, dry sample	15.7
Content of Fe_2O_3 in the ashes	11.0
Content of SiO_2 in the ashes	7.7

Source: Gengas 1950, Table 2-9

contain 20% ash or more. During flaming pyrolysis of the original biomass, the organic molecules break down to form a very finely divided soot (carbon black), such as that seen in oil or candle flames. Soot particles are much smaller than char-ash particles (ordinarily less than 1 μm). The soot is so fine that it can be expected to pass harmlessly with the gas and burn in the engine without harm.

Carbon monoxide is unstable below 700°C and, given enough time, will decompose in the presence of certain catalytic metal surfaces to form carbon and carbon dioxide, according to the reaction

$$2\,CO \rightarrow C + CO_2 \,. \tag{7-1}$$

This carbon, known as Boudouard carbon, is slippery to the touch and nonabrasive. Below about 500°C, the reaction is very slow. Normally, Boudouard carbon does not form in gasifiers because the gas cools quickly through this temperature range.

Char-ash, because of its high mineral content and abrasive potential, is the main cause of engine wear in engine systems and understandably is a main focus in gas cleanup. Similarly, soot and Boudouard carbon are inherently ash free, nonabrasive, and possibly lubricating. Despite their small size and difficulty of capture, they are not seen as a significant factor in engine wear.

Typically, the largest particles that pass through the grate can be extracted mechanically, for instance, with an auger. Subsequent to that step, removing the suspended small particles is the principal problem in gas cleanup. The particulates in Table 7-1 were collected after the cyclone separator, which caught 50% of 50-μm particles. These smaller particles are composed of very fine char-ash, soot, and tar mists.

The type of gas cleanup equipment required is determined by the particle sizes that must be removed, and therefore it is important to determine the particle-size distribution as well as the total quantity of particles. Relatively little information is available in the literature (Gengas 1950), so complete gas cleanup design necessitates measurement or knowledge of particle-size distribution. Particle-size measurement is discussed further in Section 7.8.

7.3.3 Tars

Initial biomass pyrolysis can produce up to 60% "wood oil," composed of the monomers, oligomers, and fragments of the biomass polymers cellulose, hemicellulose, and lignin. Subsequent high-temperature cracking (over 700°C) of these large molecules results mostly in gas, but also polymerization to form 5% to 10% of heavier polynuclear aromatic molecules that are similar to coal tars. Up to 20% of these tars and oils can be carried through with the gas from updraft gasifiers.

In downdraft gasification, oxygen is available to burn these oils during pyrolysis. Although flaming pyrolysis burns most of the tars and oils, 0.1% to 1% (depending on the gasifier design) can be expected to survive. These tars and oils are troublesome in the gas-processing system and the engine, so they must be thoroughly removed by scrubbing.

Tars occur mostly as a mist or fog composed of fine droplets that may be less than 1 μm in diameter (see Fig. 7-6). Tar mists continually agglomerate into larger droplets and tend to saturate and coat solid particles. If not removed, tar mist forms deposits that cause engine intake valves and other moving parts to stick. Before a gasifier is considered suitable for operating an engine, it is imperative that one test the producer gas for tars and particulates.

7.4 Gas Sampling

7.4.1 Sample Ports

A temporary or permanent port must be provided at each point on the gasifier where samples are desired, as shown in Fig. 7-7, such as downstream from the cyclone and before the burner or engine, as well as at each stage along the gas cleanup train when it is desired to determine the effectiveness of each system component. It is important that the gas sample is representative of the gas at each point. The port and tap may need to be heat traced to prevent premature condensation (see below).

Permanent sampling ports should be closed off with gate or ball valves, which provide a straight through passage. Needle valves and sill cock type water valves should be avoided, since much of the material being sampled will deposit within the twisted passages of these types of valves.

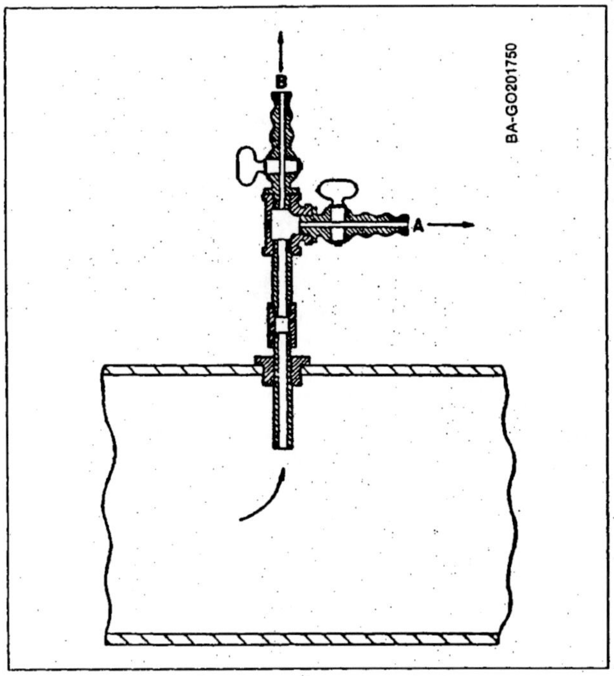

Fig. 7-7. Static tap sampling port (Source: ASME 1969, Fig. 1)

The temperature, pressure, and moisture content of gas at the nozzle must be accounted for when designing a sampling train, and measurements should be reduced to standard conditions. For average samples, the test duration should be long enough to average the reading over at least one cycle of the equipment being tested—for instance, the fuel-feed cycle, the scrubber service cycle, or the shaking cycle. For snap samples or where a transient phenomenon is being observed, then the most rapid sampling method and small samples should be used.

Hot, raw gas emerging from any gasifier will contain tar, char-ash, soot, and water vapor, and it is relatively simple to measure these quantities in a small sample of raw gas. After the gas has been cleaned and conditioned, the measurements become more difficult. Impurity levels are much lower, so it is necessary to handle much larger gas samples in order to accumulate a measurable-sized sample. The principles of measurement remain the same, but the measurements require more time to accomplish.

The physical analysis of producer gas is based on the weight of tar, particulates, and water in a measured quantity of gas. Therefore, a positive-displacement gas-testing meter (such as those made by Singer, Rockwell, and the American Gas Association) should be available for calibrating flowmeters, pumps, and similar components. An analytical balance capable of weighing to 0.1 mg is also required for accurate measurement of tars and particulates.

Approximate results can be obtained by comparing the volume of sample required for a particular depth of color deposit such as 50% grey measured on a standard grey scale such as that used for smoke testing (Dwyer 1960). Tests at SERI found that a color of 50% grey on a 47-mm filter disc, taken from a standard grey scale, represented approximately 0.12 to 0.5 mg of collected contaminants (Das 1985). A quantity of gas for chemical analysis can be collected at the same time as the sample to be used for physical analysis.

7.4.2 Isokinetic Sampling

Isokinetic (equal gas-velocity) conditions in the flow chamber and the sampling tube should be ensured where particle sizes exceed 10 μm. Otherwise particle-size distribution will not be the same in the chamber and sampling tube. The design of sampling-tube parts and their placement within the gas stream are shown in Figs. 7-8 and 7-9, respectively. Figure 7-10 shows velocity streamlines for a sampling tube in a flow chamber. Part "a" illustrates isokinetic conditions; that is, the streamlines are equally spaced within the duct and tube. In "b," sampling-tube velocity is less than in the duct (indicated by the wider streamline spacing in the tube), and proportionately more gas must flow around the tube than through it. However, the inertia of large particles impedes their being carried by the gas that deflects around the tube. Large particles in line with and immediately upstream of the sampling tube continue their flight into the tube. Hence, proportionately more large particles exist in the tube than flow in the chamber. The opposite holds true where tube velocity is greater than the velocity in the chamber (Fig. 7-10(c)); i.e., large particles are underrepresented in the tube.

Nonisokinetic sampling error is plotted in Fig. 7-11. We can see that for under 10 μm-particles, the concentration error is within ±10% over a wide range of sampling velocity from half to double the gas velocity for a velocity ratio u_s/u_g between 1/2 and 2.

In practice, the high efficiency cyclone separator of a gasifier system will remove most particles larger than 10 μm, so for the smaller particles remaining in the gas stream, the error due to nonisokinetic sampling can be ignored.

The sampling error for nonisokinetic sampling conditions also can be neglected for tar mists and other very fine aerosols. In fact, the sampling port of Fig. 7-7 can serve as a convenient 10-μm coarse inertial prefilter. The particle-sample probe of Fig. 7-9 tends to accumulate impacted large-particle deposits and eventually clogs, so it should be adjusted to face downstream during periods that it is not in use.

The isokinetic flow rate can be calculated as

$$Q_n = Q_p\,(D_n^2/D_p^2) \qquad (7\text{-}2)$$

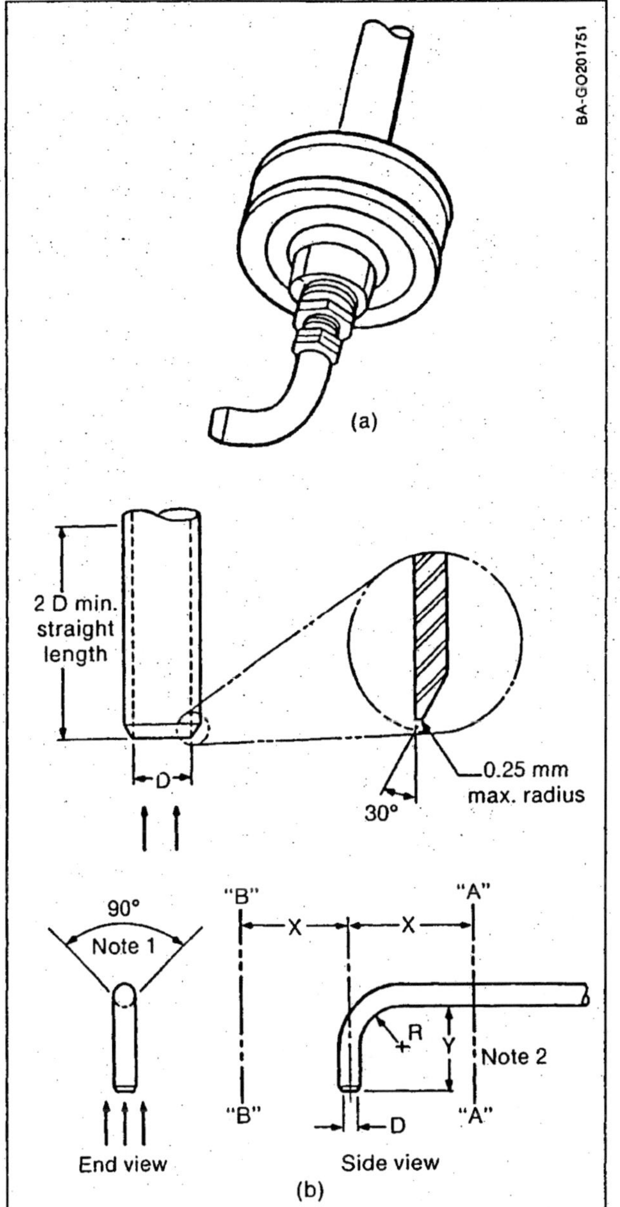

Notes:

1. The area upstream of this 90° sector shall be free of obstructions between planes A-A and B-B.
2. No portion of probe shall project upstream of nozzle entrance within a distance from nozzle centerline of 15 cm or 5 nozzle diameters, whichever is greater.

D = Nozzle diameter Y = 4 × D (minimum)
R = 2 × D (minimum) X = 6 × D (minimum)

Fig. 7-8. (a) Typical holder for flat round filters and (b) recommended design for sampling nozzle tip (Source: ASME 1980)

where

Q = flow rate,
D = diameter.

The subscripts n and p refer to nozzle and pipe, respectively.

The flow rate of a dirty-gas stream can be measured irrespective of temperature and molecular weight of the gas by using a balanced-tube, null-type apparatus. Such a system in effect measures chamber flow by collecting a portion of the gas flow through a sampling tube, cleaning it, and measuring it. The mass flow through the entire chamber is then calculated, using the ratio of chamber area to sampling tube area. Measurements are made once the velocity in the sampling tube has been adjusted (via a vacuum pump in the system) to be equal to that in the flow chamber. Velocities within the tube and chamber are equilibrated by using a differential manometer to balance the static pressures for the tube and chamber. The best placement for the probe within the chamber can be checked by testing the flow profile across the chamber. The probe should be located where the flow is average for the chamber. Where necessary, flow straighteners should be used to ensure accurate readings. Balance-tube, null-type sampling, without gas-cleanup equipment, can be used for clean gas.

7.5 Physical Gas-Composition Testing

7.5.1 Raw Gas

Sampling train options for measuring the range of levels of tar, char-ash, and water are shown in Fig. 7-12, and gasifier test-train component options are presented in Tables 7-4, 7-5, and 7-6.

Certain basic procedures must be followed whenever sampling producer gas—be it for tar, particulates, heat content, etc.:

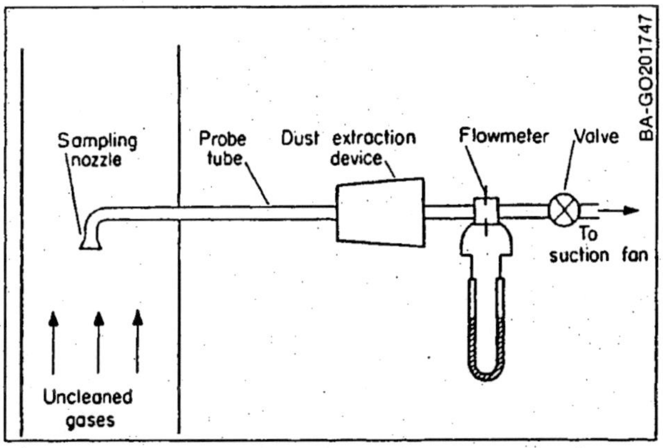

Fig. 7-9. Large particle sampling probe (Source: Strauss 1975, Fig. 2.9. © 1975. Used with permission of Pergamon Press)

- The gas, being noxious, should be either burned off or returned to the pipe downstream from the sampling point.
- The gas should be cleaned.
- The gas should be dried.

Figure 7-12(a) shows a setup for measuring particulates and tar, or moisture, or gas composition, or gas production rate.

Figure 7-12(b) is a setup for continuously measuring change in water content and change in heat content. It

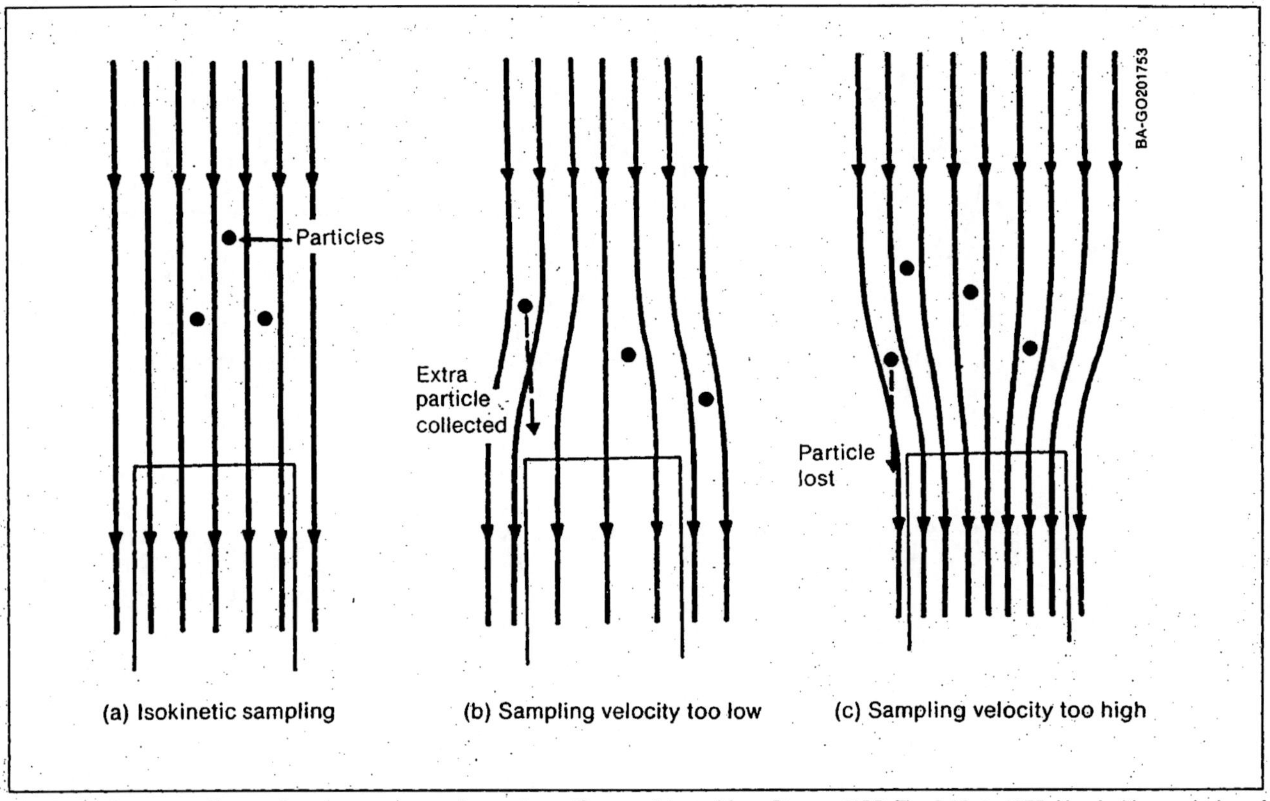

Fig. 7-10. Gas stream lines at the entrance to sampling probes (Source: Adapted from Strauss 1975, Fig. 2.12. © 1975. Used with permission of Pergamon Press)

Table 7-4. Components for Raw Gas Contaminant Test Train

Item	Requirements	Price Rank (1=Lowest)
Sample Probe	Tar-only measurements permit any suitable fitting with pipe thread access	1
	Tar and particulates require tubing with 90° bend facing the gas stream	2
Shut Off Valve	Ball Valve 1/4 in.	—
Filter Holder	47 mm reusable holder	
	Polycarbonate	1
	Aluminum	2
	Stainless	3
Filter Discs	Glass fibers 99.9% efficiency at 0.3 μm	—
Gas Sample Pump	Hand operated rubber bulb	1
	Plastic piston pumps (36 cm/stroke)	3
	Water-powered aspirator pump	2
	Motor driven vacuum pump	4
Gas Sample Flow	Flow meter, 2 in. scale, 4% full scale	2
	Gas flow indicator	1
Gas Test Meter	Positive displacement meter to indicate accumulated sample volume	1

also can be used to measure gas flow and to collect moisture and tars and particulates over a measured time.

Figure 7-12(c) shows a system train for measuring gas flow and tars and particulates.

The ball valve at the sample port (Fig. 7-12(a) and (c)) permits changing the filter disc without danger of releasing gas or admitting air. The filter holder in Fig. 7-12(a) and (c) must be maintained, by electric heating or locating close to the hot gas pipe to keep the filter hot enough, above the water dew point, typically 80°C, to avoid water condensation in the filter disc.

The desiccant-drying section (Fig. 7-12(a)-(c)) should be constructed so that it can be disconnected for weighing. We have found that an indicating desiccant assembly can be fabricated by containing the desiccant between two glass-wool plugs in Tygon or glass "U"-tubes (Fig. 7-12(b)). (Drierite is a commercial form of anhydrous $CaSO_4$ containing cobalt sulfate, which changes from blue to pink when it becomes hydrated.)

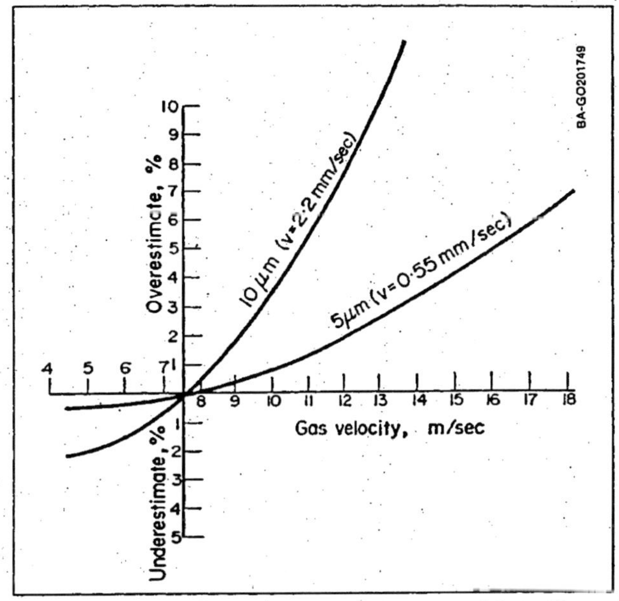

Fig. 7-11. Errors in concentrations of 5 and 10 μm particles (Source: Strauss 1975, Fig. 2.13. © 1975. Used with permission of Pergamon Press)

Table 7-5. Additional Components for Continuous Gas Quality Test Train

Function	Specifications	Manufacturer
Sample Port	1/4 in. FPT Fitting 1/4 in. MPT →1/4 in. tube compression	Common compression fittings (e.g., Swagelock)
Hygrometer Dry Bulb Wet Bulb	Cross fitting bushed down to accept 1/4 in. tubing Water reservoir in bottom 2 hole stopper with thermometers Dry bulb bare Wet bulb wrapped with wicking that dips into water reservoir	User fabricated
Dryer	Indicating desiccant Drierite or silica gel in a container large enough for several hours sample time 1.5 cm dia x 40 cm U-tube	User fabricated
Pump	Aquarium pump modified for suction and pressure service capable of 50 in. WG, 0.01 scfm	Whisper (User modified)
Flow Meter	Floating ball rotameter—RMA	Dwyer
Burner	Diffusion flame non-mixed	See Fig. 7-16 or user fabricated
	1/4 in. tubing with 1/8 in. ID opening	
Readout	Chromel-alumel thermocouple	Omega, etc.

Table 7-6. Additional Components for Condensible Collection Test Train

Item	Purpose	Price Rank
Gas Dryer	For small sample moisture determination	1
	8 in. length of 3/8 in. tubing with fittings filled with indicating desiccant weighed before and after each test	2
	For large sample pump protection 250 ml flask filled with indicating desiccant or ice bath, bubbler, impinger, condenser	3
Filter Heater	To prevent condensation at probe Heater tape around sample lines and filter holder heated chamber	1

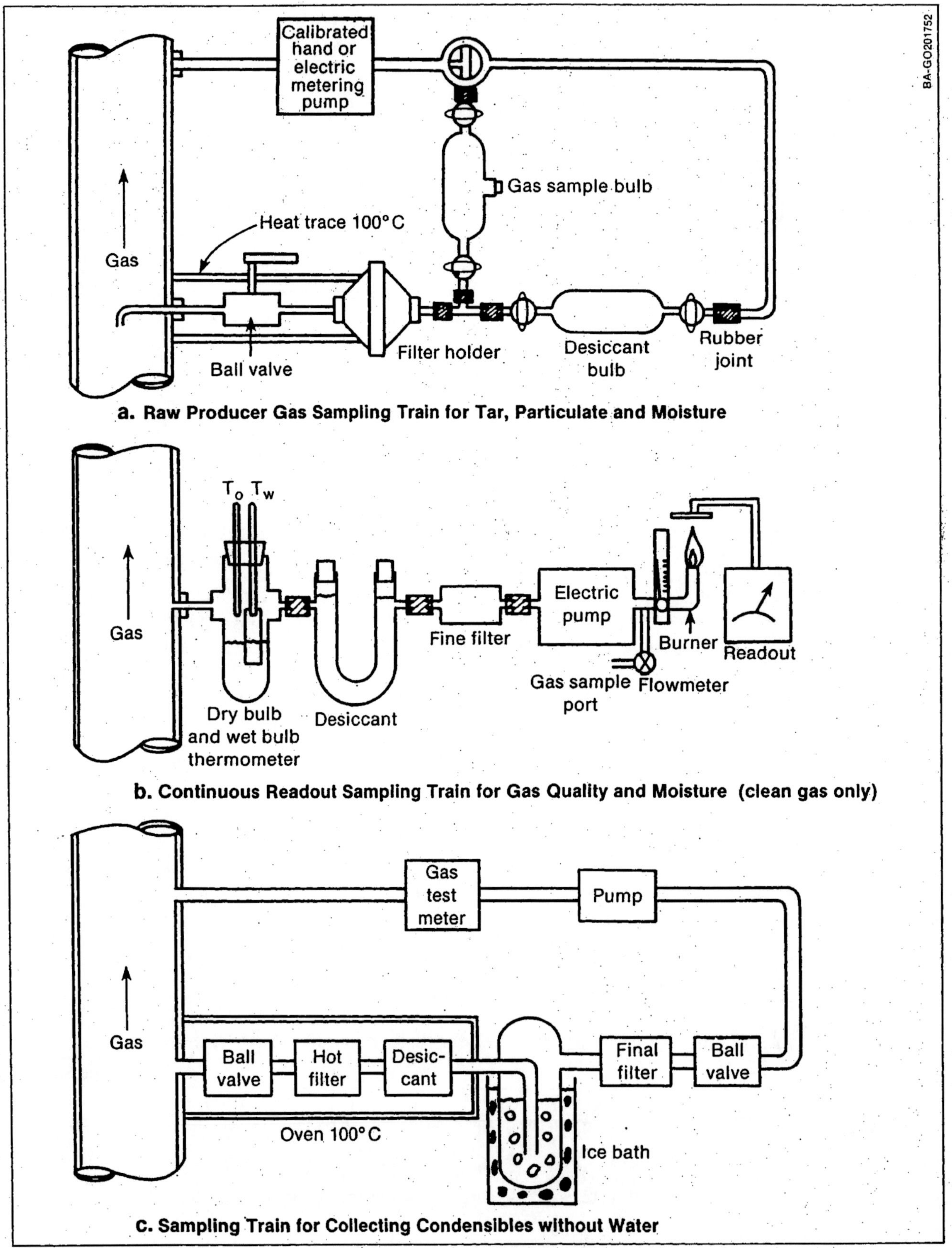

Fig. 7-12. Sampling train configurations

The desiccant container should be sealed for transport and weighing. Raw-gas moisture measurement is essential to mass balance calculations.

Finally, where volume is measured (Fig. 7-12(a) and (c)), means must be provided to pull a known quantity of gas through this train. Hand-held positive displacement vacuum pumps are made by a number of suppliers (e.g., Mine Safety, Draeger, and Gelman). We have also used a hand-held rubber aspirator bulb and found that 70 strokes collected 3 L of gas (0.1 ft^3). We also have used a Dwyer smoke test pump. The gas meter is required only for initially calibrating the sampling train and pump, since counting strokes yields adequate precision for measuring the test-gas volume. A decision on the amount of gas to be sampled should be based on the anticipated impurities in the gas and the contaminant sample quantity required for the specific analysis methods available. For instance, 50% grey scale analysis requires a 0.5 mg sample on a 47 mm filter disc.

Weighed samples require a 5 to 30 mg sample size for analytical balances with 0.1 mg readability.

7.5.2 Cleaned Gas

If the gas is cleaned sufficiently for engine use, it will be necessary to pass a much larger sample (usually 1 m^3) through the filter. A mechanical pump capable of pulling a moderate vacuum, such as a motor-driven vacuum pump or a calibrated air-sample pump, is recommended. The positive-displacement meter can also be located in the collection train between the pump and the gas return if the system pressure is close to atmospheric pressure. It is imperative to protect the pump and meter with a large absolute filter because any tar or particulates entering the pump or meter will rapidly affect their performance.

7.6 Chemical Gas Composition

7.6.1 Gas Samples for Chemical Analysis

The gas composition can be measured either continuously (on-line) or through discrete samples taken periodically from the gas stream. These methods will be discussed separately. Before the gas is analyzed, it must be drawn from the system and cleansed of tar and particulate contaminants, as described previously.

Batch-sampling requires collecting a sample of gas in a suitable container (e.g., glass cylinder, metal cylinder, Tedlar gas sample bag or syringe), as shown in Fig. 7-13. The subsequent analysis is only as good as the sample, and it is easy for gas leaks to spoil a sample after it has been taken. Therefore, it is important to use extra care to avoid leaks either into the sampling train while the sample is being taken or out of the sample bulb before the analysis is made.

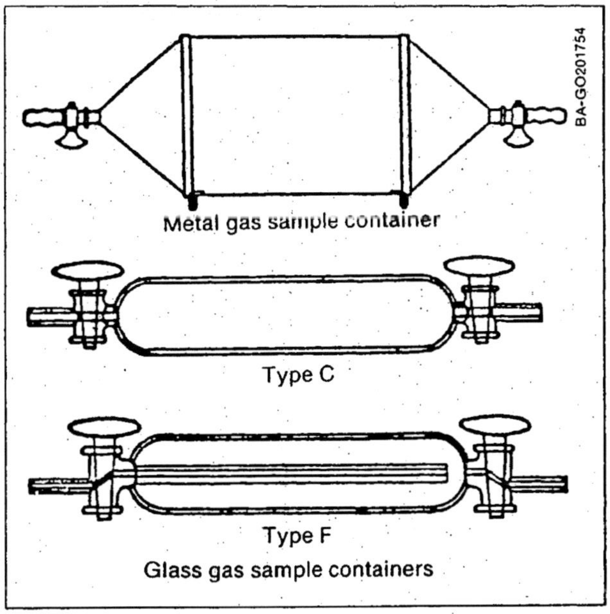

Fig. 7-13. Gas sample containers (Source: Strauss 1975, p. 13. © 1975. Used with permission of Pergamon Press)

When possible, the sample cylinder should be evacuated or, alternatively, should be very thoroughly flushed. The cylinder should be filled to at least a small positive pressure from the pump (Fig. 7-14), so that air cannot leak in before analysis. A positive pressure sample can be collected without a pump by chilling the cylinder before the gas is taken, so that a positive pressure develops as the gas in the cylinder warms to room temperature. Gas samples should be drawn from a point as close as practical to the gasifier outlet, in order to avoid errors due to air leaks in the gasifier piping.

Usually, any oxygen found in the gas can be attributed to air leaks, since oxygen is completely removed in the gasifier. When oxygen is found in the gas, the composition can be converted to an "air-free" basis by subtracting the oxygen and the corresponding ratio of nitrogen (the N_2/O_2 ratio in air is 79/21).

Some gas sample containers are shown in Fig. 7-13. A rubber septum is a desirable feature that permits one to extract the gas sample with a hypodermic syringe for injection into a gas-chromatograph without opening the stopcocks. The hypodermic syringe for injecting samples into the gas chromatograph should have a valve at the needle that can be closed between filling the syringe and analysis. Valved syringes are available as accessories from gas chromatograph manufacturers. The metal cylinder of Fig. 7-13 can contain gas at a much higher pressure than the glass system. It is important to use leak-proof valves rather than needle valves on this container and to avoid stopcock grease, which has a high hydrogen solubility. A syringe also can be used to collect a gas sample. If standard gas sample containers are unavailable, gas samples can be collected in glass bottles by water displacement, inserting a stopper while the bottle is submerged and sealing by dipping the stoppered opening in paraffin. Whichever container is used, the samples should be tested as soon as possible, since hydrogen can rapidly diffuse through rubber seals and stopcocks, thereby changing the gas composition in a few hours.

7.6.2 Methods of Analysis

7.6.2.1 Gas Chromatography

Gas chromatography (GC) is the most widely used method of gas analysis. It depends on the ability of certain adsorbent materials to selectively slow the rate of gas passage through a column packed with the adsorbent. Hydrogen is slowed least, CO, N_2, and O_2 are

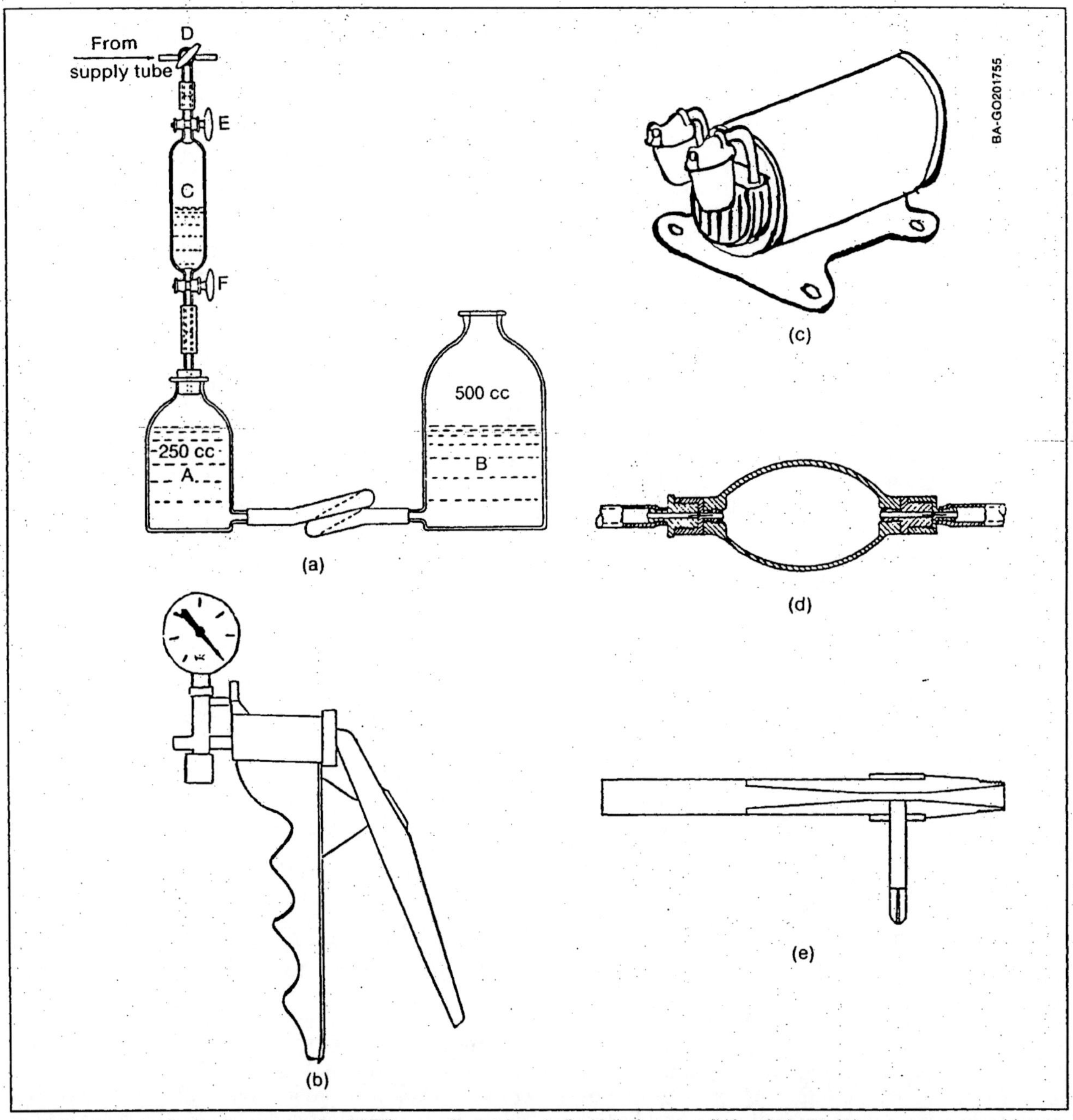

Fig. 7-14. Apparatus for drawing gas samples: (a) Filling sample containers by liquid displacement; (b) hand-operated piston vacuum pump; (c) motor-driven rotary vacuum pump; (d) rubber bulb hand aspirator; (e) Chapman filter pump (Source: (a, d, e) ASME 1969, Figs. 6 and 7)

slowed to a greater extent, and water and CO_2 are slowed to the greatest degree. The gas sample is mixed with a carrier gas; usually, helium is used because it does not occur naturally in the sample. A detector, which is inserted into the gas stream at the end of the column, records on a chart recorder both the time of passage and the quantity of each component. The presence of a particular gas component is indicated by a peak. The quantity of that gas is then determined by integrating the area under the peak in the curve and compared with that in a calibration gas of known composition. More advanced recorders include automatically controlled valving, integration of the response curves, calculation of gas quantity from calibration factors, and a printout of the composition results. Such a printout is shown in Fig. 7-15.

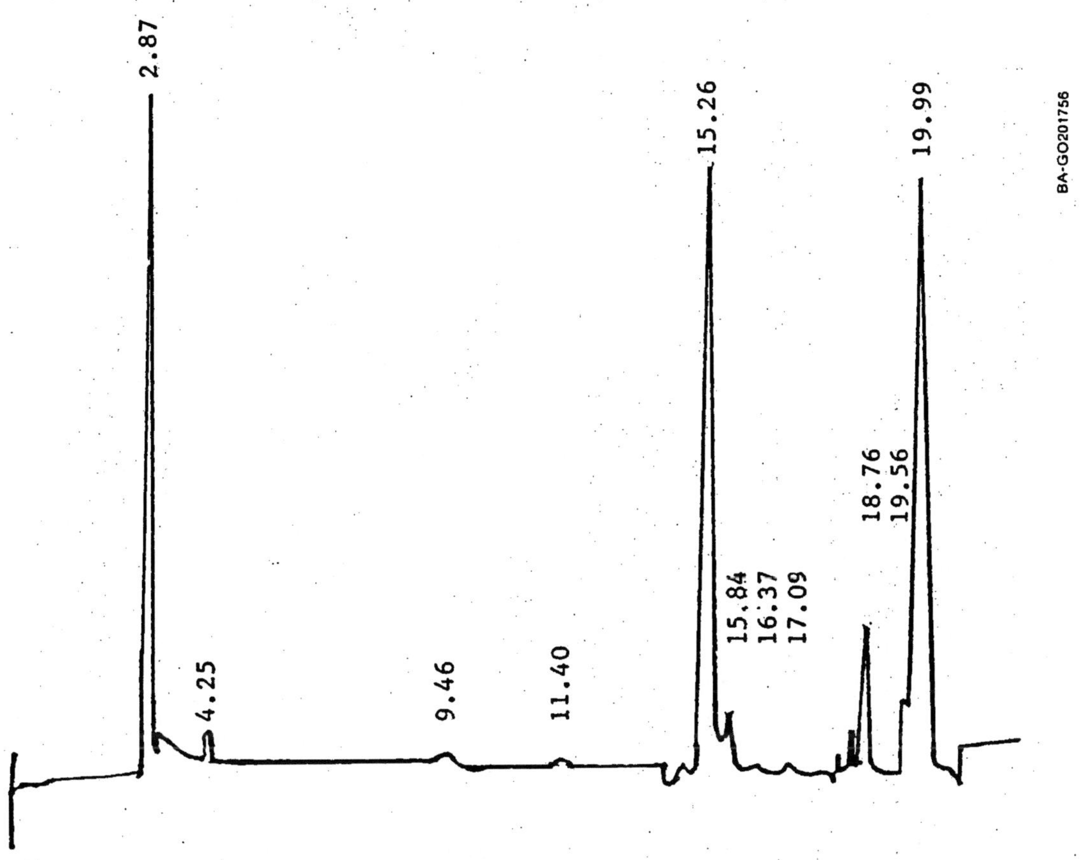

[HP] 27: MANUAL INJECTION AT 15:49 FEB 7, 1984
RUN E7 SAMPLE W6

RUN TIME	AMOUNT	NAME
2.87	25.723	HYDROGEN
4.25	0.15	PROPYLENE
9.46	0.08219	TRANSBUTENE
15.26	14.592	CO2
15.84	1.318	ETHYLENE
16.37	0.136	ETHANE
17.09	0.367	ACETYLENE
18.41	0.49	OXYGEN
18.76	1.949	NITROGEN
19.58	3.9303	METHANE
23.92	0.002827	BACKFLUSH

Fig. 7-15. Typical gas chromatography printout

The most common GC detector performs analyses by measuring the thermal conductivity of the gas (TC detector) and is the most suitable for producer gas measurement. With this type of detector, helium (or a hydrogen-helium mixture, see below) is often used because of its abnormally high thermal conductivity relative to other gases.

The flame-ionization detector (FID) measures the number of ions produced in a flame and is particularly useful for detecting hydrocarbon species. The FID is not particularly useful for producer gas, since producer gas contains few hydrocarbons other than methane.

The response of the TC detector to low levels of hydrogen in the inert carrier gas is nonlinear, and this leads to ambiguous results. There are two effective solutions to this problem. A heated palladium tube at the inlet can be used to selectively diffuse the hydrogen out of the sample into a separate nitrogen gas stream; in this secondary stream, hydrogen yields a linear response (Carle method). Alternatively, adding hydrogen to the helium carrier gas will move the baseline onto the linear region of the TC-detector response curve.

The position of a peak on the time scale of the recorder chart indicates the time of retention and is characteristic of each particular gas component. The area under the peak, obtained by analog or digital integration, indicates the volume of each gas present. Although retention times and sensitivities are listed for each adsorbent material, aging and drift are common to column packings, so it is necessary to calibrate the instrument daily to obtain an accuracy on the order of 1%. For this purpose, it is necessary to have a cylinder of previously analyzed standard gas. These cylinders are available from GC equipment manufacturers.

Although samples are usually collected as needed, it is possible to use automatic sampling with the GC to give a measurement of gas composition at regular intervals. The GC analysis cycle time depends on both the retention time of the columns used and the number of species analyzed. This time is typically 30 minutes, but note that the warmup time for the GC is one day.

A number of companies, including Carle, Hewlett-Packard, and Perkin-Elmer, manufacture satisfactory units for $3000 to $30,000 and provide excellent instruction and service.

7.6.2.2 Orsat Gas Analysis

The Orsat gas analysis system was developed to measure the gases CO_2, CO, O_2, H_2, and CH_4. It was the principal measurement method used before the GC was developed in the 1950s and is more reliable and less costly than GC; however, it requires more time (typically 30 minutes of full operator attention per analysis) and more skill.

The Orsat analysis depends on the ability of certain chemicals to react selectively with each gas component of the producer gas mixture. The components are absorbed in the order of CO_2, O_2, CO, then H_2 and CH_4, and the analysis reports the volume percent of each component directly.

Orsat analysis equipment is portable, does not require AC power, has no warmup time, and can be purchased (along with the required chemicals) from scientific supply houses for $500 to $1000.

7.6.2.3 On-Line Gas Measurement

It is convenient to have continuous "on-line" measurement of all the gas components to show instantaneous changes in composition that otherwise would not be shown by batch sampling. Methods for on-line gas analysis include flame observation, combustion calorimetry, infrared absorption, thermal conductivity, and mass spectrometry.

The heat content of the gas is a measure of a gasifier's performance and can be calculated from the gas composition (see Fig. 7-1 and Table 7-2). Most gasifier facilities, if they have gas analysis equipment, use an Orsat analyzer or a gas chromatograph, so that normally a value is available only after a considerable time delay (10-30 min). It is desirable to have a continuous indication of gas quality.

Continuous immediate readout of producer gas composition, however, has been achieved in two ways. One method, used at U. C. Davis, uses infrared (IR) absorption for continuous CO, CO_2, and CH_4 analysis with a thermal conductivity detector for continuous H_2 determination. The second method uses a mass spectrometer to give immediate on-line digital readout of all gases present, CO, CO_2, H_2, O_2, H_2O, CH_4, and high hydrocarbons (Graboski 1986).

The calorimeter shown in Fig. 7-16 is a precise primary standard for measuring HHV of the gas. Combustion air, fuel rates, delivery temperatures, and pressures are carefully measured. Heat-transfer air is also metered for inlet flow, temperature, and pressure. A counterflow heat exchanger cools the combustion products to the air inlet temperature (60°F) and simultaneously condenses water vapor to a liquid. The temperature rise of the heat-transfer air is directly proportional to the HHV of the fuel gas. The equipment pictured in Fig. 7-16 was designed for gas with a HHV of 1000 Btu/scf and may require modification of the burner to use producer gas with a HHV of 150 Btu/scf.

Other simpler, more relative methods are available and may be sufficient for many applications. It is informative simply to observe the gas flame during operation.

Flame length tends to increase with the gas heating value; flame luminance increases with hydrocarbon and tar content. After the operator has gained

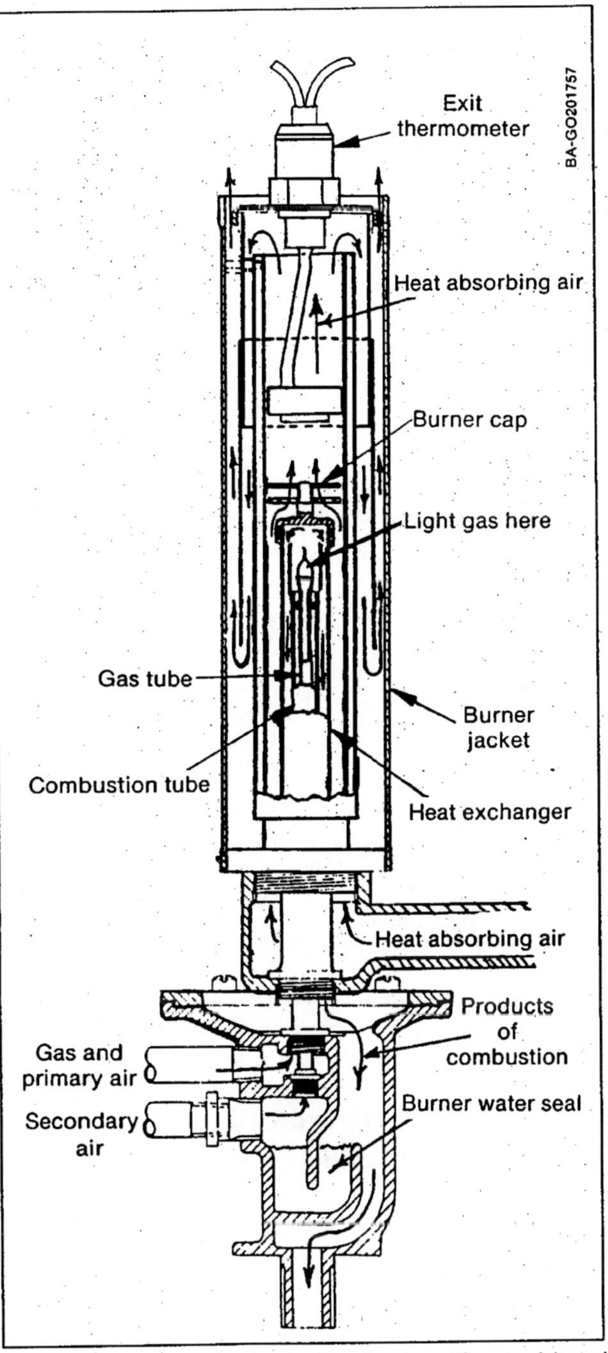

Fig. 7-16. Gas calorimeter combustion chamber (Source: Adapted from ASTM 1977)

experience, the flame can reveal a good deal on the functioning or malfunctioning system.

One author (Das) has used a small burner/thermocouple monitor shown in Fig. 7-12(b), which produces a temperature signal roughly proportional to the heat content of the clean gas.

The accuracy of continuous sampling equipment is subject to accumulation of gas contaminants, so prefiltration should be used as in Fig. 7-12(b) of adequate capacity for prolonged use and adequate efficiency for equipment protection.

7.6.3 Water Vapor Analysis

Water vapor can be determined by many methods. The three most suited to producer gas are psychrometry, condenser outlet temperature, and gravimetric methods.

7.6.3.1 Psychrometry

Water content can be determined by measuring the wet- and dry-bulb temperature of the gas as in Fig. 7-17. The moisture content is then calculated from a psychrometric chart (Fig. 7-18 or 7-19, depending on gas temperature) to find the moisture as absolute humidity.

$$\text{Moisture wt \%} = \text{Absolute humidity} \times 100 \qquad (7\text{-}3)$$

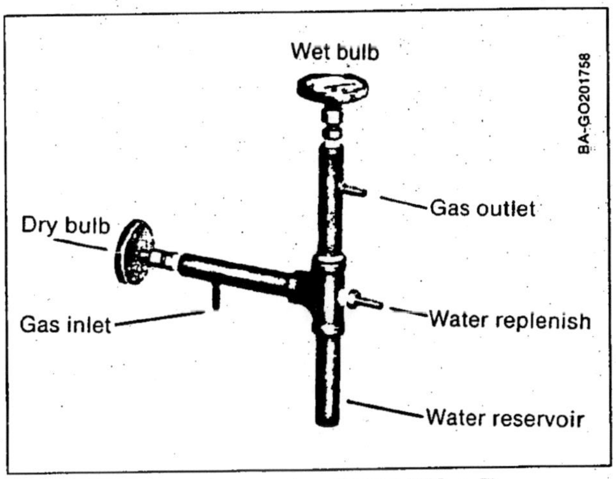

Fig. 7-17. Gas stream hygrometer (Source: RAC, p. 7)

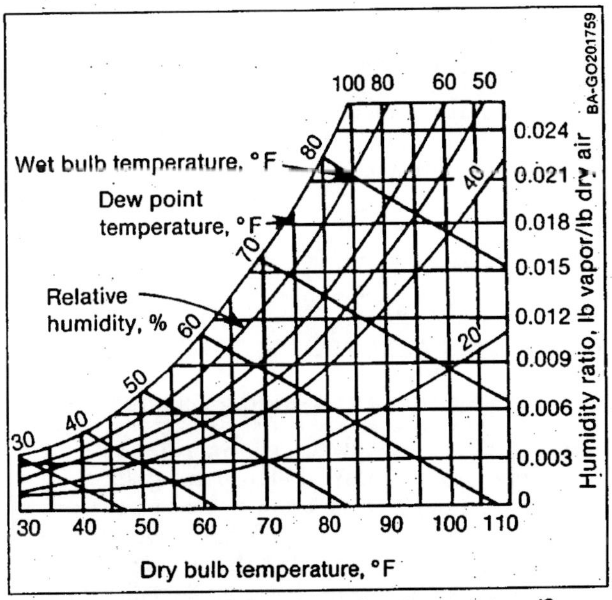

Fig. 7-18. Psychrometric chart for medium temperatures (Source: Adapted from ASHRAE 1981)

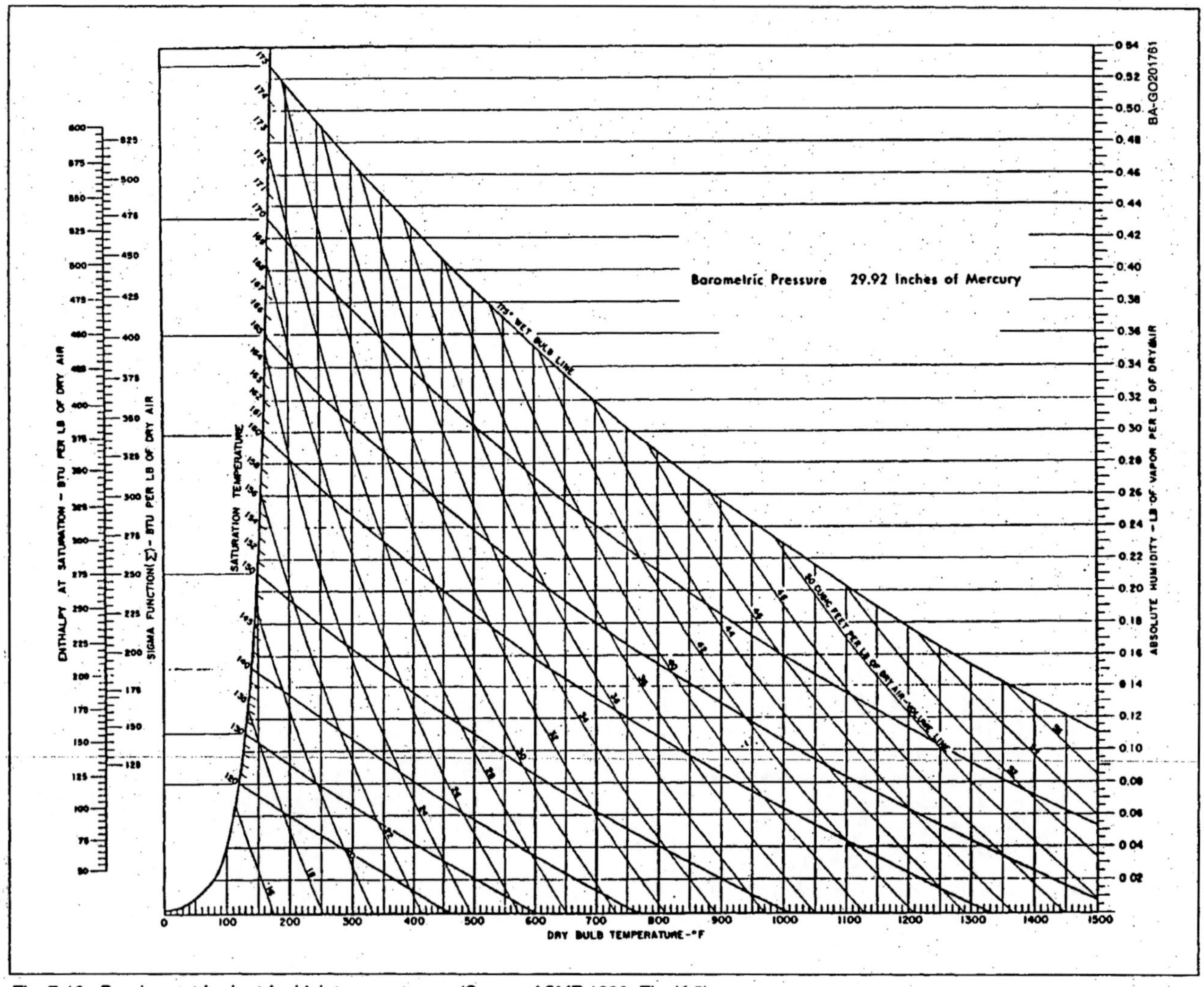

Fig. 7-19. Psychrometric chart for high temperatures (Source: ASME 1980, Fig. K-3)

7.6.3.2 Gravimetric

Gas moisture also can be determined by passing a measured volume of gas through a preweighed dryer assembly containing desiccant. Moisture can be calculated as

$$\text{Moisture wt \%} = \frac{100 \times \text{Weight gain}}{\text{Sample vol.} \times \text{Gas density}} \quad (7\text{-}4)$$

7.7 Analysis of Test Data

7.7.1 Mass Balances and Energy Balances

A typical gasifier mass balance, shown in Table 7-7, is an accounting of all mass inputs to the gasifier (or gasifier system) and all mass ouputs over a given time. Since the law of conservation of mass requires that mass be conserved in any process, the total mass input must equal the total mass output. Any deviation from an exact balance indicates either an error in measurement, or that some important flow has been overlooked.

The total inlet and outlet mass flows must not only balance each other, but also the inlet and outlet mass flows of each element (in this case, carbon, hydrogen, and oxygen) must balance. This elemental mass balance is a more rigorous test of measurement procedure; the sources of error in the global (overall inlet and outlet) mass balance may be pinpointed by the elemental mass balance.

Table 7-8 shows an energy balance, which is obtained by tabulating the energy associated with all input and output streams. The law of conservation of energy, which requires that energy be conserved, therefore provides a means for evaluating efficiency, finding instrumental errors, or calculating quantities that cannot be measured directly.

Table 7-7. Mass Balance

Run	Inputs (kg/h) Wet Chips	Dry Air	H_2O	Total	Outputs (kg/h) Dry Gas	Char	Tar	H_2O	Total	Closure (%)
98	32.0	43.1	0.5	75.6	66.3	0.9	0.09	7.4	74.7	98.9
910	32.0	45.2	0.5	77.7	68.0	0.9	0.14	7.4	76.4	98.3
920	35.7	62.1	0.4	98.2	92.9	1.4	0.09	7.1	101.5	96.6
101	52.4	76.9	0.8	130.1	116.9	1.7	0.09	13.3	132.0	98.6
929	52.7	74.0	1.0	127.7	113.0	1.8	0.14	12.2	127.1	99.5
106	58.1	76.8	0.5	135.4	117.0	1.2	0.09	10.8	129.1	95.4
1119	89.1	112.1	1.0	202.2	173.8	3.0	0.27	19.7	196.8	97.3
1117	96.2	140.4	0.7	237.3	218.9	2.5	0.18	22.9	244.5	97.0
1221	140.5	202.1	1.1	343.7	302.1	4.1	0.54	43.3	350.0	98.2

Source: Walawender 1985, p. 918.

Mass and energy balances have only been applied occasionally to gasifier development because of the difficulty and expense of measuring all flow streams. Detailed mass and energy balances usually can be performed at universities in chemical engineering laboratories or at major research laboratories, and only a few have been performed on air gasifiers. If gasification is to become a developed field, it is necessary to perform mass and energy balances.

7.7.2 Flow Rate Characterization

The variation of gas quality of a gasifier with flow rate helps determine optimum sizing parameters. Note that in Fig. 7-20 total hydrocarbons and tars steadily decrease with increased flow. In Fig. 7-21 we see that maximum heating value, peak CO and H_2, and maximum efficiency do not coincide. Peak efficiency occurs at more than twice the flow rate for maximum heating value.

Actual hearth load (see Section 5.7.3) for sizing depends on the application. Heating applications that need maximum flame temperature should be sized with lower hearth rate load to provide maximum heating value. Heating applications that need maximum fuel economy should be sized with higher hearth load to coincide with the peak efficiency curve. Engine applications should be sized for maximum hearth load to coincide with peak of the efficiency curve in order to allow maximum room for turndown.

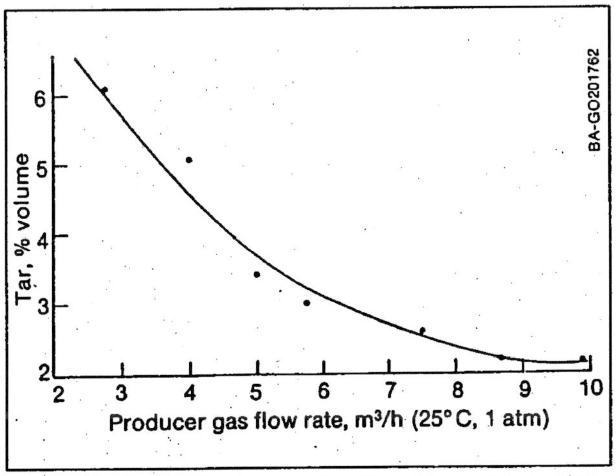

Fig. 7-20. Tar versus flow for rice hull gasifier (Source: Kaupp 1984b, Fig. 10-88)

7.8 Particle-Size Measurement

Knowledge of the size distribution and other characteristics of gas contaminants is helpful for cleanup design. Table 7-9 presents particle-size analysis methods and examples of equipment available for characterizing particle size.

7.8.1 Typical Particle-Size Distributions

The particle-size distribution of solid char and ash for raw gas shown in Fig. 7-12 was produced by mechanical screen separation for both the Imbert (tuyere and hearth constriction) and the SERI unconstricted gas producers. The difference between the two gasifiers is caused by the grate design. Gas from the Imbert gasifier exits upward through a settling space that retains larger particles. On the other hand, the SERI oxygen gasifier passes all solids through the gas outlet. Note that the overall slope is the same for both size distributions.

7.8.2 Sieve Analysis

The distribution of large solid particles greater than 40 µm may be determined using sieves. Table 7-10 lists actual sieve size for various mesh size.

7.8.3 Microscopic Size Analysis

Particles captured on a filter disc can be counted by microscopic examination. However, particles smaller than 10 µm are difficult to see under a light microscope, and liquid-droplet sizes cannot be determined by this method because droplets, once captured, coalesce, leaving no evidence of their original size.

7.8.4 Aerodynamic Size Analysis

Aerodynamic sizing can be accomplished with either a cascade impactor or a cascade cyclone. Particles and

Table 7-8. Energy Balance

Run	Dry chip rate (kg/h)	Chip moisture (% wet basis)	Matl. balance closure (%)	Dry gas HHV (MJ/m^3)	Gas yield (m^3/kg DC[a])	Char yield (kg/100 kg DC)	Tar yield (kg/1000 kg DC)	Tar in dry gas (ppm)	Dry air in (kg/kg DC)	H_2O out (kg/kg DC)	Energy yield (MJ/kg DC)	Total energy out (MJ/h)	Mass conversion efficiency (%)	Cold gas efficiency[b] (%)
98	26.7	16.52	98.9	5.33	2.22	3.28	2.55	1370	1.61	0.277	11.82	315	87.7	70.2
910	26.7	16.52	98.3	5.18	2.28	3.21	4.59	2003	1.69	0.277	11.82	315	87.5	70.2
920	33.6	6.01	96.6	5.33	2.48	4.06	2.71	977	1.85	0.212	13.14	441	94.7	78.1
927	41.2	8.11	85.4	5.63	2.54	3.08	3.30	1180	1.75	0.438	14.31	590	98.1	85.0
922	47.8	5.75	85.1	5.44	1.43	2.76	1.90	1199	1.05	0.186	7.76	370	74.9	46.1
101	48.7	6.99	98.6	5.59	2.18	3.45	1.86	777	1.56	0.273	12.19	594	89.8	72.5
929	48.8	7.41	99.5	5.63	2.10	3.72	2.79	1205	1.52	0.250	11.82	576	88.5	70.2
1011	49.5	7.31	91.6	5.44	2.55	3.94	4.59	1607	1.88	0.321	13.84	685	96.1	82.2
1015	52.3	7.95	88.8	5.33	2.73	4.69	5.21	1679	2.08	0.381	14.52	759	97.5	86.3
924	53.7	5.50	91.8	5.66	1.62	2.79	2.54	1426	1.13	0.209	9.11	490	80.9	54.2
106	54.3	6.67	95.4	5.48	1.95	2.18	1.67	776	1.42	0.199	10.68	579	86.5	63.5
123	54.3	14.61	94.2	5.51	1.90	4.27	3.35	1594	1.35	0.236	10.44	567	83.0	62.1
104	63.7	7.72	92.6	5.55	2.61	2.71	2.85	989	1.90	0.307	14.50	924	96.2	86.2
1129	68.9	14.27	91.9	5.59	2.85	3.82	3.96	1254	2.04	0.281	15.89	1094	98.1	94.4
121	72.5	15.07	80.4	5.63	1.39	3.57	3.76	2445	0.99	0.169	7.81	567	70.8	46.5
1119	74.6	16.32	97.3	5.74	2.11	3.96	3.65	1567	1.50	0.264	12.08	901	86.0	71.8
1117	85.4	11.26	97.0	5.78	2.30	2.92	2.13	830	1.64	0.268	13.31	1136	92.2	79.1
1110	119.7	11.56	91.0	5.74	1.79	1.67	3.03	1544	1.27	0.206	10.23	1224	81.6	60.8
1221	125.5	10.67	98.2	5.44	2.10	3.25	4.34	1803	1.61	0.345	11.42	1434	87.9	67.9
all	—	—	—	5.51	2.17	3.33	3.20	1380	1.57	0.268	11.93	—	88.3	70.9
mean	—	—	—	0.15	0.41	0.72	0.99	424	0.31	0.067	2.25	—	7.5	13.0
data														
<95%	—	—	—	5.51	2.19	3.34	2.92	1256	1.60	0.263	12.03	—	87.9	71.5
mean	—	—	—	0.19	0.14	0.54	0.99	433	0.12	0.039	0.76	—	2.1	4.5

[a]DC = dry chips

[b]Energy in gas/energy in wood

Source: Walawender 1985, p. 917

droplets are collected inertially as a function of their aerodynamic size. Once they have been separated by size, there is no need to prevent the droplets from coalescing. Quantities, and therefore distributions, are subsequently determined by the relative masses represented in each size grade.

7.8.5 Graphic Analysis of Size Distribution

The cumulative particle-size distribution shown in Fig. 7-4 plots as a straight line on probability paper, thereby indicating log normal distribution about a mean particle diameter, d_p at 50%, with a geometric

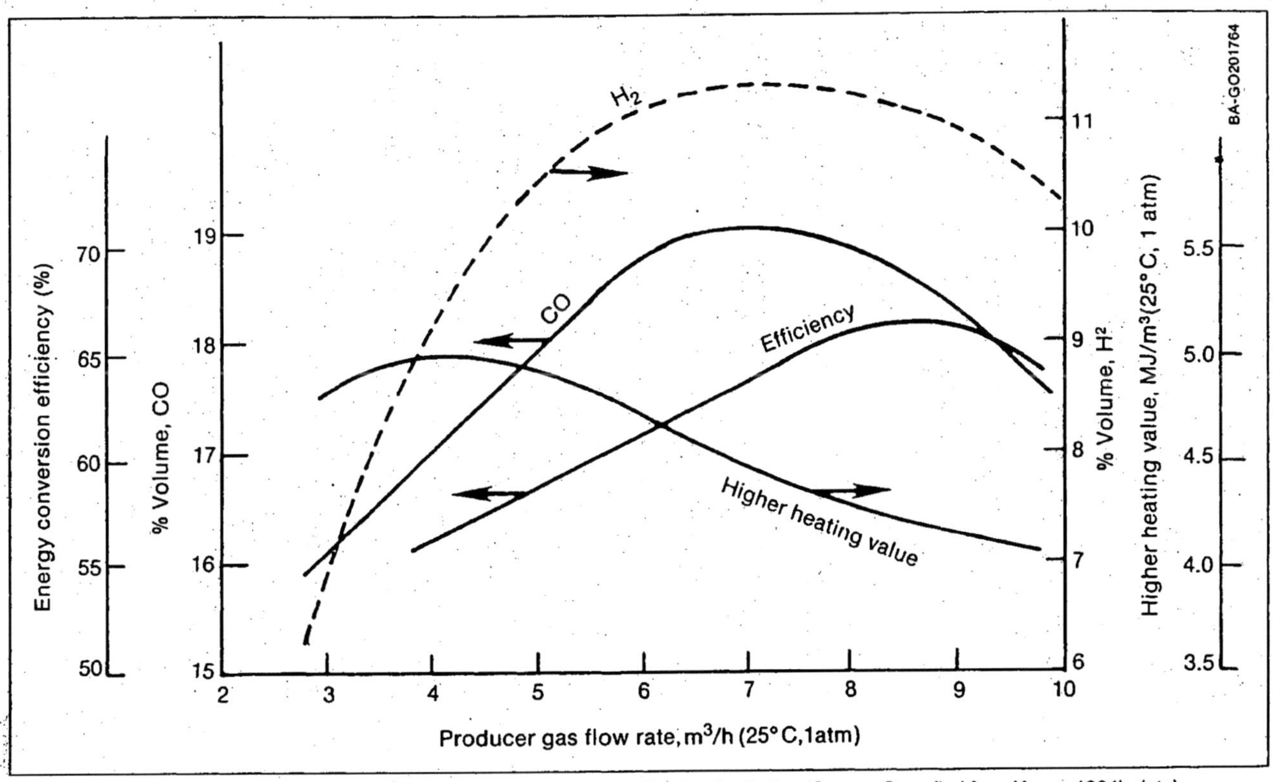

Fig. 7-21. Flow rate effects on efficiency, heating value, and gas composition for rice hulls (Source: Compiled from Kaupp 1984b data)

Table 7-9. Examples of Size-Analysis Methods and Equipment

Particle Size, µm	General Method	Examples of Specific instruments*
37 and larger	Dry-sieve analysis	Tyler Ro-Tap, Alpine Jet sieve
10 and larger	Wet-sieve analysis	Buckbee-Mears sieves
1-100	Optical microscope	Zeiss, Bausch & Lomb, Nikon microscopes
	Microscope with scanner and counter	Millipore IIMC system
	Dry gravity sedimentation	Roller analyzer, Sharples Micromerograph
	Wet gravity sedimentation	Andreasen pipet
	Electrolyte resistivity change	Coulter counter
0.2-20	Light scattering	Royco
	Cascade impactor	Brink, Anderson, Casella, Lundgren impactors
	Wet centrifugal sedimentation	M.S.A.-Whitby analyzer
0.01-10	Ultracentrifuge	Goetz aerosol spectrometer
	Transmission electron microscope	Phillips, RCA, Hitachi, Zeiss, Metropolitan-Vickers, Siemens microscopes
	Scanning electron microscope	Reist & Burgess system

*This table gives examples of specific equipment. It is not intended to be a complete listing, nor is it intended to be an endorsement of any instrument.
Source: Perry 1973, Table 20-33.

Table 7-10. Sieve Number Versus Mesh Size

Sieve Number	Mesh Size, μm
80	180
100	150
120	125
140	106
170	90
200	75
230	63
270	53
325	45
400	38

standard deviation σ as indicated by the slope. Note that both distributions shown in Fig. 7-4 have the same overall slope. This slope is typical of large materials that have been broken up into a wide range of smaller particles. It will be helpful for us to express this slope as the geometric standard deviation, σ_g

$$\sigma_g = d_{p84}/d_{p50} = d_{p50}/d_{p16} \qquad (7\text{-}5)$$

where d_{p50} is the diameter for which 50% of the total particles are captured. The other subscripts denote similar cumulative percentages of particles smaller than the respective particle diameter d.

The solid particles that pass through a cyclone can be expected to have a mean particle diameter near the cyclone cut diameter, d_{p50}, with a standard deviation, σ_g equal to 2.5, which is characteristic of cyclones, gravity separators, and all Stokes' law particle movement.

7.8.6 Physical Size Analysis

The major methods for particle-size measurement are shown in Table 7-9. Screening and microscopy are used to determine linear dimensions. The Stokes' radius, r_g, is the radius of a hypothetical spherical particle with the same falling velocity and bulk density as the particle. The aerodynamic diameter, d_{pa}, is the diameter of a hypothetical sphere of density 1 g/cm³ that will attain the same falling velocity as the particle in question.

The number mass and area distributions all have the same geometric standard deviation, σ_g.

Scrubber performance can be characterized similarly by the size particle diameter which is captured at 50%. The preciseness of the size cutoff point is characterized by the value of σ_g. Various scrubbers and separators are compared in Table 8-1 for cut diameter and sharpness. Note how many have a standard deviation near 2.5. Consider those that have a sharper or broader deviation and why.

Chapter 8
Gas Cleaning and Conditioning

8.1 Introduction

If the gas is to be used in a burner application, an updraft gasifier can be used, and no cleanup will be needed. However, if the fuel gas will be fed to an engine, then a downdraft or other tar cracking gasifier must be used; and the gas must be cleaned and conditioned before it is fed to the engine.

The gas emerging from a downdraft gasifier is usually hot and laden with dust, containing up to 1% tars and particulates. If these materials are not removed properly, they can cause maintenance, repair, and reliability problems much more costly and troublesome than operation of the gasifier itself. In fact, it is likely that more gasifier engine systems have failed because of improper cleanup systems than for any other cause. In particular, the gas is very dirty during startup and should be burned at the gasifier until the system is fully operational. (See Sections 9.3.3, 9.3.4 and 9.4.1 for blowers, ejectors, and flares.)

In 1983, the Minneapolis Moline Engine Company became the first contemporary engine manufacturer to offer a 5000-h warranty on its engine—based on a fuel gas at the engine containing less than 5 mg/Nm^3 of total combined tars and particulates (Mahin, June 1983). This amounts to 99% removal of all dust particles.

Prior to 1950, manufactured gas was widely distributed to homes, and the technology for gas cleanup was used extensively and well understood at that time. The chemical and energy industries of today routinely use the methods that will be described in this chapter.

In order to design effective gas cleanup systems, one must determine the magnitude, size, and nature of the contaminants, and then couple that information with knowledge of methods available for their removal. This chapter presents the principles of gas cleanup, the available types of separation equipment and their respective capabilities and suitability, and some approaches for overall cleanup systems.

The basic cleanup system design strategy should be based on the required cleanliness goals (determined by the application), the order of removal, temperature, and the intended deposit site for collected materials. In addition, size, weight, cost, reliability, the need for exotic materials, water consumption, effluents disposal, the time between cleaning cycles, and the ease of equipment servicing must be considered.

The first step toward producing clean gas is to choose a gasifier design that minimizes production of tars and particulates to be removed, such as a downdraft or other low-tar gasifier, and to make sure that the gasifier is operated in a manner that will minimize particulate production by proper sizing (see Chapter 5). Development of cleaner gasifiers is proceeding in the United States and Europe at a good pace (see Chapter 5).

The next step, which simplifies the handling of captured contaminants, is to remove particulates, tars, and water in the proper order and at the right temperature. If the gas is immediately cooled and quenched in one operation, then char, tars, and water all are removed at one location to form a sticky, tarry mess. If particulates are removed first at a temperature above the dewpoint of the tars (~300°C), tars are removed next at intermediate temperatures (above 100°C), and water is removed last at 30°-60°C, then each separated contaminant can be handled much more easily. The relation between gas temperature and each operation is shown in Fig. 8-1.

The final step of effective gas cleanup is to wisely choose a site for depositing the collected materials. Devices can be classified as either "in-line" or "off-line." In-line devices, such as fabric bags and packed-fiber filters, cut off the gas flow as they become filled with the tar or particulate material that they have

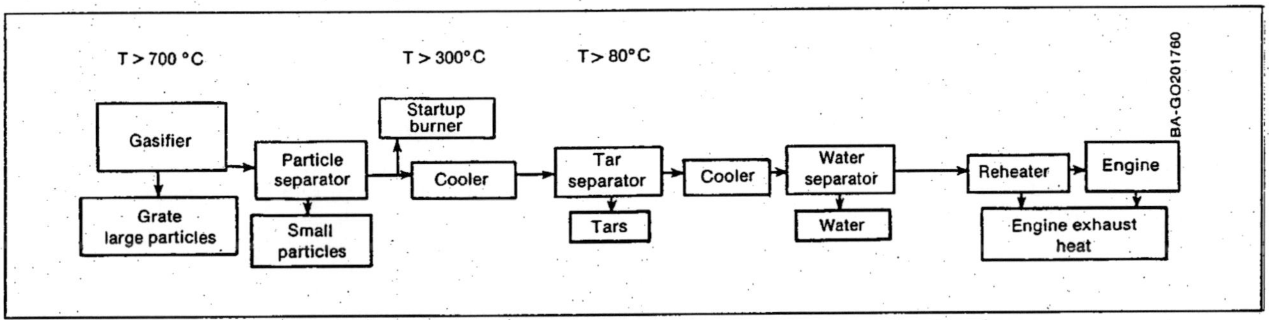

Fig. 8-1. Schematic relationship of gas temperature to contaminant removal

captured. The pressure drop across the cleanup system steadily rises with the accumulation of captured materials, requiring frequent or automatic cleaning or replacement. Collection efficiency is low for a clean, in-line filter but climbs steadily with the increasing pressure drop as the filter becomes plugged. Collection efficiency measurements of in-line filters should clearly indicate loading effects or be averaged over a full cleaning cycle in order to be meaningful.

Off-line devices, such as cyclone separators, wet scrubbers, and electrostatic precipitators, deposit captured materials outside of the flow path. These devices separate the contaminants into one stream and the gas into another stream. The pressure drops and efficiencies associated with these devices are predictable and independent of the amount of captured materials, eliminating the slow buildup of pressure drop with use. Off-line methods are preferable in applications where they can be used.

8.2 The Power Theory of Gas Cleanup

According to the contact-power theory of gas cleanup (Perry 1973), for a given power consumption, as measured by the gas pressure drop or water flow rate, all cleaning devices give substantially the same collection efficiency, and the collection efficiency increases with increasing power. Some improvement over contact-power theory limitations can be gained by using designs for reduced power consumption that use small parallel streams, multiple stages in series, diffusion, mass transfer, or condensation.

The performance and sharpness of size separation of various cleanup components are compared in Table 8-1 and Fig. 8-2. Particle cut diameter (as written d_{pc} or d_{p50}) is the particle diameter at which 50% of particles are captured. A capture rate other than 50% may be noted as a different subscript. A convenient relationship exists for particle collectors with standard deviation 2.5. Particles with diameters that are double

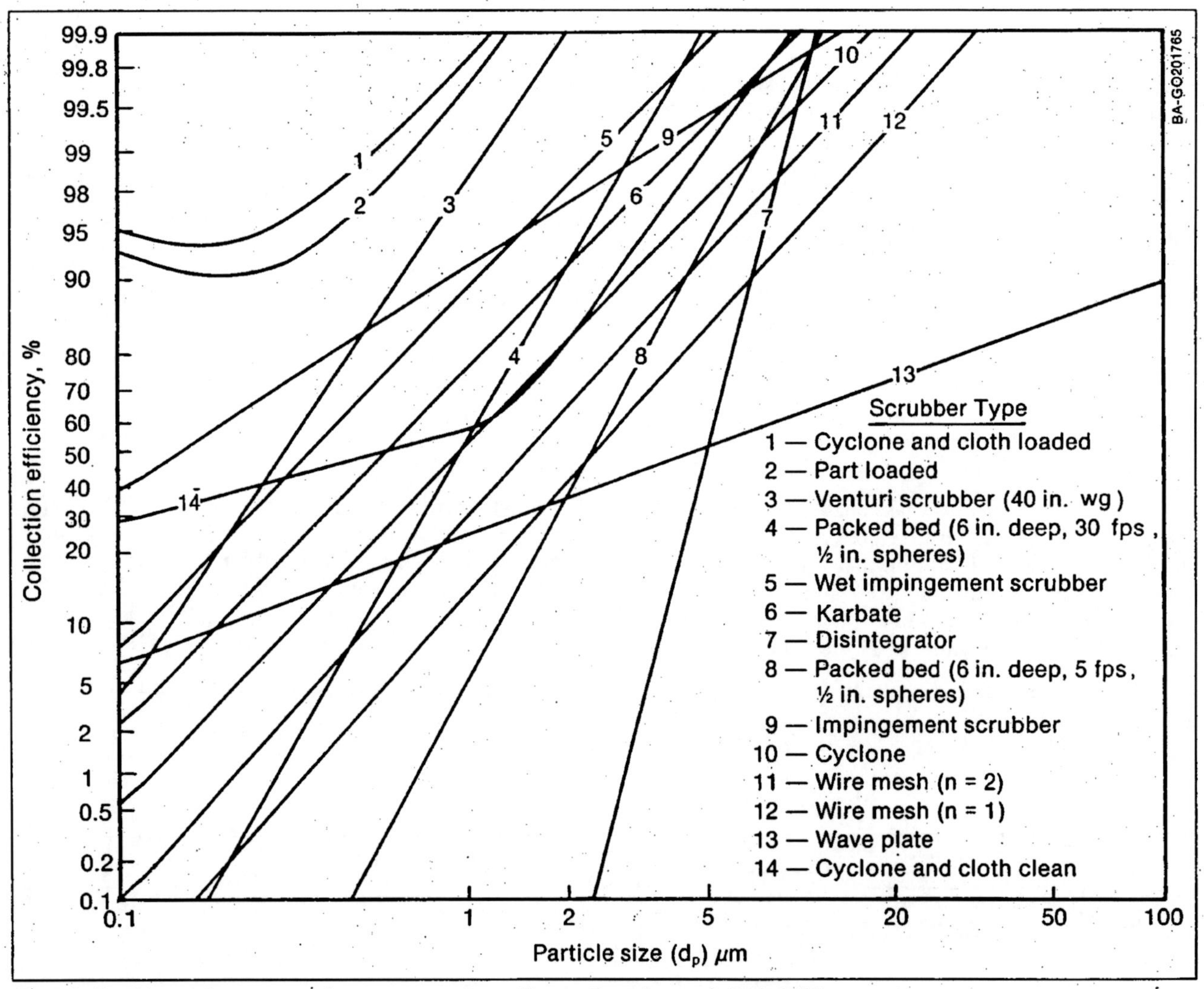

Fig. 8-2. Scrubber performance and sharpness comparison (Source: See reference in Table 8-1)

Table 8-1. Sharpness of Particle Collection Methods

Scrubber Type	Cut Diameter (d_{p50}) μm	Sharpness of Size Separation Standard Deviation $\sigma_g = d_{p84}/d_{p50}$	Reference
Gravity Settling Chamber	50	2.5	Perry 1973
Single Cyclone	1	2.5	Kaupp 1984a
Cascade Cyclones			Das 1986
n in series identical			
n = 2	0.67*	1.8	
n = 3	0.53*	0.65	
n = 4	0.48*	0.55	
Disintegrator	0.6	2.5	Perry 1973
Karbate 1 in. Δp	5	1.3	Perry 1973
Wire Mesh 2 layers	2.4	2.5	Perry 1973
(0.011 in. wire) 3 layers	1.5	2.5	
Packed Bed 6 in. deep			Perry 1973
1/2 Spheres + Saddles			
Velocity 5 fps	2.3	1.65	
Velocity 30 fps	0.95	1.8	
Venturi Scrubber 40 in. ΔpWG	0.3	1.8	Perry 1973
Sieve Plate 1.5 in. ΔpWG			Calvert 1972
Hole Velocity V_h = 75 fps	0.625	2	
Impingement Scrubber			Kaupp 1984a
4 mbar/Stage			
1 Stage	1.5	6.7	
2 Stage	0.076	8.5	
3 Stage	0.015	20	
Fabric Filter	**	1.8	Peterson 1965
Spray Cyclone Dia. 24 in.	2	3	Perry 1973
Δp = 2-10 in. H_2O V_i = 57 fps			
Droplet 40-200 μm			
Spray Tower 2-4 fps	5	2.5	Perry 1973
Droplet 500-1000 μm			
53 ft high			
Wave Plate 90°			Perry 1973
7 waves 7/16 in. radius	5	11	
0.15 in. spacing			

*Cyclone cut diameter is determined by design, down to 1 μm.

**Fabric filter collection efficiency is above 90% for all particle sizes.

the d_{pc} value are captured with 80% efficiency, and those with diameters triple the d_{pc} value are captured with 90% efficiency. In other words,

$$d_{p80} = 2\, d_{p50}, \tag{8-1}$$

and

$$d_{p90} = 3\, d_{p50}\,. \tag{8-2}$$

For example, a cyclone rated at d_{p50} = 10 μm can be expected to capture 50% of particles having 10 μm aerodynamic diameter, 80% of particles d_p = 20 μm, and 90% of particles d_p = 30 μm.

Note that gravity settling chamber, cyclone, disintegrator, wire mesh separator, and spray tower all have the same standard deviation—characteristic of inertial and gravitational separation mechanisms. Sharper particle separation as indicated by smaller standard deviation involves the benefit of other or additional capture mechanisms such as condensation, cascading diffusion, or mass transfer. Similarly, poorer separation sharpness indicates that particle separation is degraded by some mechanism such as reentrainment or fragmentation.

The minimum particle size and typical pressure drops are shown for various scrubbers in Table 8-2. The scrubber selection parameters are the particle size to be removed, the desired collection efficiency, and the maximum pressure drop.

8.3 Gas Cleanup Goals

8.3.1 Gas Contaminant Characteristics

Gas cleanup goals should be based on the degree of contamination, the size, distribution, and nature of the contaminants, as well as the degree of cleanliness required by the equipment. Both solid and liquid contaminants are present in producer gas. The solids are char, ash, and soot, and they cover a wide range of sizes. The liquid is initially a fine mist or fog composed of droplets smaller than 1 μm, but the droplets agglomerate to increase in size as the gas cools.

8.3.2 Typical Dirty Gas

A typical specification for dirty gas might be 100 mg/Nm³ of particulates with mean diameter d_{p50} = 100 μm, a geometric standard deviation $\sigma_g = 3.5$, and tar contamination of 1000 mg/Nm³.

8.3.3 Gas Cleanup Goals

The solids can be quite abrasive, and the tar mist can cause the inlet valves, rings, throttle shafts, and other moving parts to stick. Therefore, both contaminants must be thoroughly removed for reliable engine operation (Gengas 1950; Freeth 1939; Goldman 1939; Kaupp 1984a; Kjellstrom 1981). Successful gasifier-engine systems have required gas cleanliness standards from 10 mg/Nm³ to less than 1 mg/Nm³.

Many gasifiers can produce very clean tar-free gas under certain conditions. However, it is best to design the gas cleanup system with adequate capability for the very dirty gas that is occasionally produced by every gasifier, especially during startup, idling, and when wet fuel is used.

Table 8-2. Minimum Particle Size for Various Types of Scrubbers

	Pressure Drop, in. water	Minimum Particle Size, μm
Spray towers	0.5-1.5	10
Cyclone spray scrubbers	2-10	2-10
Impingement scrubbers	2-50	1-5
Packed- and fluidized-bed scrubbers	2-50	1-10
Orifice scrubbers	5-100	1
Venturi scrubbers	5-100	0.8
Fibrous-bed scrubbers	5-110	0.5

Source: Perry 1973, Table 20-41.

8.3.4 Cleanup Design Target

Requirements for solid-particle removal may be determined from knowledge of average particle diameter d_p, and the worst-case char-ash dust content (c_d). This information can be gathered using isokinetic sampling techniques to collect a representative sample of all particle sizes. If c_{dmax} represents the maximum permissible dust level for engine use, then the maximum permissible dust penetration α is given by:

$$\alpha = c_{dmax} \times 100\% / c_{dmeasured} \quad (8\text{-}3)$$

On a probability plot of the size distribution of the dust shown, for example, in Fig. 7-4, we then find the particle size d_p where cumulative mass % less than d_p equals α. This d_{p50} is then the cyclone cut diameter required for solids cleanup.

A useful rule of thumb is that the cut diameter (d_{p50}) required for a cyclone or scrubber will be about the same as the diameter at the cumulative fraction corresponding to the maximum permissible penetration (Calvert 1972).

8.4 Classification of Particles

Solid particles with diameters greater than 1 μm are called **dust**, and those with diameters below 1 μm are referred to as **fume**. Liquid droplets over 10 μm in diameter are called **spray**, and droplets with diameters below 10 μm are called **mist**. **Aerosols** are solids or liquids suspended in a gas (Calvert 1972).

Dispersion aerosols are materials that begin as large particles and subsequently are broken into smaller sizes. They tend to be coarse with a wide size-range, composed of irregular particles and aggregates (i.e., char-ash dust). **Condensation aerosols** are formed from supersaturated vapors, such as tar and water mist from chemical reactions, and soot formed from cracked hydrocarbon molecules. They tend to be very fine and of uniform size.

8.5 Particle Movement and Capture Mechanisms

Methods for separating particulates from the gas stream usually depend on the mass of the particles. The simplest method allows the particles to settle under the influence of gravity, with the gas stream flowing vertically upward or horizontally. For horizontal separation, the process can be accelerated by providing multiple horizontal plates. Particles also can be separated from the gas on the basis of their mass by using the centripetal force provided by a centrifugal separator.

The details, including the mathematics, of separation can be found in Strauss (1975). However, it is interesting to note that the relation between any separator's 50% capture cut-particle diameter and the capture rate for other sized particles is:

$$d_{p50} = (1/2)\, d_{p80} = (1/3)\, d_{p90} = (1/4)\, d_{p95} \qquad (8\text{-}4)$$

All collectors using the same capture mechanism will be characterized by the same slope (Fig. 8-2) and standard deviation (Table 8-1).

8.6 Dry Collectors

8.6.1 Gravity Settling Chambers

As long as unlimited space and materials are provided, a gravity settling chamber theoretically can achieve any level of particle separation down to the Stokes' limit of about 1 µm. In fact, many of the earliest gasworks used gigantic settling chambers. However, even though it is effective, this method tends to be a bit cumbersome.

8.6.2 Cyclone Separators

Cyclones are simple and inexpensive dust and droplet separators; they are widely used on gasifiers and will be discussed in extra detail in this section.

Hot gas cyclone separators are well suited to remove solid particles larger than 10 µm as a prefilter for the gas cooler and fine particle removal, as shown in Fig. 8-3, for a vehicle gasifier of the 1939-1945 era.

Cyclone separators are also used widely in industrial processes. The principles are well-developed, and designs are easily scaled to the necessary size. High-efficiency cyclone separators can be fabricated readily by a sheet-metal or welding shop. Cyclone design parameters are presented in this section and at greater length in Perry (1973), Calvert (1972), and Strauss (1975). Unfortunately, the small cyclones required for small gasifiers are not available commercially, so they must be custom designed and fabricated.

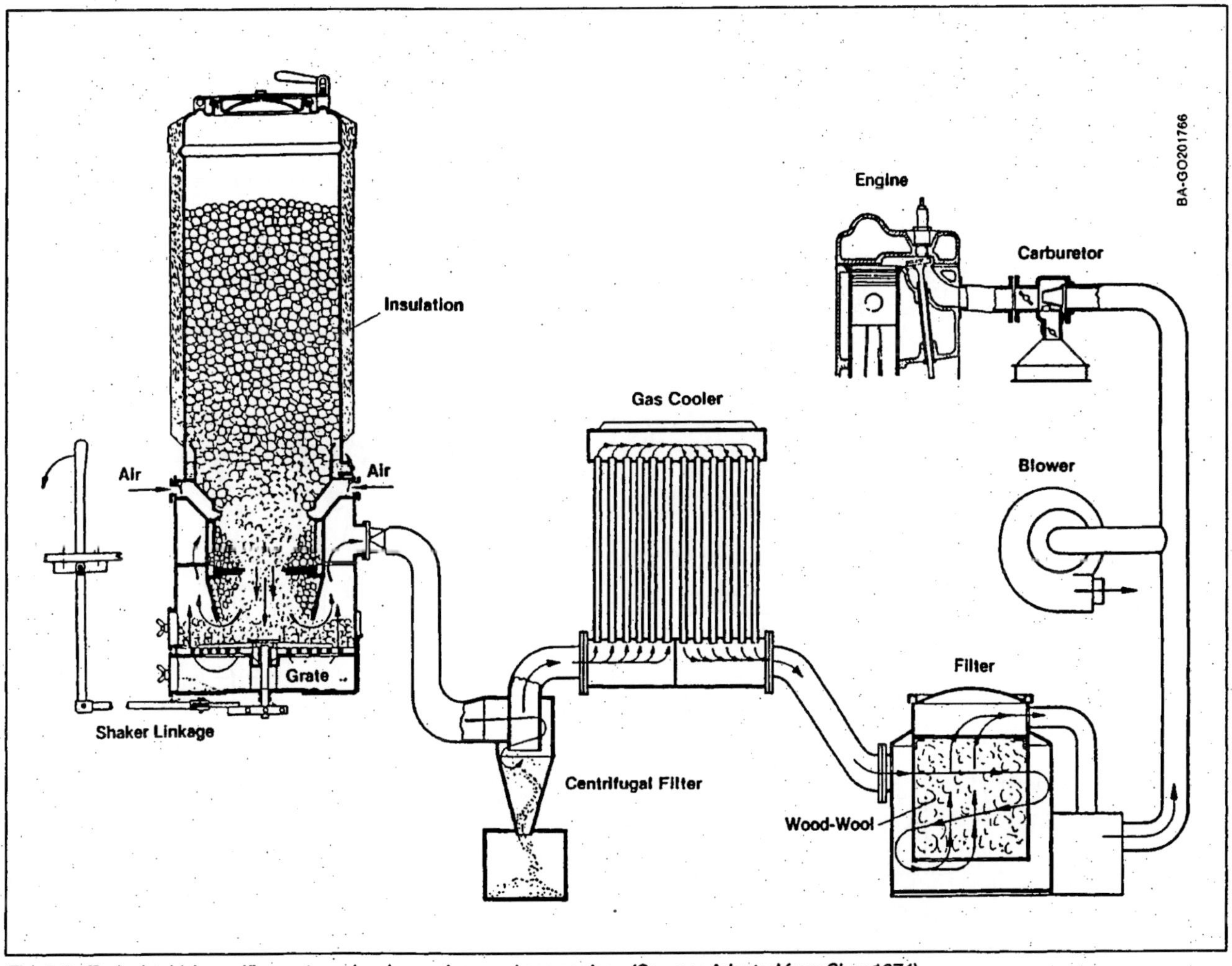

Fig. 8-3. Typical vehicle gasifier system showing cyclone and gas cooler (Source: Adapted from Skov 1974)

8.6.2.1 Cyclone Operating Principles

A cyclone separator imparts a rotary motion to the gases and thereby enhances the settling rate to many times that induced by gravity alone. A cyclone separator is essentially a gravitational separator that has been enhanced by a centrifugal force component. The cyclone separator grade efficiency curve, Fig. 8-4, applies to all cyclone separators, as well as to inertial and gravitational collectors.

Cyclone performance is rated in terms of particle cut diameter or cut size. The cut size, d_{p50}, is the particle size which is captured 50%.

The relationship between particle cut diameters for this type separator is given by Eq. (8-4), where d_p is the particle diameter and the numerical subscript denotes the collection efficiency of that size particle.

8.6.2.2 Cyclone Design Principles

The proportions for high-efficiency cyclone separators are shown in Fig. 8-5.

The particle size that can be separated with 50% efficiency is predicted for general cyclones and for the high-efficiency cyclone proportions of Fig. 8-6 by

$$d_{pc} = \sqrt{9\,\mu_G\, b/[2\pi N_e V_i (\rho_p - \rho_G)]} \qquad (8\text{-}5)$$

From Eq. (8-5) we can derive the relationship between the cyclone separator's particle cut size d_{p50} and the cyclone inlet width b for a given pressure drop as shown in Fig. 8-6. Notice that the effect of temperature

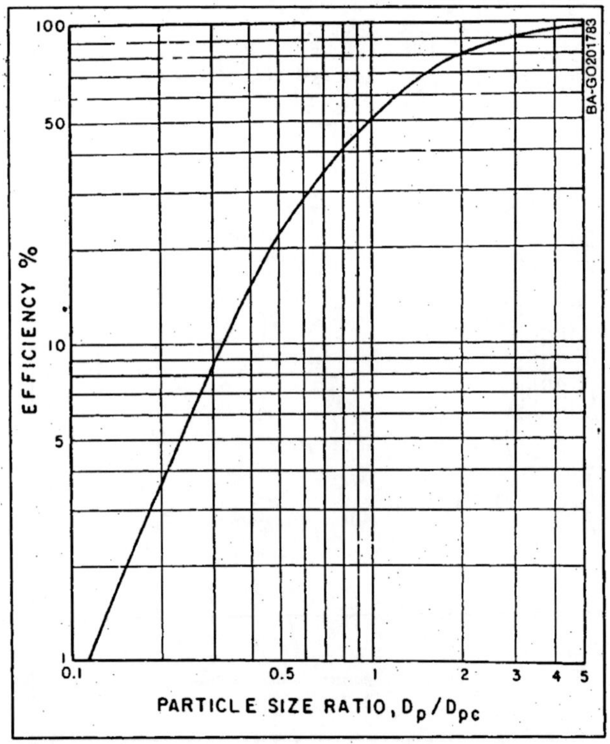

Fig. 8-4. Cyclone grade efficiency curve (Source: Kaupp 1984a, Fig. 138)

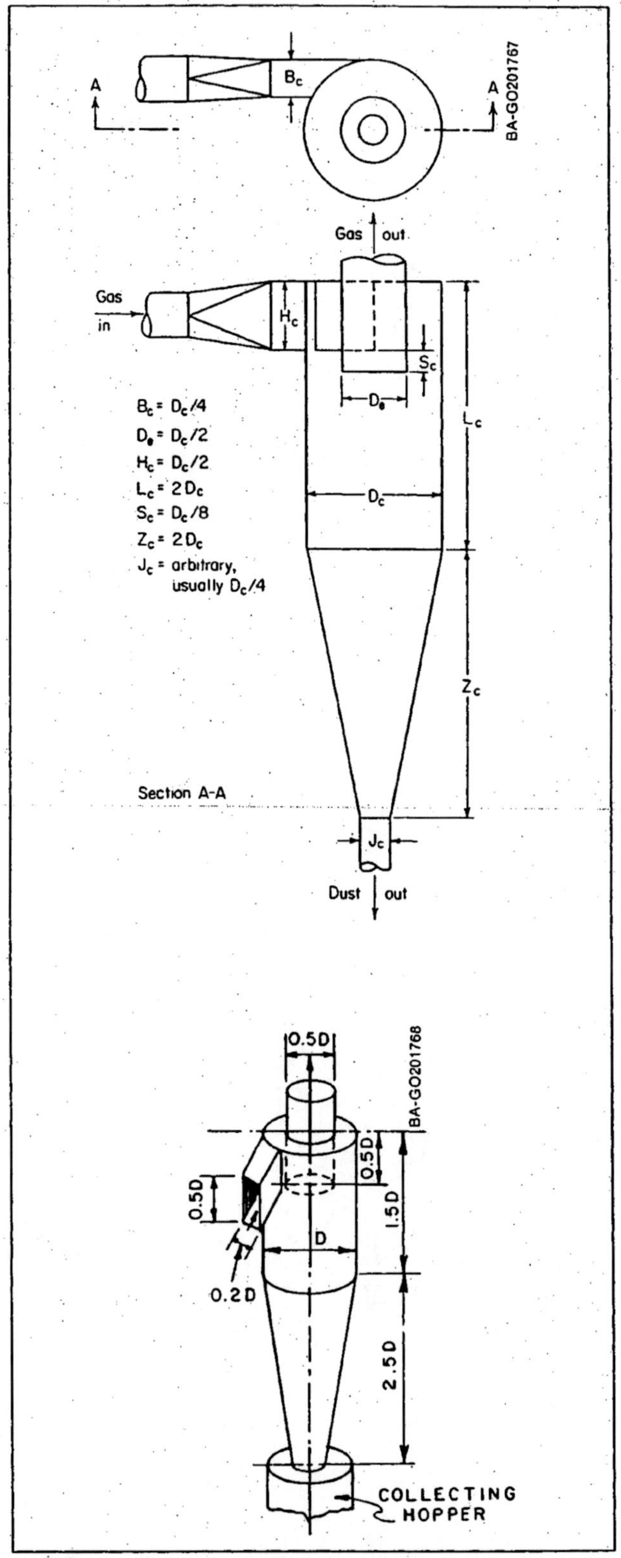

Fig. 8-5. High-efficiency cyclone proportions (Source: Perry 1973, Fig. 20-96. © 1973. Used with permission of McGraw Hill Book Co.; Kaupp 1984a, Fig. 134)

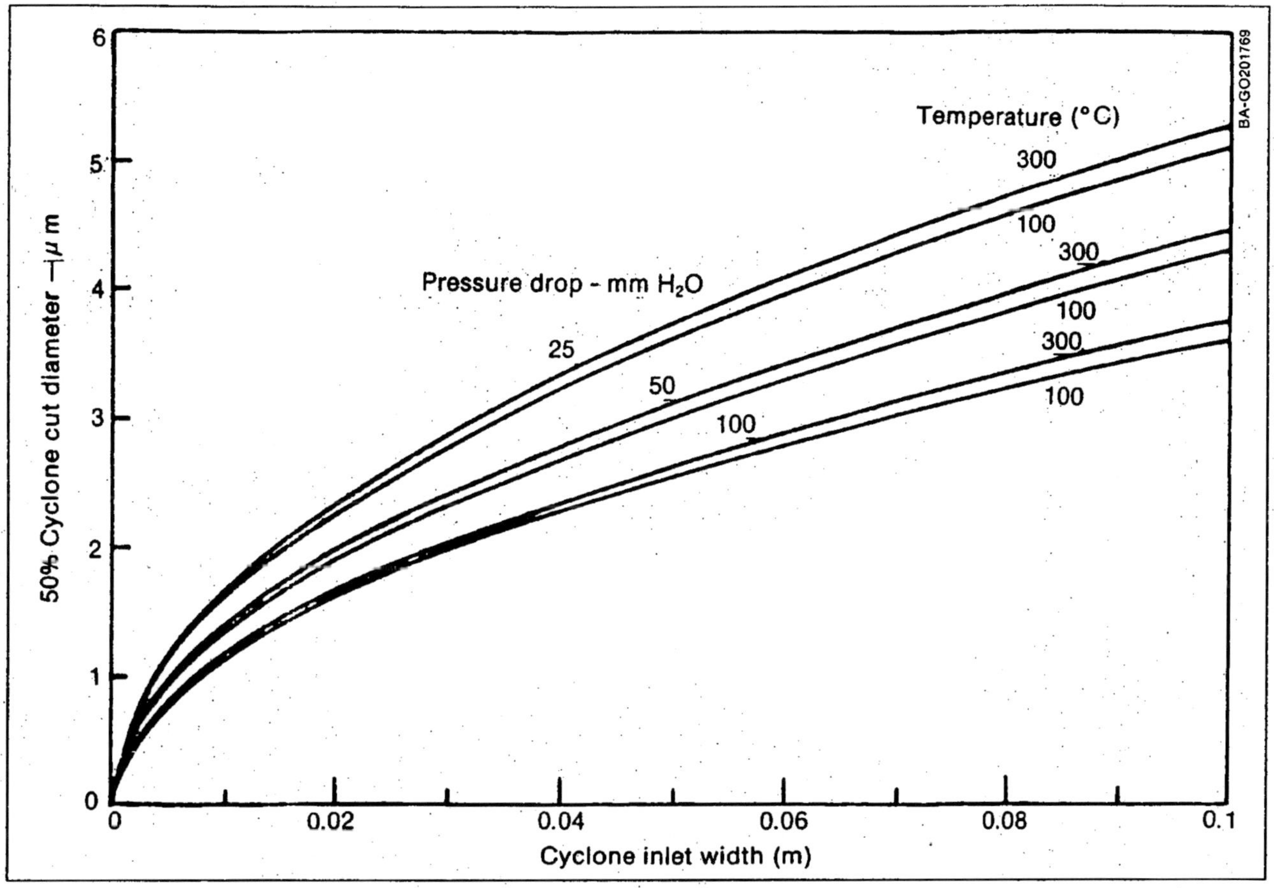

Fig. 8-6. High-efficiency cyclone — cyclone cut size versus inlet width

on cyclone cut size is minimal. This is due to the counterbalancing variations of density and viscosity with temperature.

If coarse particles are introduced into a fine-particle cyclone separator, then two detrimental effects may occur. First, the large particles may block the small inlet. Second, the high velocities within the separator may break up the coarse particles by erosion, impact, and attrition. This latter effect can generate fine particles that may be harder to capture than anything previously present in the gas (Perry 1973). For this reason it is preferable to provide a gas-disengagement space for gravity settling within the gasifier rather than two stages of cyclones. This will allow the coarser particles to settle out prior to the cyclone separator. Settling velocities are related to particle size as shown in Figs. 7-6 and 8-7.

8.6.2.3 Cyclone Design Strategy

In our experience, the cyclones fitted to gasifiers are too large for optimum particle removal. Therefore, we present here an example of detailed cyclone design.

The diameter of pipe leading from the gasifier outlet to the cyclone inlet should be selected to allow an adequate solids-conveying velocity within the pipe. Typical solids-conveying velocities for light materials range from 10 to 15 m/s (30-50 ft/s), as shown in Table 6-1. One should select the cyclone inlet pipe width (b) equal to the gasifier outlet pipe diameter or set the cyclone inlet velocity equal to the pipe velocity and design the cyclone according to the proportions in Fig. 8-5. The equations that appeared previously then can be used to predict particle cut size and pressure drop.

8.6.2.4 Cyclone Design Example

For example, let us design a high efficiency cyclone for a 10 kW (13.4 hp) gasifier engine system.

First we must determine the gas flow rate for a typical 22% engine efficiency and an assumed heating value of 1300 kcal/Nm^3 (5.44 MJ/Nm^3, 157 Btu/scf). Then specific gas consumption is 2.2 Nm^3/hp-h (1.43 scfm/hp) per horsepower or 3 Nm^3/kWh (25 cfm/kW) (Gengas 1950).

A 10 kW engine will require 30 Nm^3/h (20 scfm) of producer gas, which corresponds to a gas energy output of 163 MJ/h (188 kBtu/h). The volume of gas at the cyclone inlet temperature of 300°C will be

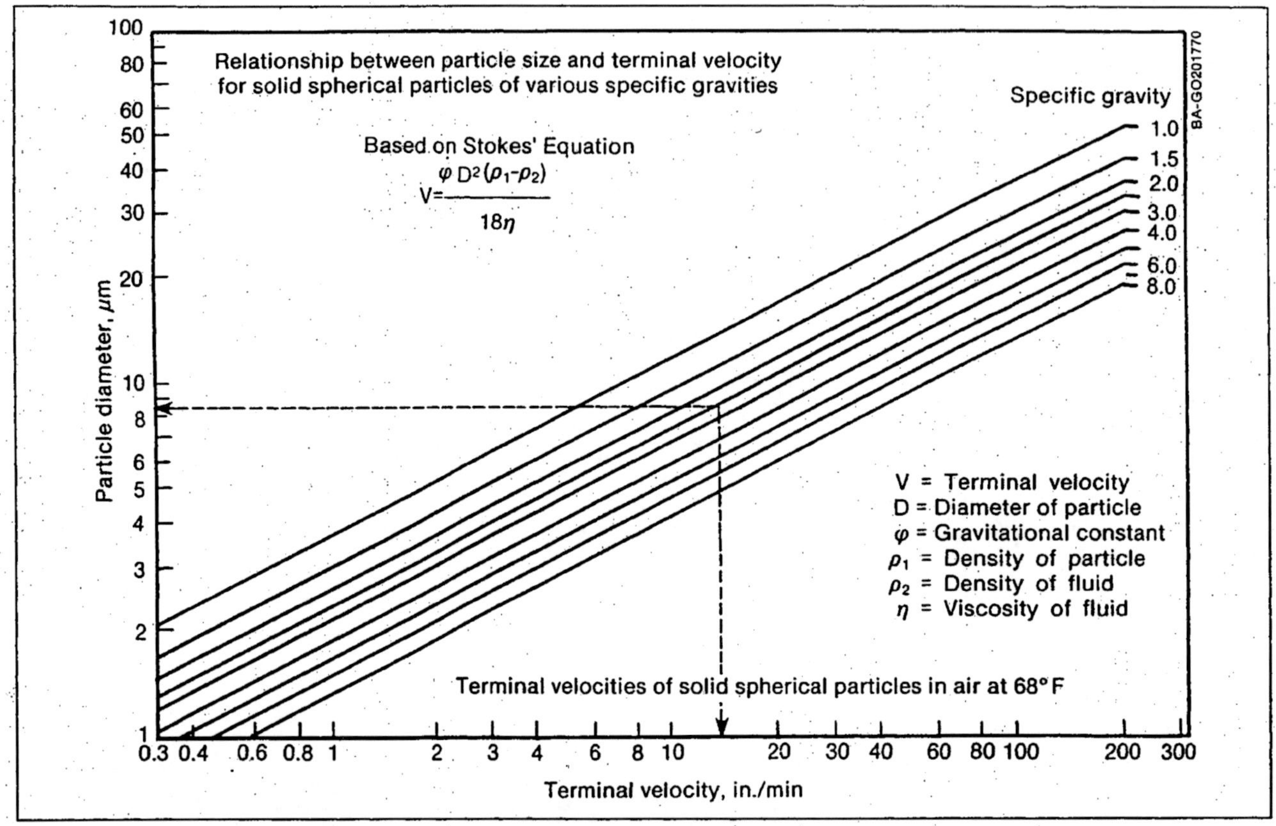

Fig. 8-7. Particle settling velocity versus size and density (Source: ASME 1965, Fig. 6)

$$30\ Nm^3/h\ \frac{573.16\ K}{273.16\ K} = 63\ m^3/h\ (37\ ft^3/min) \quad (8\text{-}6)$$

We wish to reduce solid particulates to 10 mg/Nm^3 from raw gas that exits the gasifier at 700°C. The gas cools within a few feet of pipe to 300°C, which we will consider as the cyclone inlet temperature. Screening analysis of airborne char (Fig. 7-4) shows a mass mean particle diameter $\bar{d}_p$ = 100 μm with a geometric standard deviation σ_g = 2.5. Particulate sampling indicates that total dust load at peak flow is 5000 mg/Nm^3. For a turndown ratio requirement of 4:1, the maximum dust penetration at peak flow rate is $P'_D = 100 \times [(10\ mg/Nm^3)/(5000\ mg/Nm^3)] = 0.2\%$. Derating by square root of turndown, the maximum penetration for turndown is $P_D = 0.2\%/\sqrt{4} = 0.1\%$.

On Fig. 7-4 we follow the particle size distribution line for an Imbert gasifier to the data point where the cumulative fraction equals the maximum penetration allowing for turndown P_D (0.1%). The corresponding particle diameter d_p = 3 μm is the cut point we will require. d_{p50} = 3 μm, as shown by the dotted line.

The recommended minimum gas velocity for conveying medium density dust is 15 m/s, and for heavy dust (metal turnings) is 25 m/s.

A pipe 2.5 cm (1 in.) inside diameter should provide a gas velocity of

$$V = \frac{4Q}{\pi D^2}$$

$$= \frac{4(63\ m^3/h)}{3.14\ (2.5\ cm/100\ cm/m)^2\ (3600\ s/h)}$$

$$= 35\ m/s\ (7000\ fpm) \quad (8\text{-}7)$$

which is well above the minimum.

Selecting the cyclone inlet width equal to the gas pipe diameter, the cyclone is designed by the proportions from Fig. 8-5. For inlet width 2.5 cm (1 in.) and inlet height 5 cm (2 in.) then the cyclone inlet velocity will be

$$V = \frac{Q}{A}$$

$$= \frac{(63\ m^3/h)(10^4\ cm^2/m^2)}{(2.5\ cm)(5\ cm)(3600\ s/h)}$$

$$= 14\ m/s\ (2755\ fpm) \quad (8\text{-}8)$$

Cyclone cut size is the size particle that will be collected with 50% efficiency. Then we calculate, using the viscosity and density of producer gas from Fig. 8-8 and assuming ash density 2.0 (2000 kg/m^3) and char

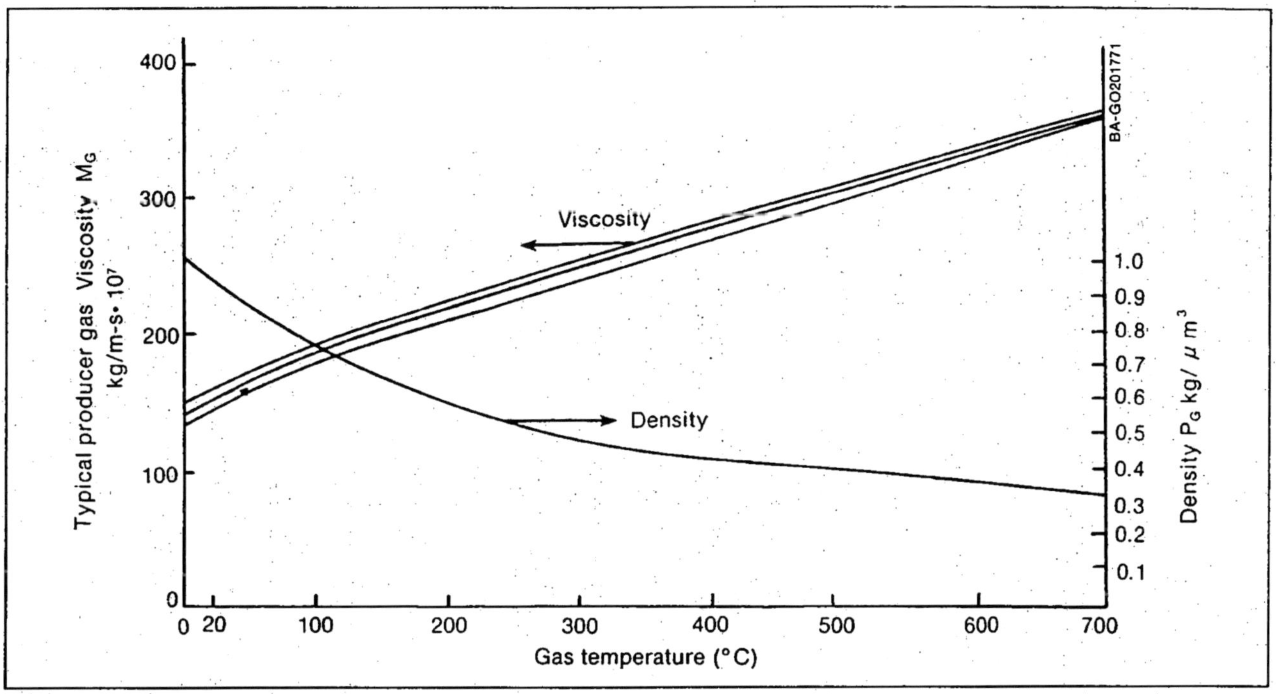

Fig. 8-8. Gas viscosity versus temperature

density 0.2 (200 kg/m³), that the cyclone cut size will be

$$d_{pc} = \sqrt{\frac{9\,\mu b}{2N_eV_i\,(\rho_p - \rho_G)\,\pi}}$$

$$= \sqrt{\frac{9(255 \times 10^{-7}\ \text{kg/m-s})(0.025\ \text{m})}{2(5)(14\ \text{m/s})(2000 - 0.489\ \text{kg/m}^3)\ (3.14)}}$$

$$= 2.5\ \mu\text{m} \qquad (8\text{-}9)$$

for ash, and

$$d_{pc} = \sqrt{\frac{9(259 \times 10^{-7}\ \text{kg/m-s})(0.025\ \text{m})}{2(5)(14\ \text{m/s})(200 - 0.489\ \text{kg/m}^3)\ (3.14)}}$$

$$= 8\ \mu\text{m} \qquad (8\text{-}10)$$

for char.

The cyclone pressure drop will be

$$\Delta p = \frac{(.065)(\rho_G)(V_i^2)A_d}{De^2}$$

$$= \frac{(.065)(0.489\ \text{kg/m}^3)(14\ \text{m/s})^2(0.025\ \text{m})(0.05\ \text{m})}{(0.05\ \text{m})^2}$$

$$= 31\ \text{mm}\ (1.22\ \text{in.})\ H_2O \qquad (8\text{-}11)$$

We can see that this cyclone should achieve the desired particulate removal without excessive pressure drop.

However, there is still a finite possibility of large particles passing through a cyclone, so it is not advisable to use a cyclone as the sole method of particulate removal. Also, any tars in the gas stream must still be removed by other means.

8.6.2.5 Other Factors in Cyclone Performance

The most common errors encountered in cyclone design are a low intake velocity caused by an oversized cyclone, and reentrainment of solid particles caused by improper cone design or faulty design of the discharge receiver.

Reducing the flow rate decreases the separator's performance, but it has only a slight effect on gas cleanliness because the dust load entrained in the inlet gas to the separator is also lower at the reduced flow rate. The resulting effect is that this outlet dust load slowly increases with a decreased flow rate by the inverse square root of gas flow rate (Gengas 1950; Calvert 1972; Perry 1973).

It is tempting to reduce the cyclone inlet width with an inlet vane; however, the upset cyclone proportions have been found to reduce the effective number of gas rotations N_t to as low as two and to increase reentrainment (Perry 1973). Therefore, to remove finer particles using cyclones at a lower pressure drop, it is preferable to reduce the individual cyclone diameter using multiple parallel cyclones (multi-clones) if necessary.

Air leaking into the char-ash hopper at the bottom of a cyclone separator deteriorates performance substantially. Similarly, removing the gas through the bottom improves the efficiency of the cyclone separator.

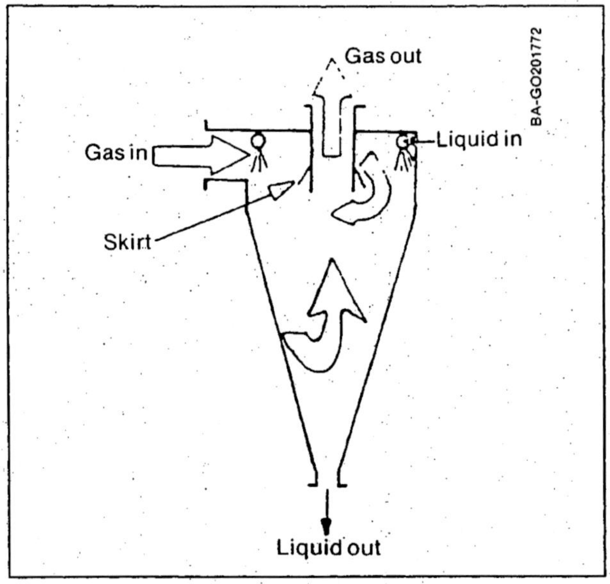

Fig. 8-9. Wet cyclone (Source: Calvert 1972, Fig. 3.1-6)

Collecting entrained droplets from wet scrubbers with a cyclone separator requires an outlet skirt to prevent reentrainment of liquids that have impinged on the outlet tube, as shown in Fig. 8-9.

8.6.2.6 Recent Cyclone Development

Recent work has been done on cyclone design specifically applied to gasifiers by LePori (1983).

8.6.3 Baghouse Filter

8.6.3.1 Principle of Baghouse Filters

Baghouse filters (such as shown in Fig. 8-10) are used widely today to capture fine dust particles and to separate flyash from combustion gases. A baghouse filter consists of one or more fibrous filter bags supported on metal cages enclosed in a chamber through which the gases must pass. A deposit of the separated particles soon builds up on the bag and establishes a dust cake of appropriate pore size through which additional particles cannot pass. As more dust is accumulated, the pressure drop increases. When the cake is an optimal thickness for removal, the bag is agitated either by gas pressure or by mechanical means, causing the excess cake to drop to the bottom of the housing where it is eventually removed.

8.6.3.2 Action of the Filter Cake in the Operation of Fabric Filters

Fibrous bag filters have been found to be outstanding in the removal of particles down to submicron sizes, as shown in the grade efficiency curve of Figs. 8-2 and 8-11. High-efficiency capture of small particles is surprisingly independent of the size of openings in the filter weave. The reason for this is that the primary capture element is the dynamic cake that forms on the filter surface. This cake, which consists of captured particles, presents a circuitous path that effectively captures fine particles, while coarser captured particles maintain an open cake structure to promote high gas permeability. When a new filter fabric is inserted, the main mechanism of particle collection is physical sizing as determined by the openings in the weave. At first, small particles may pass uncaptured until some buildup accumulates on the filter. From this point on, the gas must effectively pass through a packed bed of micrometer-sized particles. Interception and impaction then emerge as significant collection mechanisms.

8.6.3.3 Application of Baghouse Filters

Baghouse filters have been used with good success in many of the more successful and reliable engine gasifier systems (Breag 1982; Kjellstrom 1981). The use of fabric filters has virtually eliminated the corrosive

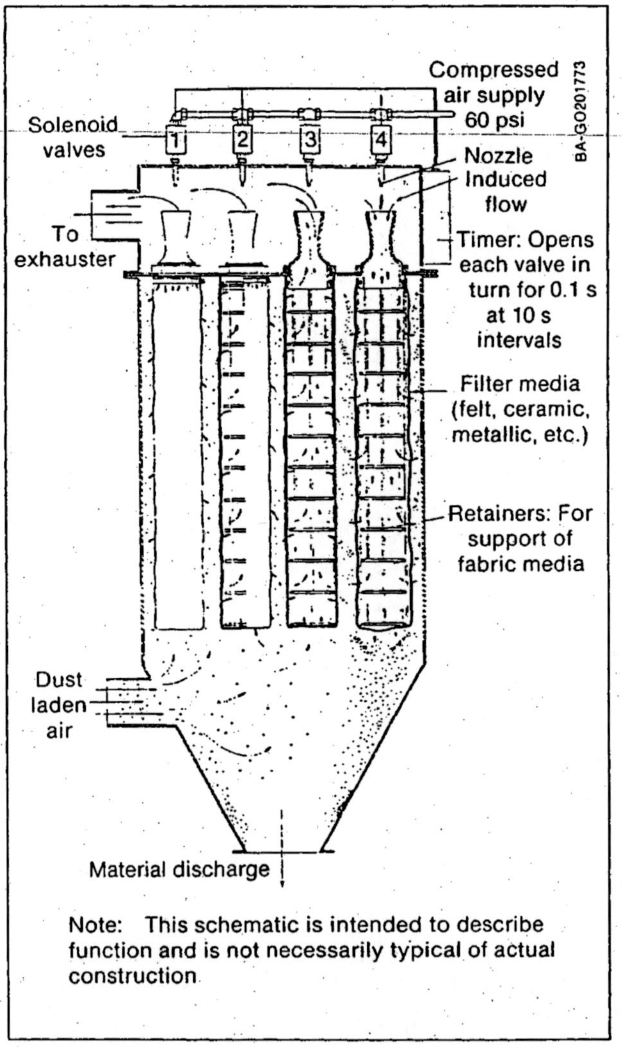

Fig. 8-10. Cloth bag filter with intermittent reverse pulse cleaning (Source: Work 1955, p. 483)

ash that otherwise was present in condensate or scrub water. The fabric filter is no doubt the most efficient device for fine cleaning; but for wood gas, extensive precautions against condensation of tar or water are necessary (Gengas 1950).

Note in Table 8-3 that the cylinder wear is less for producer gas with a fabric filter than for diesel oil alone.

During operation, the previously described filter cake grows steadily in thickness, collection efficiency climbs, and the pressure drop across the filter rises. When the filter cake has reached optimal thickness for removal, the filter cake must be removed by one of the following methods: momentary flow reversal to collapse the bag and dislodge the cake as shown in Fig. 8-10, a pulse jet of compressed gas or air to create the momentary bag collapse, or dismantling and manually shaking the bag (Breag 1982).

After cleaning, the filter efficiency is lower until the filter cake reforms. It is wise to use a conservatively designed fabric filter (5-10 cfm/ft^2), or even larger bag area, to maximize the interval between bag cleanings so as to maintain clean gas flow.

Bag filters are suitable only for removing dry particulates; sticky or tacky materials do not release from filter bags. Therefore, special provisions and precautions are required to maintain the bag filter temperature in order to prevent water vapor or tars from condensing on the filter bag. In particular, since tar-laden start-up gas should not be drawn through a cold bag filter, the design should locate the flare outlet upstream from the bag filter and provide means for preheating the bag filter assembly.

Materials that have been used for bag filters include natural and synthetic organic fiber, glass fiber, ceramic fiber, and stainless steel. The properties of these materials are outlined in Table 8-4. Organic-fiber bag

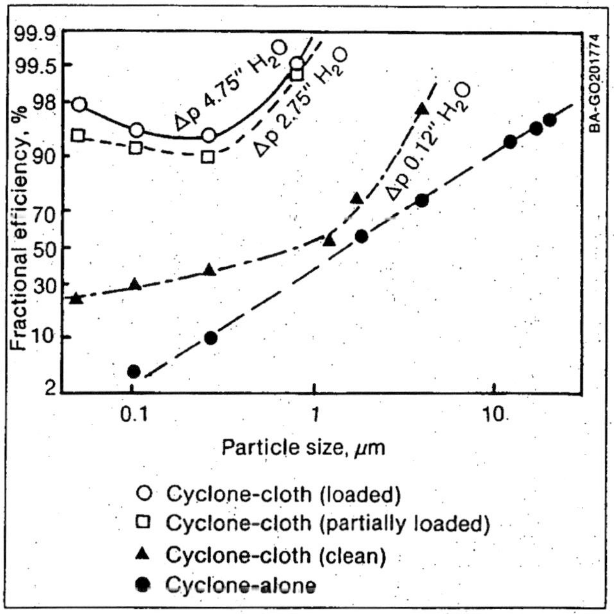

Fig. 8-11. Fractional efficiency curves of cyclone-alone and cyclone-cloth collectors (Source: Peterson 1965, p. 48)

filters are limited to low-temperature operations requiring accurately controlled gas cooling. Various cloth filters were used on gasifiers between 1939 and 1945, but these proved to be a continual source of difficulty because they could catch fire if they became too hot, or would get wet (from condensate) if they became too cold. Polyester felt bags, the most widely available, are rated for 135°C continuous service temperature. Stainless steel, glass-fiber, and ceramic-fiber bag filters have been used successfully at higher temperatures and show good promise (Johansson 1980). Unfortunately, the abrasion and flexing resistance of glass and ceramic fibers can be low; after installation, and especially once they have been heated, these materials should be handled as little as possible. High-temperature bag materials are not as widely available as other materials.

Table 8-3. Cylinder Wear for Gas Cleaned with Wet Cleaning and Fabrics Filters

		Dual Fuel — Diesel Oil/Producer Gas	
	Diesel Oil	Wet Cleaning System	Fabric Filter
Cylinder wear			
1000 h			
Tractor 01	0.015 mm	0.05	
02	0.028	0.05	
03	0.031	0.06	0.007
06	0.005-0.010		0.019
08	0.020		0.011
Oil contamination expressed as amount of insoluble products in benzene after 100 h	0.2% - 0.3%	0.54% - 1.97% average 0.75% (9 tests)	0.12% - 0.25% (2 tests)

Source: Kjellstrom 1981, Table 2.4.

Table 8-4. Filter Fabric Characteristics

Fiber	Operating Exposure (°F)		Supports Combustion	Air Permeability[a] (cfm/ft^2)	Composition	Resistance[b] to				Cost Rank[c]
	Long	Short				Abrasion	Mineral Acids	Organic Acids	Alkali	
Cotton	180	225	Yes	10-20	Cellulose	G	P	G	G	1
Wool	200	250	No	20-60	Protein	G	F	F	P	7
Nylon[d]	200	250	Yes	15-30	Polyamide	E	P	F	G	2
Orlon	240	275	Yes	20-45	Polyacrylonitrile	G	G	G	F	3
Dacron[d]	275	325	Yes	10-60	Polyester	E	G	G	G	4
Polypropylene	200	250	Yes	7-30	Olefin	E	E	E	E	6
Nomex[d]	425	500	No	25-54	Polyamide	E	F	E	G	8
Fiberglass	550	600	No	10-70	Glass	P-F	E	E	P	5
Teflon[d]	450	500	No	15-65	Polyfluoroethylene	F	E	E	E	9

[a] cfm/ft^2 at 0.5 in. water gauge

[b] P = poor, F = fair, G = good, E = excellent.

[c] Cost rank, 1 = lowest cost, 9 = highest cost.

[d] Du Pont registered trademark.

Source: After Kaupp 1984a, Table 51.

8.6.3.4 Safety Filter

If the filter bag ruptures, contaminants harmful to the engine will be released. Therefore, a safety filter or other effective warning means should always be used in conjunction with bag filters. The safety filter acts by plugging quickly and shutting down the system in the event of an upstream equipment failure. A 200-mesh screen is suitable for a safety filter, as shown in Fig. 8-12.

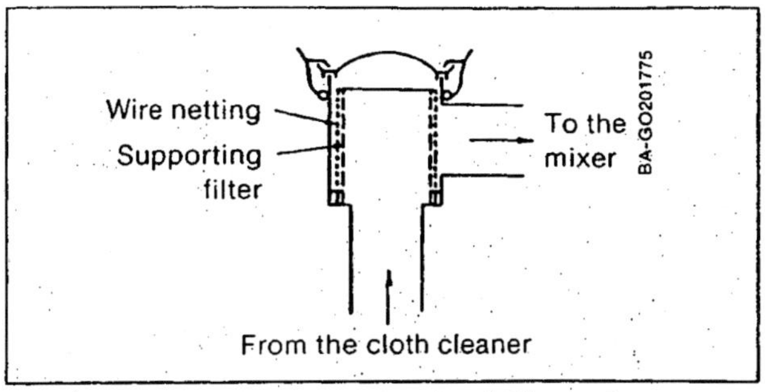

Fig. 8-12. Flame arrestor and safety filter (Source: Gengas 1950, Fig. 166)

8.6.4 Electrostatic (Cottrell) Precipitators

Electrostatic precipitators have a long history of industrial use to produce exceptionally clean gas. During operation, the gas passes through a chamber (as shown in Fig. 8-13) containing a central high-voltage (10-30 kV) negative electrode. A corona discharge forms around the central electrode, which imparts a negative charge to all particles and droplets. The negatively charged particles then migrate to the positive electrode, which may be washed by a continuous water stream to remove these particles. The electrostatic precipitator is effective for all drop and particle sizes.

A small precipitator (20 cm in diameter and 1 m in length) was operated at SERI to clean gas produced by a 75-hp Hesselman gas generator powering a 15-kW electric generator. The initial results were very dramatic, and the tar mist at the flare could be seen to disappear instantaneously when the voltage was applied. However, the electrodes and insulators soon became coated with soot and tar, and formed a short-circuit path that supported an arc. A means for cleaning the electrodes must be provided, along with a means to warm the insulators to prevent a water-condensation short-circuit. These problems are being investigated.

For tar-mist removal, wire and tube electrostatic precipitators are preferred over the plate-type electrostatic precipitator (Strauss 1975). Typical performance

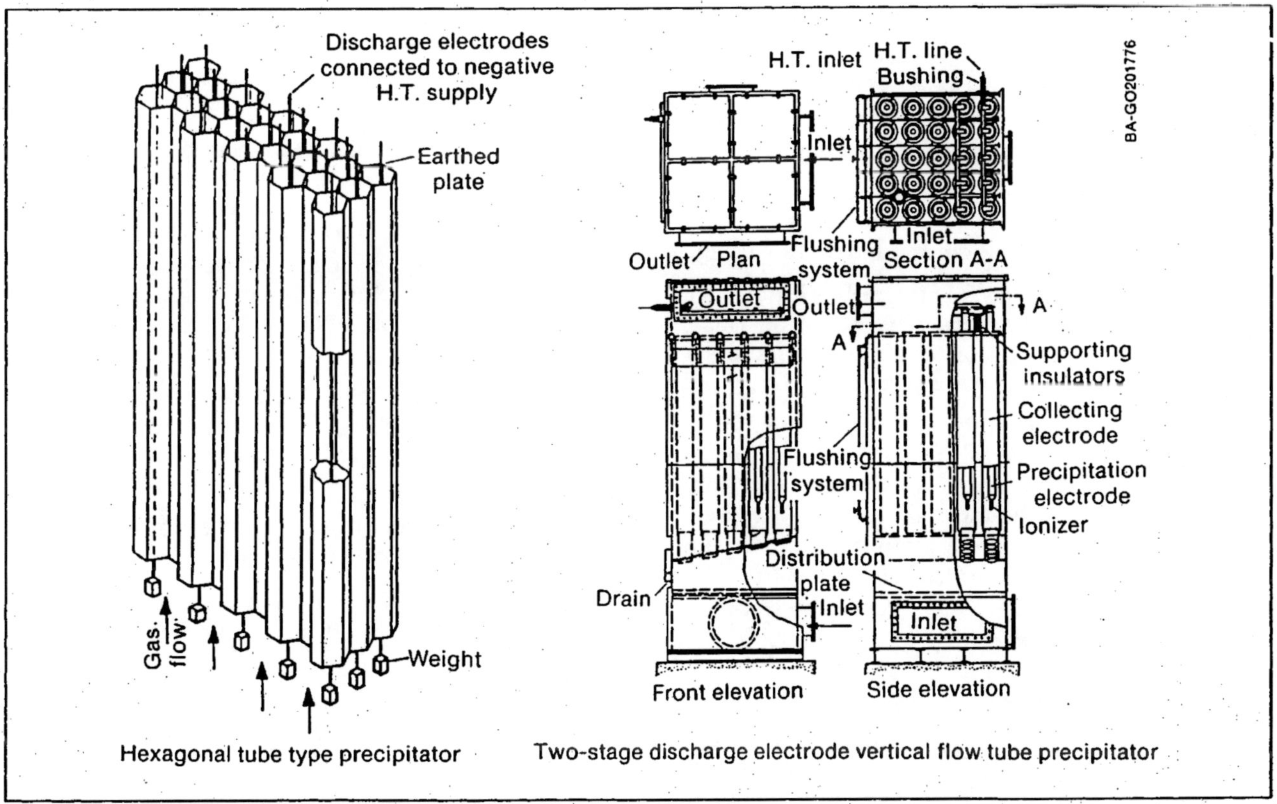

Fig. 8-13. Electrostatic precipitator examples (Source: Strauss 1975, Figs. 10-19, 10-20. © 1975. Used with permission of Pergamon Press)

Table 8-5. Typical Performance Data for Electrostatic Precipitators

Type of Plant	Dust Concentrations g/m^3 at Operating Temp.		Collecting Efficiency (%)	Power Consumption W/1000 m^3/h
	Inlet	Outlet		
Coal Gas Industry				
Peat gas producer	5.34	0.008	99.85	702
Cracking plant for natural gas	0.224	0.002	99.20	120.4
Producer gas from lignite briquettes	37.7	0.20	99.47	652
Producer gas from sembituminous lignite	28.7	0.10	99.7	602
Shale-gas cleaning plant	40.0	0.006	99.9	903
Coke oven town gas cleaning	24.15	0.010	99.9	903
Coke oven town gas cleaning	17.0	0.003	99.9	1605
Coke oven gas cleaning	28.0	0.078	99.8	752
Oil carburetted water gas cleaning	4.73	0.039	99.2	1404
Tar carburetted water gas cleaning	10.0	0.050	99.5	1805

Source: Perry 1973, Table 20-45.

characteristics for electrostatic precipitators are shown in Table 8-5, indicating high capture efficiency.

The precipitator tube diameter should be small enough to allow the corona discharge to be established at a reasonable voltage and large enough so that its volume will provide the necessary residence time with a reasonable length. Low flow rates result in a higher residence time and higher collection efficiency.

Multiple parallel precipitator tubes, as shown in Fig. 8-13, permit a more compact precipitator design and use a lower voltage than a single larger tube. The sparking voltage for tube precipitators is shown in Table 8-6. In practice, precipitators are operated at the highest operating voltage without excessive sparking (Perry 1973). One-second delays between sparks have been found to result in effective precipitator operation.

Application of a negative (rather than positive) voltage on the center electrode is favored because this arrangement results in a more stable corona and less sparking.

The typical current through the electrode is low: 0.1 to 0.5 mA/m^2 of collecting surface (Perry 1973). Half-wave rectification of a 50 to 60 Hz electric supply provides adequate time for extinguishing sparks.

The power consumed by an electrostatic precipitator is very low, typically 1.5 W/hp (Strauss 1975), and the pressure drop also is very low, at considerably less than 1 in. of water. High-voltage equipment requires rigorous safety measures. In addition, unforeseen power failures may cause a loss of the electrostatic precipitator's cleaning ability, with a subsequent release of tars to the engine.

8.7 Wet Scrubbers

8.7.1 Principles of Wet Scrubbers

As we have previously stated, particles with diameters larger than 1 μm settle by gravity and inertia. They follow Stokes' law and can be captured by impaction, gravitational, or centrifugal means. For particles smaller than 0.1 μm, motion is dominated by molecular collisions. They follow Brownian motion principles, behave more like a gas, and may be collected by diffusion onto a liquid surface. In this section we will look at the basic mechanisms of particle movement and capture for wet scrubber systems.

Particles with diameters between 0.1 and 1 μm fall within the so-called "open window." They are the most difficult particles to capture, either by diffusion or inertial mechanisms. They are too large to diffuse well but too small to settle. However, they can be made to grow in size, since small particles collide naturally and agglomerate into larger particles that are easier to capture.

One method of high-efficiency collection uses primary collection of large particles by inertia and diffusion, followed by an increase in fine particle size by agglomeration, and finally by collection and entrainment separation. The rate of agglomeration is proportional to the total number of particles present. Agglomeration is also assisted by the presence of droplets that act as nuclei.

Particles tend to move toward a surface on which condensation is taking place. This phenomenon is referred to as "Stefan motion." Particles tend to migrate away

Table 8-6. Electrostatic Precipitator Sparking Potentials

Pipe Diameter, in.	Sparking Potential,[a] volts	
	Peak	Root Mean Square
4	59,000	45,000
6	76,000	58,000
9	90,000	69,000
12	100,000	77,000

[a]For gases at atmospheric pressure, 100°F, containing water vapor, air, CO_2 and mist, and negative-discharge-electrode polarity.

Source: Perry 1973, Table 20-43.

from a hot surface and toward a cold surface. This phenomenon is called "thermophoresis."

Wetted particles tend to stick together better when they collide, thereby assisting agglomeration. Wet scrubbers have been used widely, especially in stationary applications for cleaning and cooling the gas. A scrubber operates by creating conditions for maximum contact between the gas to be cleaned and a scrubbing liquid medium.

Basic scrubber types and performance characteristics are summarized in Tables 8-1, 8-2, and 8-7, and grade efficiency curves are shown in Fig. 8-2. Scrubbers can be divided into impingement-plate, packed-bed, sieve-plate, spray tower, and Venturi scrubbers.

Small, difficult-to-capture droplets may be made to grow in size with time until they are large enough to be captured by simply providing adequate residence time in the scrubber volume. Particles grow in size by agglomeration and condensation. Agglomeration is particle growth through particle collision. Almost all high-concentration clouds tend to have the same particle concentration within 1 min after formation.

A novel method to capture 0.2-μm mist is to provide water fog nuclei and ample residence time. Water fog is introduced at a concentration of 0.25 L/Nm^3, at 30 psig spray pressure, and in a high intensity sonic field of frequency 600 to 800 Hz. A 12-s residence time permits the particles to agglomerate to a size large enough to be captured with 94% efficiency by a 5-μm cyclone (Calvert 1972).

If a condensation nucleus is absent, but the degree of supersaturation (S) exceeds 200% to 400%, then homogeneous self-nucleation occurs. Self-nucleation produces extremely small droplet sizes. The droplet growth rate is inversely proportional to the droplet radius, so it proceeds slowly at first, accelerating with droplet size.

Nucleated condensation dominates over homogeneous self-nucleation when nucleation sites are present. Vapor condenses more readily within concave surfaces, filling the void fraction of solid particles. Soluble aerosol particles nucleate even more readily by boiling point depression in solution. A small droplet grows slowly by chance agglomeration until it reaches its critical size; after that, it grows rapidly by acting as a nucleation site. Soluble particles behave as nucleation sites without having to achieve critical size. The char-ash dust particles present in the producer gas stream at temperatures below the tar dew point will act as nuclei

Table 8-7. Scrubber Types and Performance

Method	Size Limit Particle Cut Dia. (dp_{50}) μm	Pressure Drop Water Column cm (in.)	Comments
Gravity Settling	>30	Low	Coarse separator; very large and bulky
Massive Packing	>5	1 cm/10cm of column height	Free draining coarse demister
Fiber Packing	1	10 cm	Viscous materials can cause plugging
Preformed Spray	>5	Low	High water consumption
Gas Atomized Spray	>5	0.002 (0.001)	Good for sticky materials
Venturi and Sieve	>2	5.7 (2.2)	
Plate Scrubbers	>1	22.6 (8.9)	
	Submicron	91 (35.9)	
Centrifugal	1-2	7.5-20 (3-8)	Compact; good for preliminary cleanup
Baffle Plate	20 (solids) 5 (mist)	2.5-7.5 (1-3)	Large coarse collector
Impingement	2-3	10-50 (4-20)	Recoil bounce; can reentrain
Entrainment Separator	5	32 (5)	Can be clogged
Mechanically Aided	1-2	Acts as a blower	High power and maintenance
Moving Bed	1	7.5-15 (3-6)	Good mass transfer
Fabric Filter	0.3	13 (5)	Excellent; can be clogged

Source: Compiled from data in Calvert 1972.

for tar condensation, thereby reducing the amount of very fine and persistent self-nucleated tar mist (Calvert 1972).

The design of a good scrubber must maximize the gas-liquid contact area while minimizing the pressure drop through the scrubber. For instance, the gas-liquid contact area for a foam is much greater than for a spray, given equal energy inputs.

If a gas stream enters a liquid-filled chamber at high velocity through a small hole at the bottom of the chamber, then all of the entering gas must experience the subsequent impaction and diffusion environments. When water enters the gas stream as a high pressure spray, only a small fraction of the gas is close enough to the nozzle to receive the benefit of impaction with the high-velocity droplets. Spray droplet agglomeration proceeds rapidly, causing the gas-liquid contact area to drop off sharply within a short distance from the nozzle. This effect seriously limits the collection ability of spray scrubbers.

8.7.2 Scrubber Equipment

8.7.2.1 Spray Towers

The simplest type of scrubber is the spray tower (shown in Fig. 8-14), which is composed of an empty cylinder with spray nozzles. The optimum spray droplet size is 500 to 1000 μm. Typical upward superficial gas velocity for a gravity spray tower is 2 to 4 ft/s, and particle collection is accomplished when particles rising with the gas stream impact with droplets falling through the chamber at their terminal settling velocity. The spray tower is especially well-suited as a prefilter for extremely heavy dust loads (over 50 g/Nm^3), which would plug other less-open types of scrubbers. Full-cone spray nozzles produce 500 to 1000 μm droplets, which fall with a settling velocity of 13 ft/s. For a spray tower 53 ft high, the value of d_{p50} is 5 μm.

8.7.2.2 Cyclone Spray Scrubbers

The cyclone spray scrubber combines the virtues of the spray tower and dry cyclone separator. It improves the particle-capture efficiency of the spray droplets in ordinary spray scrubbers by increasing spray-droplet impact. The cyclone spray scrubber also has the advantage, compared with the spray scrubber, of being self cleaning, of collecting more particles regardless of size, and operating at smaller pressure drops. A basic design is shown in Fig. 8-9; others are described in Strauss (1975). Commercial cyclone scrubbers are better than 97% efficient at removing particles with diameters greater than 1 μm. The cut diameter for a cyclone spray scrubber is about an order of magnitude less than that for either a dry cyclone or spray scrubber.

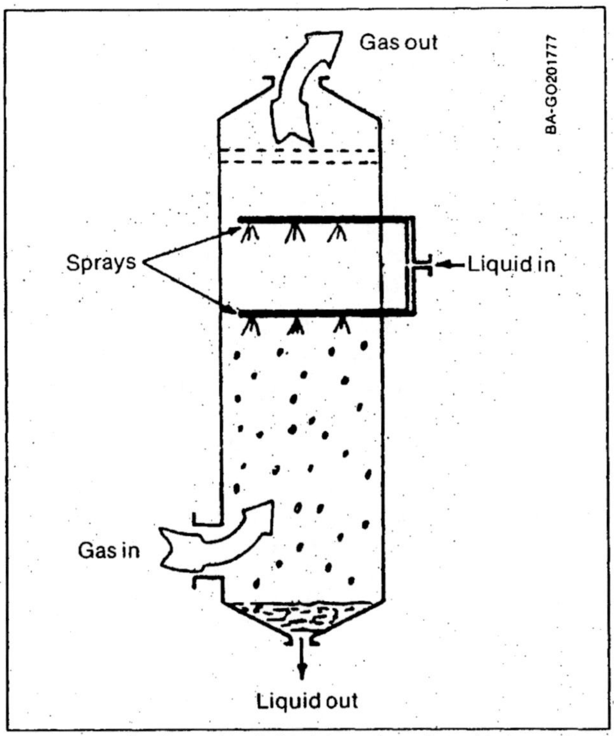

Fig. 8-14. Spray tower scrubber (Source: Calvert 1972, Fig. 3.1-4)

8.7.2.3 Sieve-Plate Scrubbers

A sieve-plate scrubber (Fig. 8-15) consists of a vertical tower with a series of horizontal perforated sieve plates. The scrubbing liquid is fed into the top of the column and flows downward via downcomers from plate to plate; the gas to be scrubbed is introduced at the bottom of the column and passes upward through the sieve holes counter to the liquid. Contact between the liquid and gas is enhanced by using plates with bubble caps, impingement plates, or sieve plates.

The sieve-plate scrubber captures large particles by impingement and impaction, and small particles by diffusion. Gas passes upward into the water layer through holes in the sieve plate. The high gas velocity through the sieve holes atomizes the scrubber liquid into fine droplets, and most inertial particle collection takes place just as the bubble is being formed, by impaction on the inner surface of the bubble. Diffusive particle collection dominates as the bubble rises. Here, surface-active agents can reduce the collection efficiency because of Stefan motion, but a cold water scrubbing liquid receiving a hot aerosol increases the collection efficiency. A deeper foam reduces inertial effects and increases collection by diffusion. Inertial collection is only slightly increased by adding plates or increasing the pressure drop.

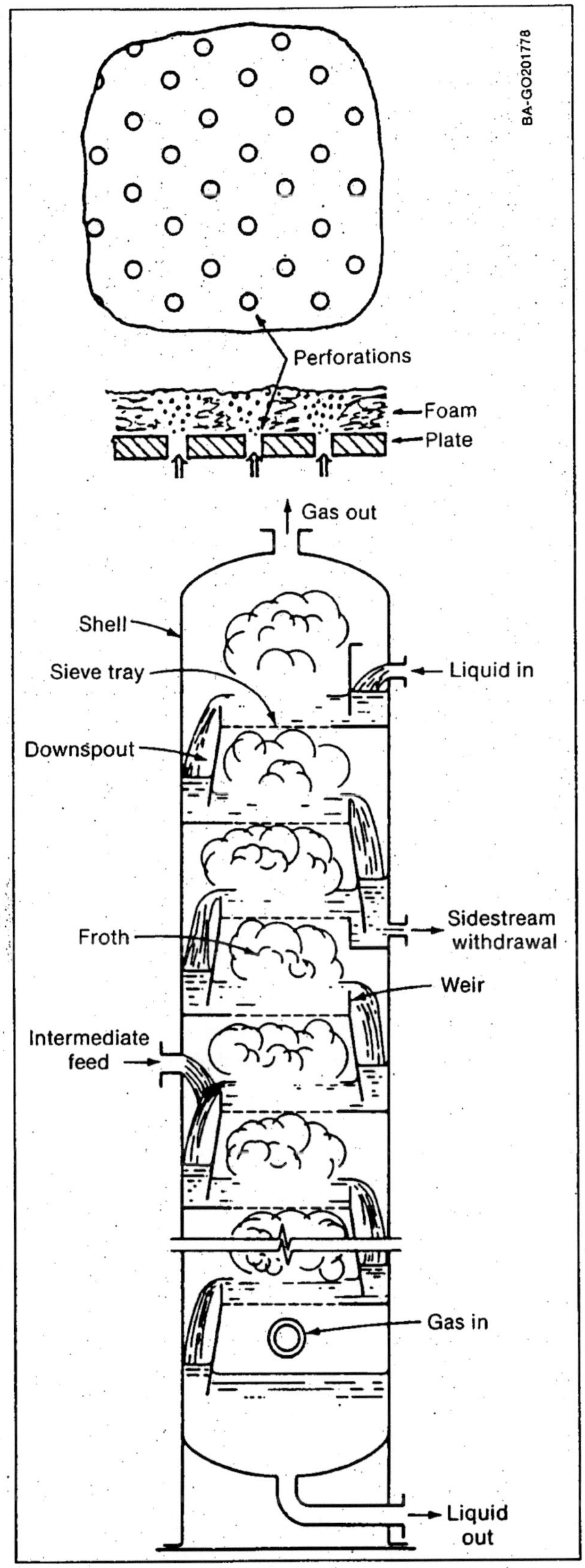

Fig. 8-15. Sieve plate scrubber (Source: Kaupp 1984a, Figs. 141, 146)

A typical sieve-plate scrubber can attain 90% efficiency for 1-µm particles using 3/16-in. sieve holes, at a specific velocity of 15 m/s (50 ft/s). Typical performance characteristics of sieve-plate scrubbers are discussed in Kaupp (1984a).

8.7.2.4 Impingement Plate Scrubbers

The impingement-plate scrubber shown in Fig. 8-16 is similar to a sieve-plate scrubber, except impingement plates are arranged so that each hole has an impingement target one hole diameter away from the hole. Gas flow past the edge of the orifice produces spray droplets that, when formed, are at rest, resulting in a large relative velocity between dust particles and these droplets. The gas velocity usually is above 15 m/s (50 ft/s), and the typical operating pressure drop is 1.5 in. water gauge (4 mbar) per plate. An increased pressure drop raises the collection efficiency. The required water flow rate is 1 to 2 gpm per 1000 cfm of gas flow.

8.7.2.5 Venturi Scrubbers

The Venturi scrubber (Fig. 8-17) captures large particles by impaction and impingement, and also rinses away any deposits that might otherwise form. Some fine particles are also captured here by diffusion. High-velocity flow through the low-pressure throat area atomizes the droplets. The low pressure at the throat causes condensation, and the high relative velocity of the droplets with respect to the gas captures most larger particles by impaction.

The atomized droplets present a considerable surface area for fine particles to be captured by diffusion. Furthermore, condensation in the throat improves capture through diffusion because of the phenomenon of Stefan motion. The atomized droplets rapidly agglomerate in the diffuser section, where collection through diffusion continues. Entrained droplets containing captured contaminants are separated inertially from the cleaned gas. Liquid recycle requires cooling and removal of captured materials, or disposal and replenishment.

The collection efficiency and droplet size are determined by the pressure drop: efficiencies may be increased by reducing the throat area to raise the pressure drop. The efficiencies of Venturi scrubbers are discussed in Calvert (1972).

8.7.2.6 Ejector Venturi Scrubbers

The velocity of the contacting liquid both pumps and scrubs the entrained gas in an ejector Venturi scrubber, as shown in Fig. 8-18. Spiral spray nozzles impart axial and tangential velocities to the liquid jet. The contacting liquid must be removed after the scrubber by a suitable entrainment separator. Compared to a Venturi scrubber, the ejector Venturi scrubber requires both more liquid and more power to achieve the same particle collection and gas movement. Ejector Venturi

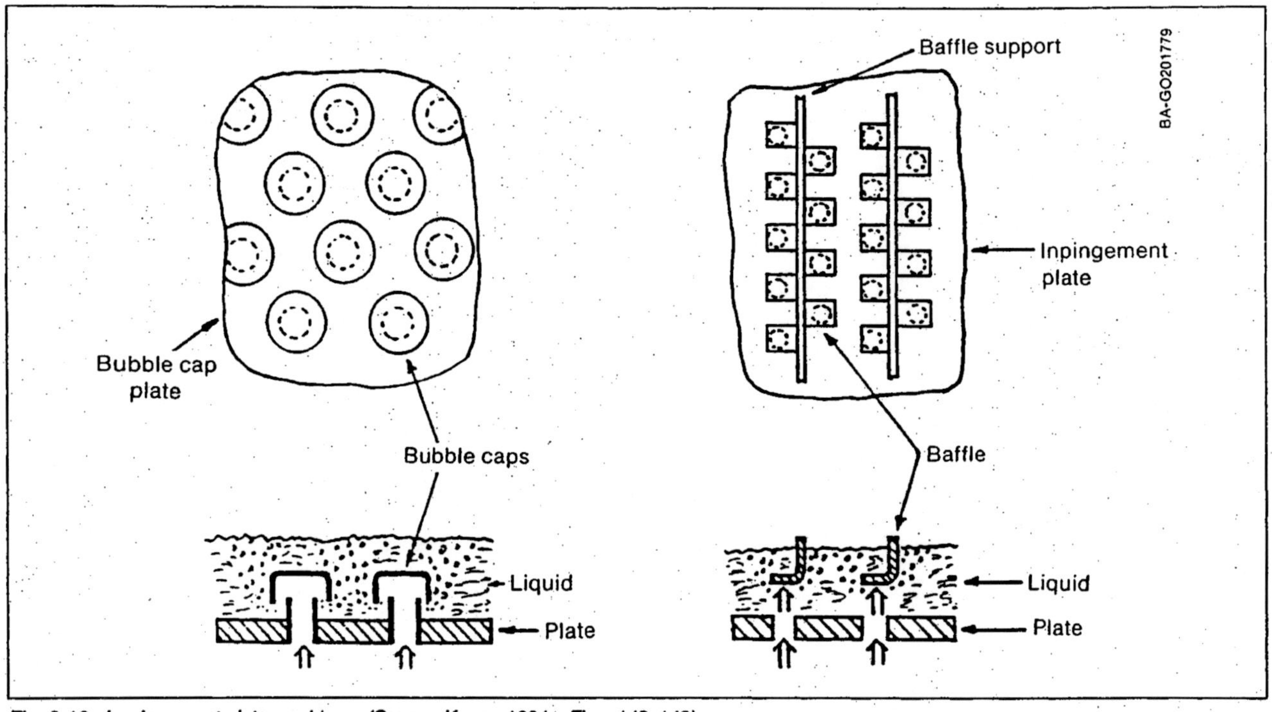

Fig. 8-16. Impingement plate scrubber (Source: Kaupp 1984a, Figs. 142, 143)

scrubbers have no moving parts and are especially well-suited for very dirty, corrosive, or abrasive materials that might otherwise damage a blower impeller (Calvert 1972).

8.7.2.7 Packed-Bed Scrubbers

The packed-bed scrubber (Fig. 8-19) is simple and open in design, and uses spheres, rings, or saddles as random packings to enhance the gas-liquid contact area. Packed beds are more effective for both gas absorption and liquid-gas heat exchange than they are for particle collection. However, packed beds are excellent for capturing entrained liquids. For entrainment separation, the optimum superficial gas velocity for packed-bed scrubbers using 1/2-in. spheres is 10 to 12 ft/s. Flooding and reentrainment occur above a gas velocity of 12 ft/s. The pressure drop is 7.5 to 8.5 in. water gauge for a 6-in.-deep bed. Performance characteristics of packed beds are shown in Fig. 8-20. Packed beds are free-draining; they may be irrigated to remove accumulations with a water flow (Calvert 1972).

8.7.2.8 Entrainment Separators

Entrained liquids from the wet scrubber must be thoroughly removed from the gas stream because they carry a slurry of captured materials. Entrainment droplets are typically greater than 10 µm and may be captured using a variety of techniques, including a packed bed, a packed fiber bed, a cyclone separator, an impingement separator, a spray tower, or a settling chamber. Poor entrainment separation has been a common problem for wet scrubbers in gasifier systems. Gas contaminant testing is advisable for all unproven designs.

8.7.3 Auxiliary Equipment

8.7.3.1 Gas Cooling

Water vapor acts as an inert dilutent of producer gas, initially lowering the gas heating value and ultimately lowering engine power or burner rating, as shown in Fig. 8-21. Much of this water vapor can be removed by cooling the producer gas and subsequently condensing out the water.

The amount of water vapor remaining after the cooling and condensation steps can be determined readily from the lowest temperature to which the gas has been cooled. If condensation has occurred, then the lowest gas temperature is of course the dew point of the gas mixture. The water vapor content of the gas may be determined from Fig. 8-22, or the psychrometric chart of Fig. 7-19. Note that the moisture fraction roughly doubles with each 10°C increase in the dew point temperature. We can calculate that at the 70°C dew point, water vapor represents 25% of the gas volume. Cooling the gas to 40°C reduces the water content to less than 8%, resulting in a substantial improvement in gas quality.

Water vapor dilution will be minimized by using fuels that are as dry as possible and then condensing water vapor to remove it from the gas.

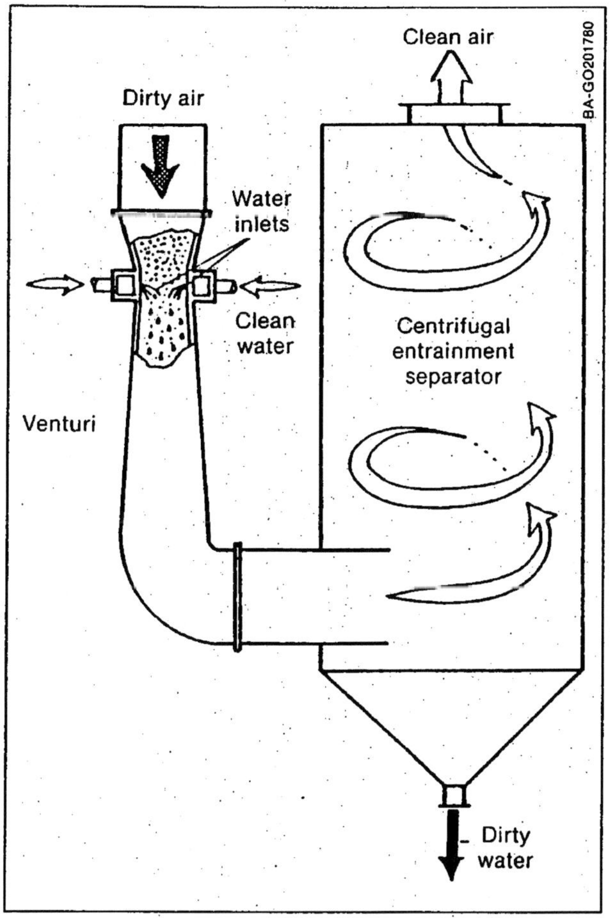

Fig. 8-17. Venturi scrubber with centrifugal entrainment separator (Source: Calvert 1972, Fig. 5.3.6-1)

When the gas leaves the char gasification zone at about 800°C, the sensible heat of the gas accounts for about 15% of the initial energy in the wood. If the gas is burned while it is still hot, then the sensible heat can be utilized. However, if the gas is to be used in an internal-combustion engine, it must be cooled to prevent preignition, to improve the engine volumetric efficiency, and to facilitate gas cleanup.

The thermal energy in the raw gas may be either dissipated, used for low-temperature applications such as drying, or recycled into the gasifier by using the energy to preheat the incoming air. Each method has been used on gasifiers. Airblast preheating was used extensively in European vehicle gasifiers to improve the gas and to permit wetter fuels to be gasified (Schläpfer 1937; Egloff 1941; Lutz 1940).

Using sensible heat to do more than preheat the air blast or heat the fuel hopper is hardly justified because the gas is dirty and only a small quantity of heat (15%) is involved. If usable heat is desired, then the engine exhaust gas and engine coolant fluid are much cleaner and more abundant sources of heat, representing 60% to 70% of the total energy contained within the initial fuel.

The heat losses from surfaces vary from 1 to 5 Btu/ft^2-h-°F, depending on the geometry of the gas cooler and the temperatures involved. Thus, a great deal of the cooling at higher temperatures can be accomplished in the pipes and at the surfaces of the gasifier itself, as well as in the cyclone separators or other cleaning equipment. However, as the gas approaches ambient temperature, additional cooling surface through some form of gas cooler is required.

Gas coolers exchange heat between the gas and the surrounding air, or between the gas and a liquid. A typical radiator used in vehicle applications is shown in Fig. 8-3. Here, the motion of the vehicle increases air flow around a gas cooler, so that more cooling air is available at the higher speeds when the heat load is greatest. In stationary applications, forced ventilation is required to move air through the gas cooler.

As the gas cools, tars begin to condense at temperatures below 350°C. As the temperature passes below the dew point of the gas (typically 40°-60°C), water also will condense. Water condensation helps to remove tar particles but yields a contaminated water condensate in the process. If tars and particulates are removed from the gas before it enters the gas cooler, then the gas cooler will be able to operate longer between cleanings. All heat-exchange and gas-cooling surfaces in contact with

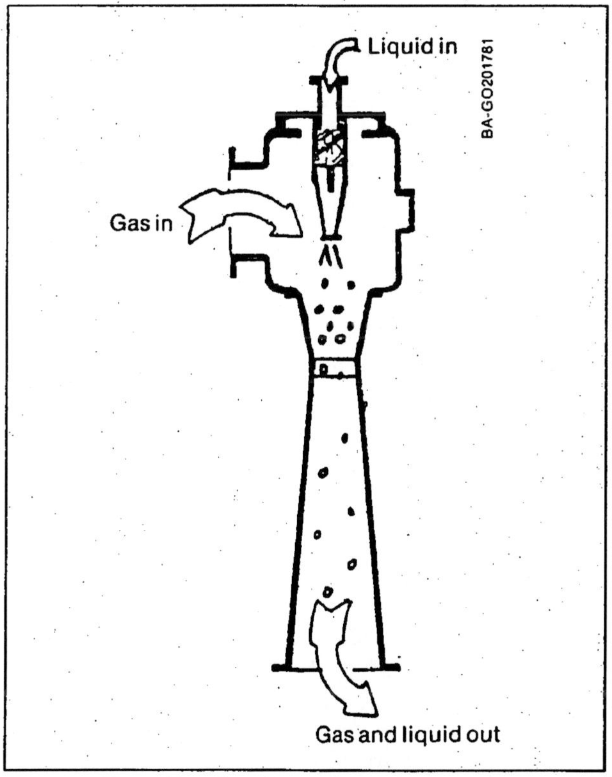

Fig. 8-18. Ejector Venturi scrubber (Source: Calvert 1972, Fig. 3.1-4)

the raw gas are subject to ash and tar buildup, so ample cleanout ports should be provided to clean these surfaces without requiring extensive disassembly.

8.7.3.2 Gas Drying

The diluting effect of water vapor on the wet gas heating value (HV_w) may be determined from the heating value of the dry gas (HV_d) and the moisture fraction (F_m) from Fig. 8-22, where F_m = water vapor partial pressure/total (absolute) gas pressure.

Then, HV_w is found from

$$HV_w = HV_d(1 - F_m) \qquad (8\text{-}12)$$

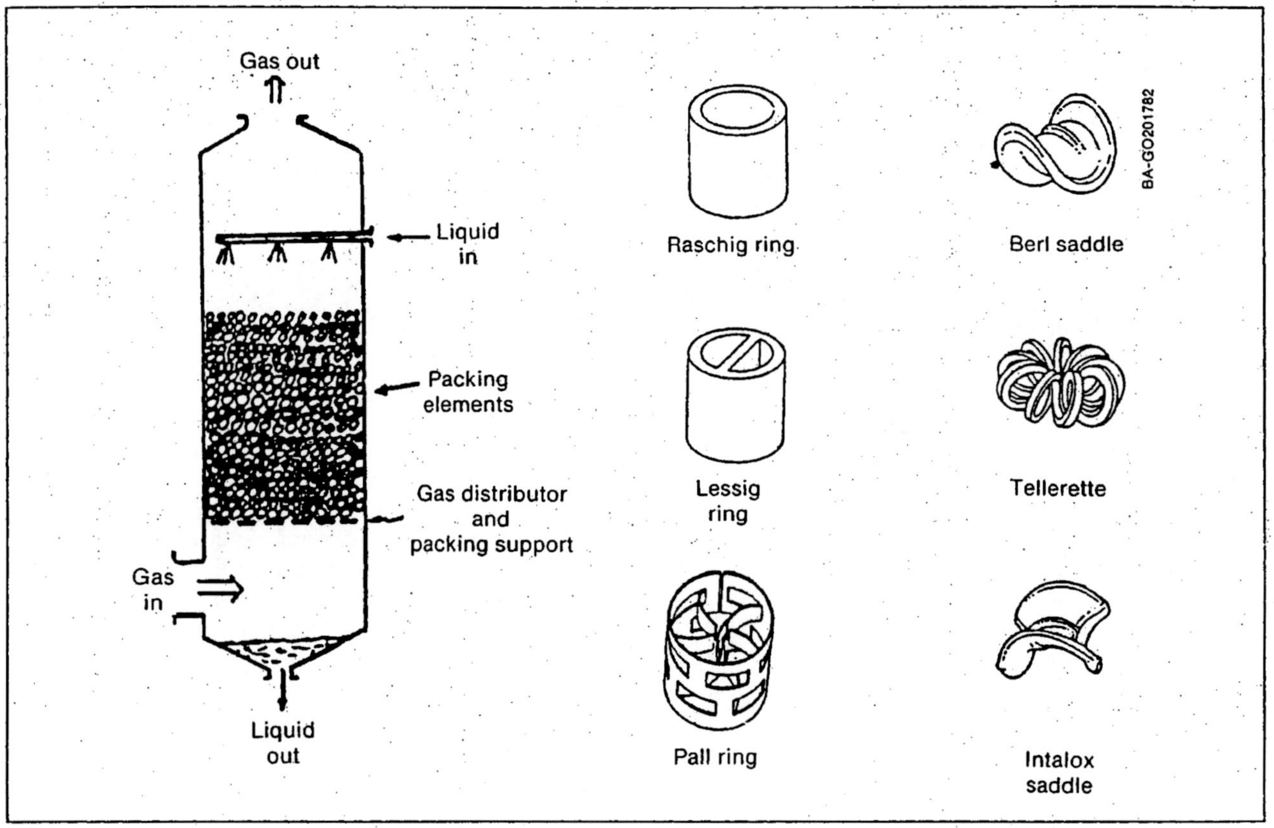

Fig. 8-19. Packed tower and packings (Source: Kaupp 1984a, Fig. 151)

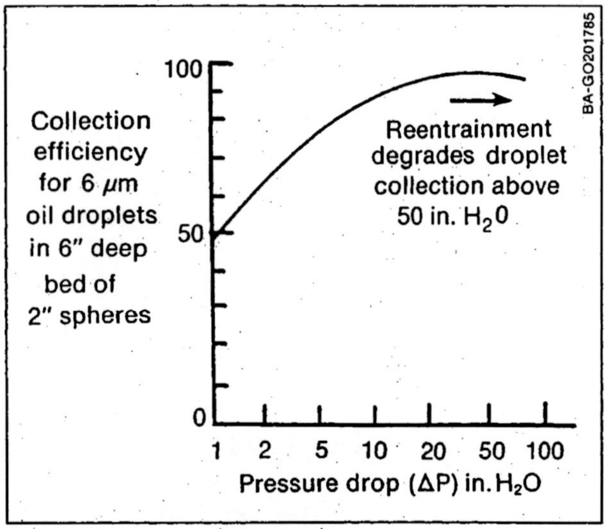

Fig. 8-20. Performance of packed bed

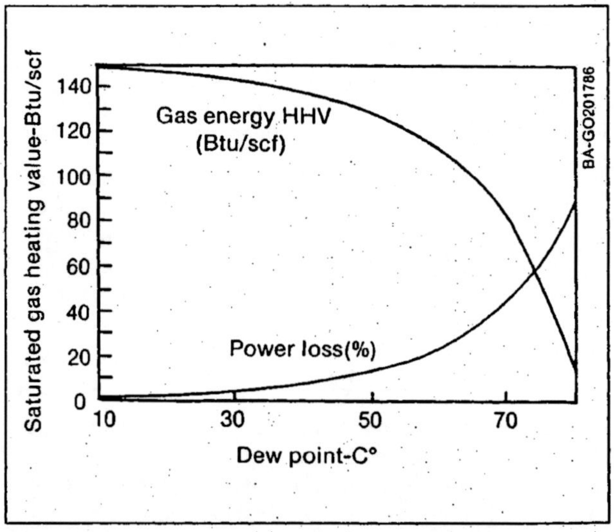

Fig. 8-21. Power loss due to moisture saturated producer gas for 150 Btu/scf dry basis gas

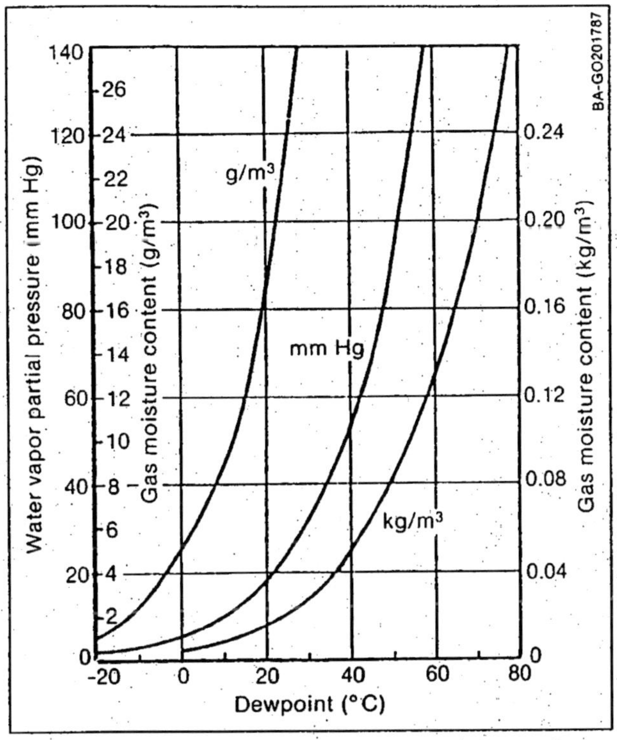

Fig. 8-22. Water content of saturated producer gas (Source: Gengas 1950, Fig. 82)

(The moisture fraction F_m is approximately the same value for either mass fraction or volume fraction of water, since 1 Nm^3 of water and 1 Nm^3 of producer gas each weighs approximately 1 kg.) The effect that water-vapor dilution has on the heating value of a 150-Btu gas is shown in Fig. 8-21.

Methods of measuring gas moisture are presented in Chapter 7. To minimize the power loss from water-vapor dilution, the necessary cooling surface can be roughly determined from Fig. 8-23. Ample ventilation must be provided to cool and condense the gas with a 60°C dew point from 700°C down to 40°C.

8.7.3.3 Demisting/Entrainment Separation

A common problem with otherwise adequate gas cleanup systems is inadequate removal of entrained scrubber liquids.

The gas emerging from the gas cooler and from wet scrubbers contains droplets of dirty water entrained in the gas stream. Most engine trouble is caused when these entrainment-borne contaminants form deposits on the engine parts. Therefore, they must be removed to finish the job of gas purification.

Wet cyclone separators (shown in Fig. 8-9), with a maximum spray velocity of 45 ft/s (15 m/s), are good for removing large mechanical entrainment drops more than 100 µm in diameter but have low efficiency for fine mist particles less than 10 µm in diameter.

Packed bed separators are good for finer droplet removal. For example, Fig. 8-24 illustrates that a 6-in.-deep bed packed with 1/2-in.-diameter spheres will capture 50% of 2.5-µm diameter droplets from a superficial gas velocity at 1.5 m/s (5 ft/s). Deeper beds and finer packings will increase collection performance; however, excessive gas velocity may cause reentrainment, deteriorating overall performance. The minimum packing size is limited by the fact that smaller packings more rapidly become plugged by viscous tar deposits. In these cases flow can be restored by stirring or replacing the packing.

It was common practice during World War II to pass the gas through wood chips, cork, or other fibrous materials to remove tars. Some of these materials subsequently can be used as fuel in the gasifier and thus dispose of the pollutants. Fiberglass filters have been used to clean gas (Johansson 1985) as has char (Humphries 1985).

Fiber-type demisters have limited applications because viscous tar deposits on fine wire mesh do not drain freely and are prone to plugging.

An electrostatic precipitator may be useful for entrainment separation. However, these units have not yet been proven reliable for continuous operation with producer gas.

8.7.3.4 Preventing Further Condensation

The scrubbed gas may have a very high humidity (from 80% humidity to the saturation point). Further condensation can be expected to occur either as the pressure drops or when the producer gas is mixed with combustion air.

To prevent further unwanted condensation, one may heat the gas or secondary air (the engine intake air) with engine exhaust heat as shown in Fig. 8-25.

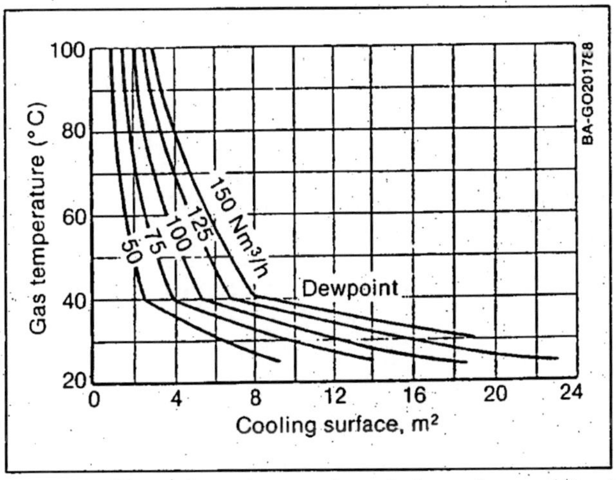

Fig. 8-23. Gas cooler surface requirements for various outputs at 700°C (Source: Gengas 1950, Fig. 99)

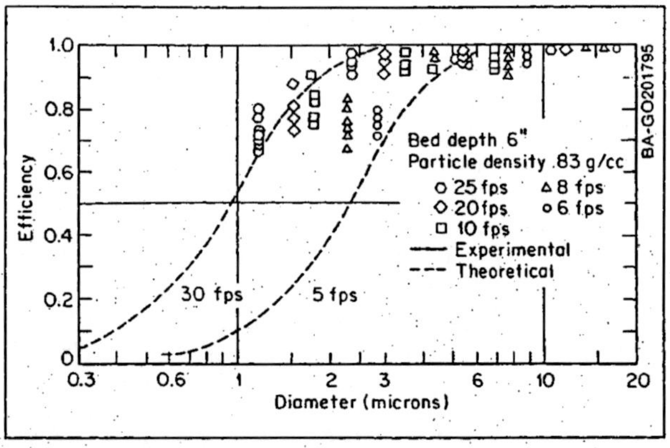

Fig. 8-24. Entrainment separator collection efficiency of packed bed for droplet removal (Source: Perry 1973, Fig. 18-142)

8.8 Disposal of Captured Contaminants

8.8.1 Char-Ash

The char-ash removed from producer gas is free of dangerous materials and can be burned or safely disposed of in a landfill. When burned to white ash, char-ash contains valuable minerals that may be beneficially returned to the soil. Charcoal is a valuable, clean-burning fuel, worth several times the value of wood. Alternate uses and possible markets definitely should be examined.

Larger char material may be saleable for further charcoal gasification, combustion, or briquetting.

In addition to valuable soil minerals, charcoal has been successfully used as a soil conditioner in Japan, resulting in improved crop yields (Kishimoto 1985). It also has been used as a livestock feed supplement to reduce digestive problems and meat hormone levels (Taylor 1986). Charcoal has long been considered a premium cooking fuel in many developing countries.

8.8.2 Tar

A gasifier that produces more than 0.5 g/Nm^3 of tar cannot be suitably cleaned for engine applications due to the large amounts of tar that must be captured and disposed of. For a worst-case tar production scenario of 2 g/Nm^3 (0.2%), up to 2 g of tar per hp-h may arise with each horsepower. Thus, a 100-hp engine would produce up to 4.8 kg of tar in 24 hours, or about 5 L. At room temperature, tar is a viscous, slow-flowing, molasses-like fluid. It may contain carcinogenic substances and should be treated with due caution. Collected tars that cannot be recycled to the gasifier or burned on site may need to be treated as hazardous wastes. They should not be dumped on the ground or released into waterways. Prevention through low-tar gasifier designs is the best cure for tar (see Chapters 4 and 5).

8.8.3 Condensate

The liquid condensate from producer gas may contain substantial amounts of tars and phenols. Phenols are well-known germicides and will kill soil bacteria if spread on the ground. If released into waterways, the condensate could damage lifeforms supported by those waterways.

For condensate water, as with tars, prevention is the best cure. Methods of minimizing condensate formation should be considered fully early in the design and selection of the system (see Chapter 5). Gas moisture content may be minimized by using dry fuel (the drier, the better) and by recycling heat back into the gasifier through an air-blast preheater. Condensation from the gas may be minimized by limiting the amount of gas cooling so as to use the gas above its water dew point (40°-60°C). The loss in engine power or the costs of larger engines may be more than offset by the savings in the cost of condensate disposal.

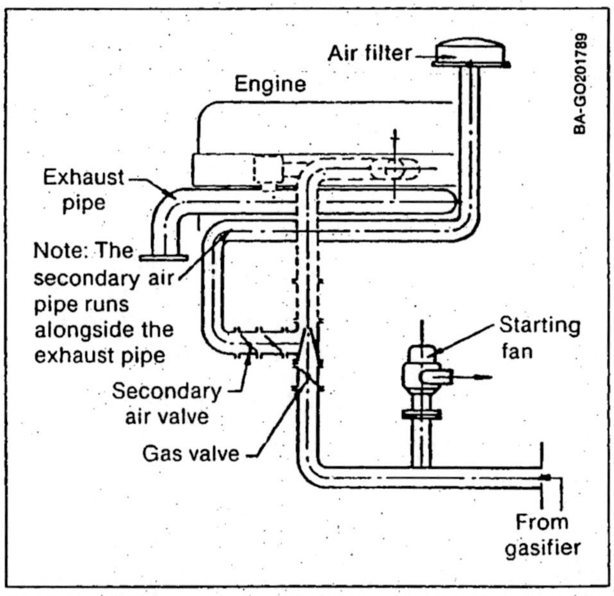

Fig. 8-25. Heating of secondary air by exhaust gas heat from the engine (Source: Gengas 1950, Fig. 85)

Chapter 9
Gasifier Systems

9.1 The Complete Gasifier System

The previous chapters have discussed the major operational components of a gasifier-engine system. However, no system is stronger than its weakest link. A complete system requires means to store and possibly dry the biomass, to feed the biomass, to remove char-ash, to push or pull the gas through the system, to clean the gas, and to burn the gas during startup, as shown in the front of this book and in Fig. 9-1.

A complete system also requires instruments to measure pressure, flow rates, and temperatures at crucial points, and controls to establish the required conditions. Instruments and controls will be discussed in Chapter 10.

9.2 Storing, Feeding, and Sealing Solids

9.2.1 Characteristics of Solids

Solids are many times more difficult to feed and seal against gas flow than liquids and gases. Because they can support a shear stress, solids can bridge and arch in cylindrical channels. An important measure of the difficulty of feeding a particular solid, the angle of repose, is the average angle from a horizontal plane assumed by individual biomass pieces when they are randomly piled up. For liquids, this angle is zero; for some solids, it can be greater than 90°! For this reason, vibrators, shakers, rakes and chains, live bottoms (on trucks), and a host of ingenious devices are used widely in industrial and agricultural solid-feeding applications. Much time and money can be wasted reinventing these devices, so the designer is advised to contact others with experience in feeding the particular form of biomass being used.

9.2.2 Storage

A closed bin, silo, or hopper must be supplied to hold the biomass feedstock (chips, cobs, pellets, etc.), to prevent it from getting wet. In many cases, industrial or agricultural containers are available in appropriate sizes at low cost.

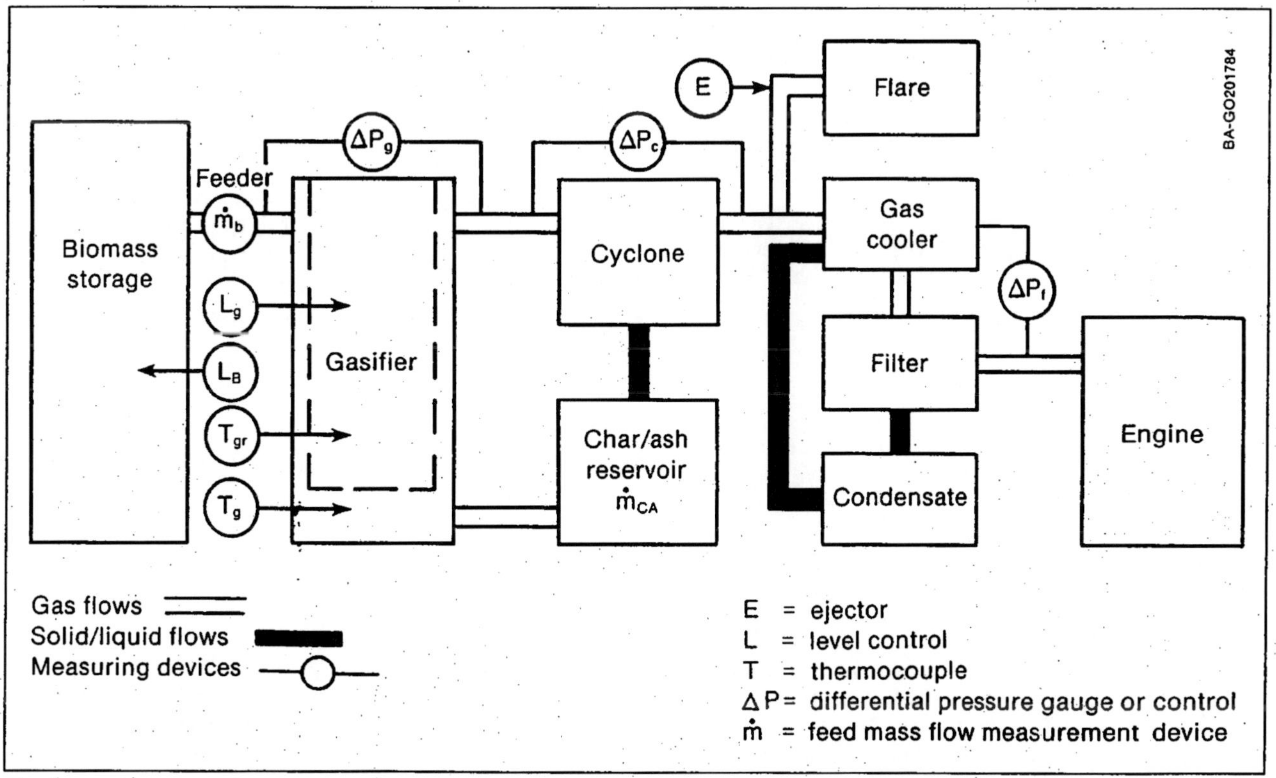

Fig. 9-1. Gasifier system showing means of moving solids and gases and positions for various instruments and controls

9.2.3 Feeding Solids

During testing, fuel can be fed manually to small gasifiers. However, there is the danger of running out of fuel, which in turn overheats the gasifier as the remaining charcoal burns. Level alarms or other controls are strongly recommended for gasifier systems intended for continuous operation (see Chapter 10).

Biomass fuels are only partially free flowing from a hopper by gravity alone so bin stirrers, vibrators, or shakers may be required for even fuel delivery. Biomass can be moved laterally and vertically by conveyor belts, chain drags, bucket elevators, augers, pneumatic blowers, and vibratory feeders (e.g., Syntron type), widely available in agricultural handling equipment. Again, those with experience with the particular biomass form should be contacted for feeding and equipment suggestions.

The flow of solids in the gasifier is also subject to irregularity and interruption and can cause great difficulty during gasification, resulting in such problems as bridging, caking, channelling, and rat-holing. The importance of uniform feeding cannot be overemphasized and experts in the field should be consulted (Guzdar 1982, Miles 1982).

Char-ash must be removed from the gasifier and stored as it is produced. An air-tight char-ash receiver should be provided, since this char material is combustible and may reignite spontaneously. In addition, it may be necessary to cool the receiver. Although the weight of the char-ash may be only 2% to 5% of the weight of the biomass fed to the gasifier, its volume may represent a larger fraction of the volume of the original biomass because of its lower density. The required receiver volume must therefore be calculated from measured char-ash bulk densities, which may range from 0.064 to 0.4 g/cm^3 (4-26 lb/ft^3).

It should be mentioned here that ash receivers can contain explosive gas even when cold. They have been known to ignite on startup, and precautions should be taken against this.

9.2.4 Sealing Solid Flows

Gasifiers may operate at pressures up to 20 in. water gauge above or below atmospheric pressure, which makes it necessary to provide a seal through which the biomass can pass without leaking air or gas. Proper seals are very important, to ensure both gas quality and safe operation (see Chapter 12).

Gasifiers from the World War II era were batch-fed through a lid that could be sealed tightly, as shown in Fig. 9-2. The spring-loaded lid would pop open in the event of an internal gas explosion. As long as the gasifier was filled quickly, the expelled smoke could be tolerated as a nuisance.

Fig. 9-2. Batch fed gasifier with lid (Source: NAS 1983, p. 68)

There are two basic types of solids feeding and sealing devices—mechanical seal type where the seal mechanically prevents gas passage, and plug seal where the fuel acts as its own plug and seal such that fuel velocity into the gasifier is greater than gas velocity out through the fuel plug. It should be noted that rotary valves and gate valves are also good firestops for flashback or explosion prevention. Inert or air purge gas should be used in pressurized gasifiers to offset leakage through rotary valves. Additionally, there is the stratified charge gasifier where air enters through the top at atmospheric pressure such that for operation above a minimum gasification rate, all smoke is drawn down into the fuel bed, and a lid is unnecessary. In many cases, the biomass feed can help to act as a seal in a long auger or vertical pipe. However, the pressure drop through the fuel is small, and the technique will not work if the gasifier fuel inlet is under a positive pressure of more than 2.5 cm (1 in.) water gauge.

Various solids feeding devices are shown in Fig. 9-3. Star valves, which rotate to feed the fuel, are available commercially.

If a gasifier is to be operated at high pressure, it becomes exceedingly difficult to feed biomass through a single seal. Lock hoppers that use two slide or bell valves supplying a metering feeder as shown in Fig. 9-4 have been used with biomass at pressures up to 30 atmospheres for making medium-energy gas for

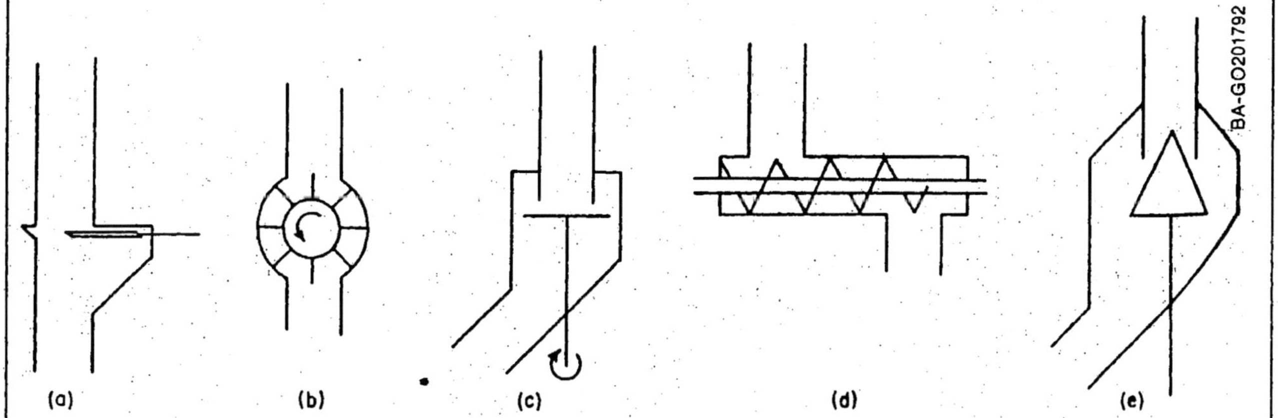

(*a*) Slide valve. (*b*) Star valve. (*c*) Table feeder. (*d*) Screw feeder. (*e*) Cone valve.

Fig. 9-3. Solids feeding devices (Source: Perry 1973, Fig. 20-79. © 1973. Used with permission of McGraw Hill Book Co.)

pipeline distribution or syngas for chemical synthesis (Reed 1982).

9.3 Fans, Blowers, Ejectors, and Compressors

9.3.1 Importance of Gas-Moving System Design

It is important to provide a suitable method for pulling or pushing the gas through the gasifier, and since the mass of gas and air being moved is much larger than the mass of fuel being fed, considerable power may be required. The engine being fueled can serve this purpose.

Every internal-combustion engine is a compressor, since it compresses the intake air and fuel to 10 to 30 times atmospheric pressure before igniting the fuel. When an engine operates on producer gas, it can also provide suction and compression for the gasifier. However, an engine is a very expensive compressor for testing new gasifiers (Arthayukti 1984; Breag 1982). It is desirable to use some other method for moving the gas that is less sensitive to tar, char, and soot during testing. Full engine power also may require some type of compressor (see Chapter 10).

The question of whether the gas should be pulled (suction operation) or pushed (pressurized operation) through the gasifier is important, and one finds strong advocates of each method. Gas leaks from gasifiers operating above atmospheric pressure can be dangerous because of the possibility of leaking carbon monoxide out of the gasifier; air leaking into gasifiers operating below atmospheric pressure can cause explosions.

9.3.2 Fans

Propeller-type fan blades usually generate under 1.25 cm (0.5 in.) of water gauge pressure and are used in gasifiers only to move air through heat exchangers and radiators. They are not suitable for moving gas against any resistance.

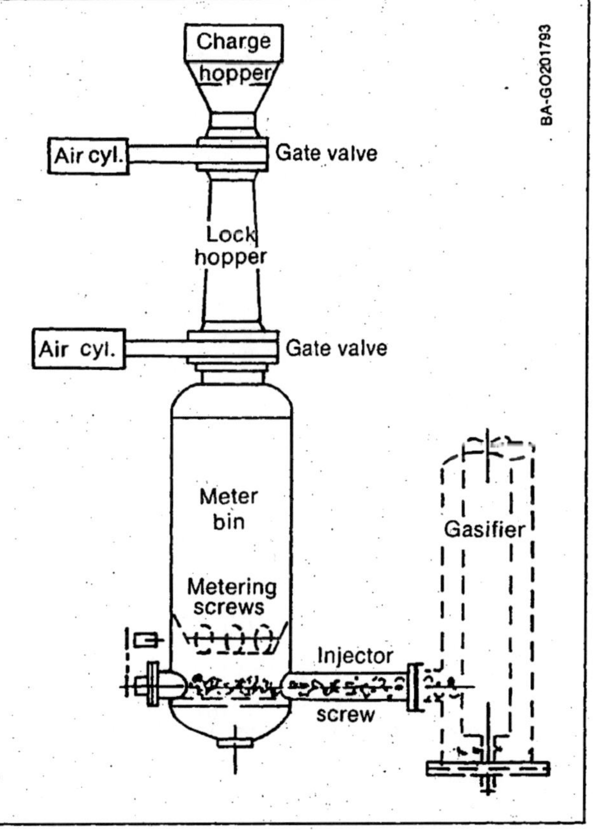

Fig. 9-4. Lock hopper for high-pressure fuel feeding (Source: Miles 1982)

9.3.3 Blowers

Centrifugal blowers (Fig. 9-5) can generate pressures on the order of 100 cm (40 in.) of water gauge pressure and are quite suitable for gasifier testing. To generate these pressures, the blowers must either rotate very fast or have a large diameter, since it is the centrifugal force that creates the pressure. The blower can tolerate, and in fact will remove, a certain amount of tar and particulates, but a means for draining and cleaning the blower should be provided. Blowers can be used either to push the air into the gasifier or to pull the hot gas through the system at negative pressure. Considerably more power is required to pull the gas through the system than to push air because there is necessarily more mass to manipulate and the gas is less dense. In addition, suction fans must be capable of handling a higher temperature than fans pushing air into the gasifier. Most blower breakdowns occur due to deposits on shaft seal and impeller or erosion of the case. Reliability is limited by deposits.

9.3.4 Ejectors

Ejectors are a very convenient and simple means for moving dirty gas. No moving parts are exposed to gas contaminants. Ejectors (Fig. 9-6) use the motion of a small amount of one gas to move larger quantities of a second gas, often at negative pressures. During startup, the gas produced initially is very tarry and may quickly clog cleanup and engine. Therefore, one should use a fan or ejector during startup to send this gas to a product-gas burner until low-tar operation is reached. Compressed air, nitrogen, or steam can be used to drive the ejector. Water jets can also be used to move, cool, and clean the gas.

Ejector design is based on the principle of the conservation of momentum of the driver gas as it aspirates

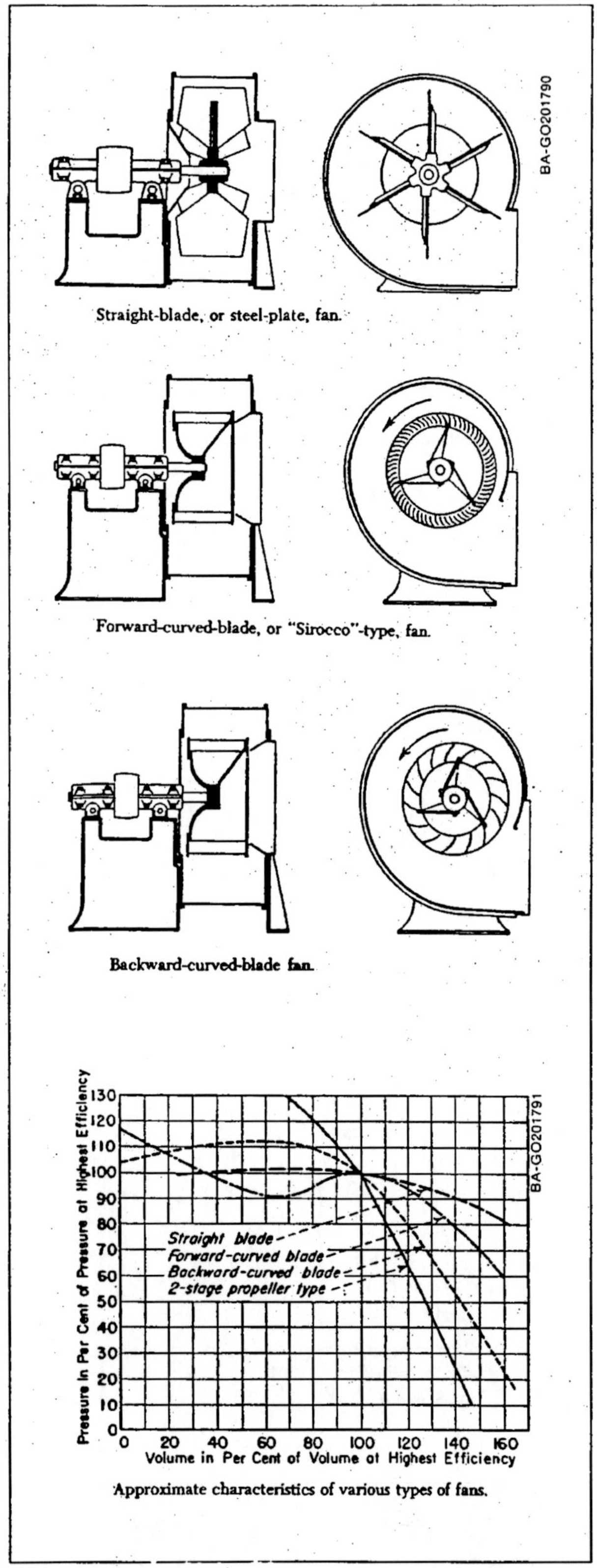

Fig. 9-5. Centrifugal blowers (Source: Perry 1973, Figs. 6-37, 6-38, 6-39, 6-41. © 1973. Used with permission of McGraw Hill Book Co.)

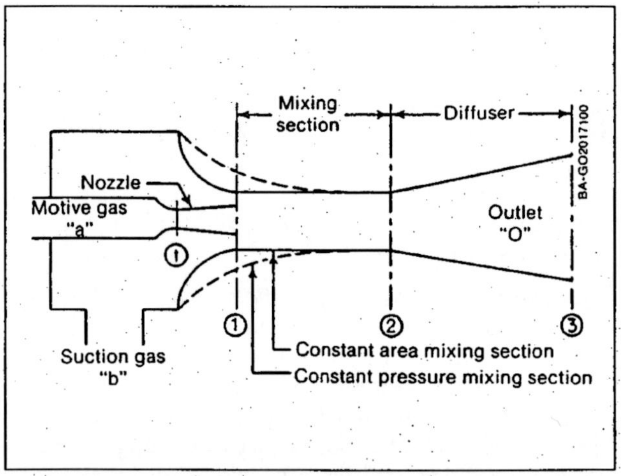

Fig. 9-6. Ejector pump for moving gas (Source: Perry 1973, Fig. 6-73. © 1973. Used with permission of McGraw Hill Book Co.)

and mixes with the driven gas. Optimum ejector dimensions are discussed in Perry (1973).

9.3.5 Turbochargers and Superchargers

The power output of internal-combustion spark and diesel engines is directly proportional to the energy content of the intake fuel-air mixture. A mixture of producer gas and air has 30% less energy than a gasoline and air mixture, resulting in a minimum 30% power loss at any given rpm. Intake pressure can be increased to overcome this power loss by a turbocharger using engine exhaust pressure to run a turbine, or a supercharger operating from the engine shaft power. This pressure boost is widely used in diesel engines and racing cars, and is coming into wider use even for spark-ignition engines.

Positive displacement rotary blowers (Fig. 9-7) and superchargers can achieve any pressure required, but they do so at higher capital, operating, maintenance, and energy costs. This increased cost must be weighed against the lower cost of using a larger engine.

9.4 Flares and Product-Gas Burners

9.4.1 Flares

Flares sometimes may be seen at oil wells or refineries in which excess gas burns with a luminous flame. In order to produce a nonluminous flame, it is necessary to provide enough air and residence time to burn the soot in a hotter, nonluminous flame. This is called a gas incinerator.

Raw producer gas contains up to 40% carbon monoxide and up to 20% volatile tars, making it absolutely essential that a reliable incinerator be available during testing to burn the gas. The incinerator must be sized to fit the gasifier. Most of the principles discussed below for incinerators also apply to developing burners for producer gas.

The three essential elements necessary to combust any gas are residence time, temperature, and turbulence (the three "T's" of gas combustion). Residence time requires a sufficently large chamber for combustion to proceed to completion. High temperature is achieved by using a refractory lining on the burner. Turbulence can be generated by high-velocity mixing of the combustion air or fuel (for instance, by passing it through a nozzle) or by tangentially mixing air with the gas. The reader is referred to books on combustion and burners for a more complete discussion (e.g., Perry 1973).

Incinerators for toxic chemicals and gases require a pilot flame to assure combustion operated on methane or propane, an ignitor to start the flame, a flame sensor to shut down the burner if the flame is extinguished, and a control system to regulate the air mixture and stack temperature.

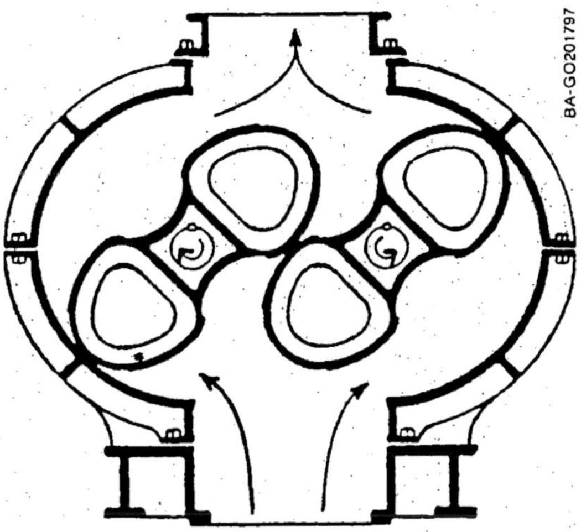

Two-impeller type of positive rotary blower.

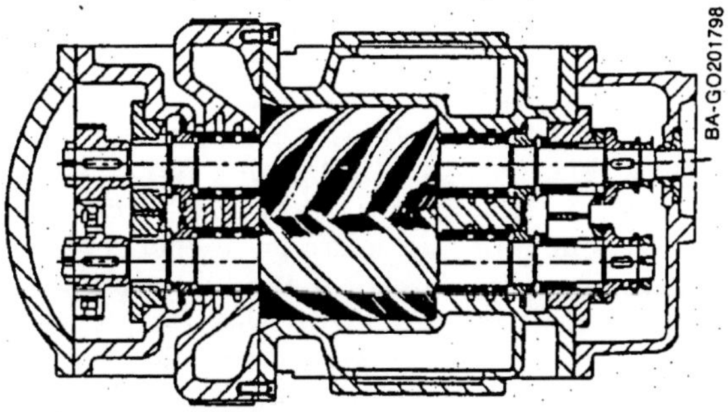

Screw-type rotary compressor.

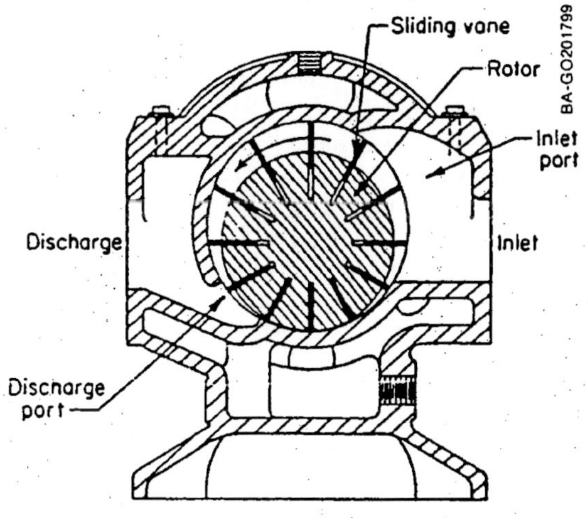

Sliding-vane type of rotary blower.

Fig. 9-7. Positive displacement rotary blowers and compressors (Source: Perry 1973, Figs. 6-49, 6-50, 6-51. © 1973. Used with permission of McGraw Hill Book Co.)

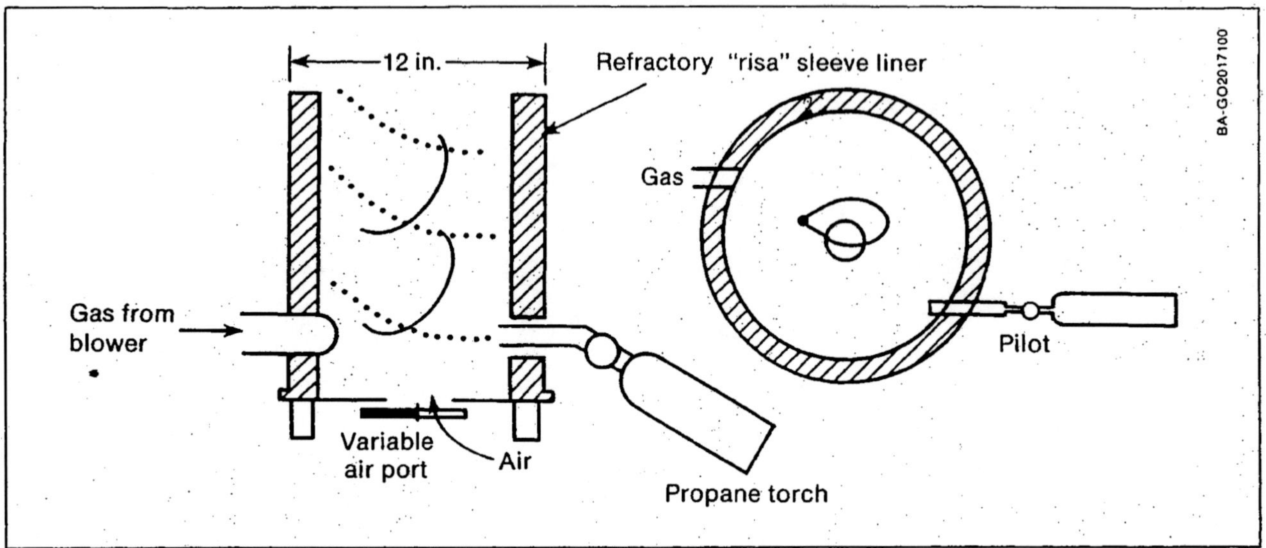

Fig. 9-8. Product-gas burner for testing small gasifiers

A small flare, suitable for burning up to 200,000 Btu/h of producer gas, is shown in Fig. 9-8. It is made from a 5-gal can lined with an 8-in.-ID ceramic fiber risa sleeve. (Risa sleeves are used in metal foundries, high-temperature flue lines, and oil-burning furnaces.) A propane torch or suitable pilot flame injected tangentially at the bottom maintains a high temperature on the wall. Producer gas is injected tangentially upstream from the propane flame and mixes with the propane flame as a source of ignition. Air is drawn through the bottom port to permit complete combustion. Smaller or larger flares can be built on this principle, or commercially available burners and gas incinerators can be used.

9.4.2 Burners

Close coupled gasifier burners offer improved clean, high efficiency burning of solid fuels compared with conventional solid burners. Gas-air mixture and mixing are more easily controlled than are conventional solid fuel burners, resulting in more complete combustion. Equipment costs for retrofitting an existing gas or oil unit for close coupled gasifying are generally less than the costs of replacing the unit with a solid fuel burner.

Close coupled gasifier burner systems are able to meet emission requirements with no expensive pollution abatement equipment.

Updraft producer gas is an excellent fuel for high-quality heat applications. The high tar content does not need to be removed, and adds to the heating value. Furthermore, the sensible heat of the gas adds to the flame temperature and overall heat output. TIPI workshop (Das 1986) has been using updraft producer gas from wood chips to fuel a melting furnace, melting copper and bronze at high efficiency using 2 to 3 lb wood chips per pound of bronze poured and a rapid heat cycle (20 to 30 min) to reach pouring temperature.

The flame inside the forge is very different from any other kind of wood burner. It has the intensity and fury of a jet engine and the clean invisible flame of a high-quality fuel. Direct combustion of producer gas lends itself to a wide range of applications (e.g., ceramics, glass, steam, drying, blacksmithing, and process heat).

Chapter 10
Instrumentation and Control

10.1 The Need for Instrumentation and Control

The gasifiers of World War II were batch-fed, and gas flowed to the engine in response to the vacuum created in the engine. As a result, usually there were no controls or instruments other than those of the engine and vehicle. The corresponding tradeoff was that a great deal of operator expertise was required in order to know when to shake the grate, stir the fuel bed, fill the fuel hopper, and clean the system, as well as how to operate the vehicle on hills and in traffic.

Just as the modern automobile has a number of instruments, sensors, warning lights, and controls, gasifiers built today should incorporate instruments and controls to improve performance, convenience, and safety.

During research and development, many measurements are required to determine operating parameters and to determine where controls should be applied. However, a production gasifier system (like a production car) should require relatively few key instruments and controls.

Ideally, gasifiers of the future will operate automatically and unattended. A working gasifier system requires the integrated operation of a number of components. It is desirable to have automatic controls and warning systems on production gasifiers in order to keep the system in balance and to warn the operator when a particular component needs attention. Although such a degree of automation would have been impossible to implement on the gasifiers of World War II, reliable, inexpensive sensors, instruments, and controls are widely available today in much of the world, due largely to the development of the modern automobile and solid-state technology.

This chapter deals with the methods and equipment that have been used to measure each physical quantity critical to optimal gasifier operation. Since these measurements are commonly used by chemical engineers, the reader is referred to chemical engineering texts, and especially to handbooks such as the *Chemical Engineers' Handbook* (Perry 1973).

10.2 Gasifier Instruments

10.2.1 Pressure Measurement

In experimental work, pressure drops may be measured routinely across the gasifier bed, the orifice plate flow meters (see below), the gas cleanup system and components, and any components that are prone to plugging, as shown in Fig. 9-1. The total pressure with respect to atmospheric pressure (known as "gauge" pressure) may also be measured at the gasifier outlet, the cleanup outlet, and (if the gasifier is operated at a pressure above atmospheric pressure) the air inlet. In locations where occasional measurements are sufficient, connections should be capable of being closed off when they are not in use.

The pressures within the gasifier will be close to atmospheric pressure (except for high-pressure gas producers) and generally will be measured in centimeters (inches) of water column.* Pressure drops and differential pressures can be measured by a U-tube manometer filled with colored liquid. Liquid-filled manometers are simple devices that use a fluid-filled (usually colored water or light oil) tube to measure pressure. For convenience and portability, they can be made from flexible, transparent plastic tubing and a meter stick. Commercial units are available in a wide range of accuracies, from 0.25% to 3% of the full-scale reading.

For more sensitive measurements, an inclined-tube manometer may be used (Dwyer 1960). Typical inclined-tube scales can be read to 0.1 in. water gauge, and curved-tube manometers (e.g., the Dwyer Mark II) provide high resolution over a wide range of readings. Commercial units are available for as low as $10, dual-range vertical inclined manometers start at $30, and stationary high-resolution manometers can cost as much as $400. Special handling is required to avoid losing or blowing out gauge fluid because of excessive pressure. Accurate leveling is important for all manometers.

Diaphragm-type differential pressure gauges give a needle-pointer reading of the differential pressure. They are available in full-scale accuracies of 2% and full-scale ranges from 0.25 to 150 in. water gauge (e.g., those manufactured by Dwyer). These gauges are rugged and can be used in all positions with positive and negative pressures.

For nondifferential pressures in excess of a few pounds per square inch, the well known Bourdon gauges should be used.

*1029 cm water = 1 atm; 406.8 in. water gauge = 1 atm; 26.4 in. water = 1 psi.

10.2.2 Gas Flow Measurement

10.2.2.1 Floating Ball Rotometers

Clean gas flow can be measured using a rotometer (ball and tube) flowmeter, which comes in a wide range of sizes up to 100 scfm. The gas must be clean at the rotometer, and therefore rotometers usually can be used only on air or oxygen streams at the gasifier inlet. In addition, rotometers must be calibrated for the specific gas to be used, and the reading also depends on the absolute gas pressure. Nevertheless, they are very convenient flow measurement devices.

10.2.2.2 Gas Flow by Differential Pressure Measurement

Gas flow may be measured by a number of methods that produce a differential pressure signal. Pitot tubes, orifice meters, Venturi meters, and flow-restriction meters all are based on measuring the pressure drop produced by gas flow across or through the device. For each technique, the velocity is proportional to the square-root of the pressure drop (Δp).

Pitot Tubes. The standard Pitot tube shown in Fig. 10-1 is recognized as a primary standard for velocity measurements. When properly designed, it has a unit calibration factor, K = 1. For pipes under 10 cm (4 in.) in diameter, the average velocity is 90% of the center-stream velocity, ±5%. For air, the gas velocity (ft/min) is given by the equation:

$$V = 1096.7\,K\,[\Delta p\, T_{test}\, P_{std}/(\rho_G\, T_{std}\, P_{test})]^{1/2} \quad (10\text{-}1)$$

where:

ρ_G = gas density at standard conditions

Δp = the differential pressure in inches of water

P = gas pressure absolute (atmospheres)

T = gas temperature (Kelvin)

K = calibration constant

The subscripts std and test refer to standard conditions and actual test conditions, respectively.

The standard Pitot tube shown in Fig. 10-1 introduces the smallest pressure drop of any pressure-sensing flow

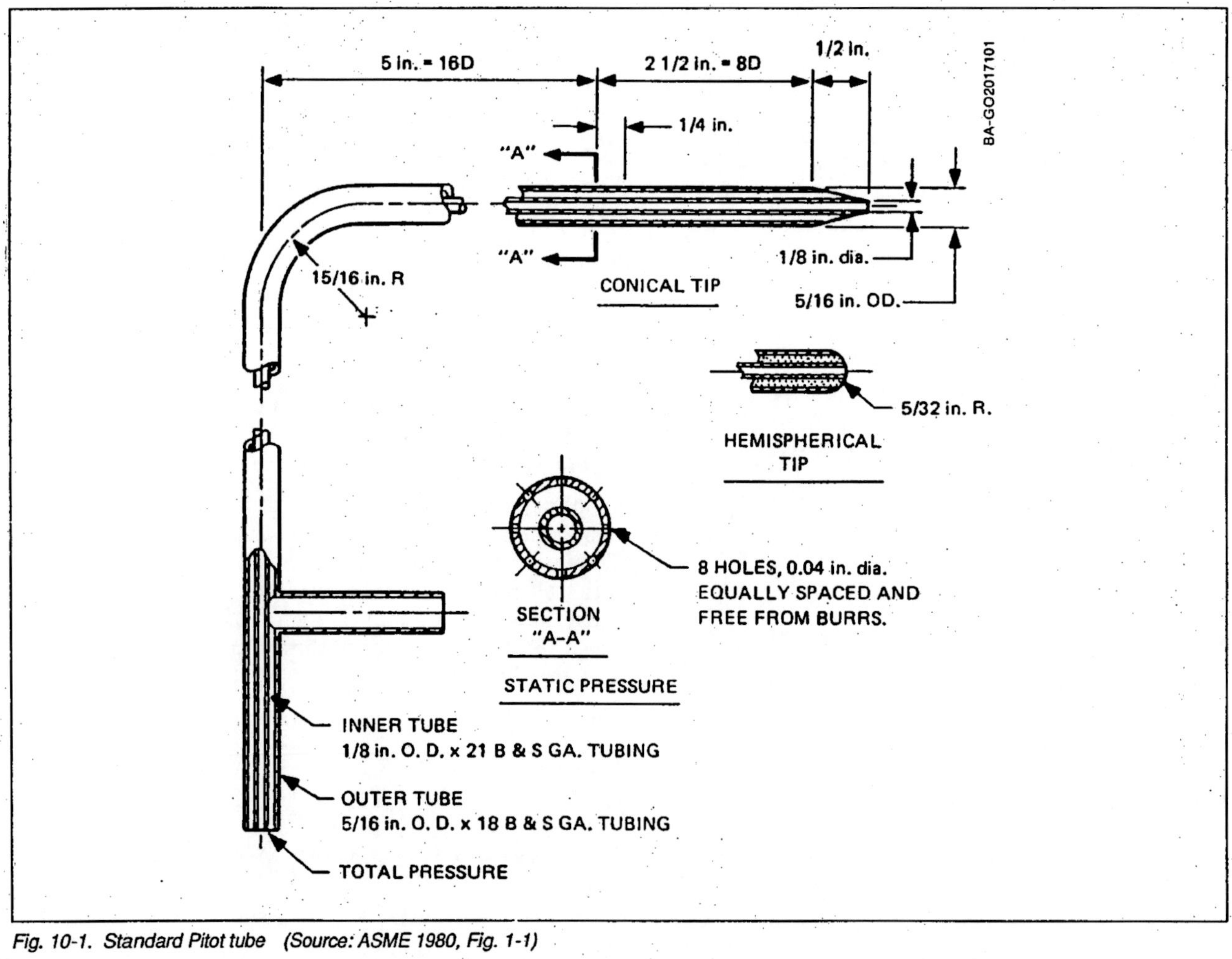

Fig. 10-1. Standard Pitot tube (Source: ASME 1980, Fig. 1-1)

method. However, the small openings are easily clogged by dirty gas. The "s" type Pitot tube shown in Fig. 10-2 is more suitable for dirty gas. Its calibration factor is approximately K = 0.83 for velocities up to 50 ft/s. For more accurate work, the individual probe should be calibrated against a primary standard.

The low pressure signal characteristic of Pitot measurements requires a sensitive and delicate readout. A typical gasifier pipe velocity of 20 ft/s will show a pressure drop of 0.09 in. water gauge. This pressure drop can be measured by a Dwyer Magnehelic® 0-0.25 in. water gauge pressure gauge, which has a resolution of 0.005 in. water gauge, a pressure readability of 5%, and a velocity readability of 2.7%.

The velocity of a gas in a pipe is highest at the center and zero at the wall. The gas velocity can be used to measure flow by traversing the velocity profile and integrating the result.

Orifice Meters. The orifice meter shown in Fig. 10-3 gives a much higher pressure reading than a Pitot tube, but it does so at the expense of creating a slightly higher

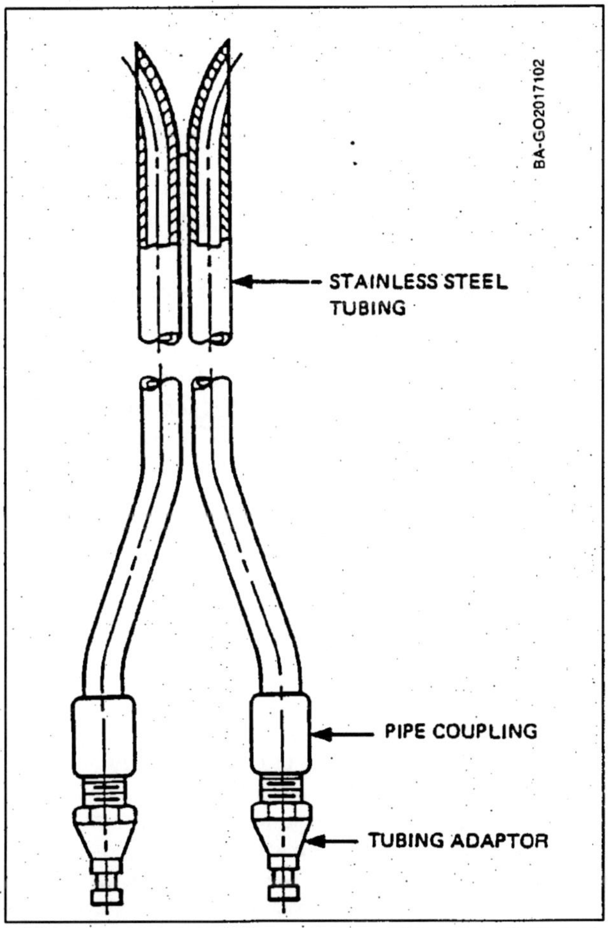

Fig. 10-2. "S"-type Pitot tube for dirty gas (Source: ASME 1980, Fig. 1-1)

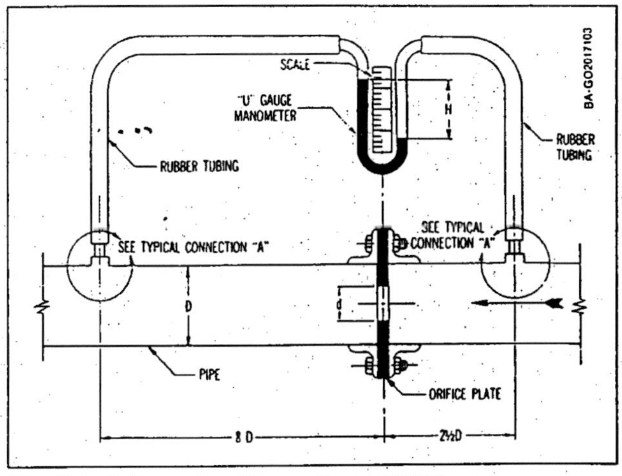

Fig. 10-3. Orifice meter (Source: Haaland 1968)

pressure drop in the system. If the added pressure drop can be tolerated, then the orifice meter is usually preferred over the other alternatives because it is low in cost, can use more rugged, less sensitive readout, and is not sensitive to small amounts of tars and particulates. An orifice meter consists of a washer-shaped plate with a hole diameter d that is placed in the gas line, with a plate thickness at the orifice edge of no more than 3% of the pipe diameter, D. If the orifice is an accurately centered, round opening of diameter d, the diameter ratio, d/D, should be between 0.3 and 0.6. A means for measuring the differential pressure drop across the plate, as shown in Fig. 10-3, must be provided.

The calibration can be predicted from the gas properties and dimensions of the pipe and orifice using formulas given here and in more detail in chemical engineering texts and handbooks, such as the *Chemical Engineers' Handbook* (Perry 1973). The location of pressure taps on an orifice meter will significantly affect the calibration. For an accurate measurement, the orifice meter should be calibrated against a primary standard such as a Pitot tube or dry gas test meter. Although small deposits of tars will not plug the meter, they may alter the calibration; frequent cleaning or recalibration may be required. To prevent fouling from char or condensate, the orifice opening may be moved to the bottom of the pipe to permit free passage of solids and liquids. Calibration is advisable against a standard orifice meter or other known standard. If the gasifier uses a cyclone, then it can be used as a convenient flow meter by locating temperature and pressure taps at the cyclone inlet and outlet. Calibration again may be predicted from gas properties and cyclone dimensions or calculated against a known standard.

Flow is related to pressure by the equation

$$Q = K\sqrt{\frac{\Delta p T_o}{T_G}} = a\, D^2 \sqrt{\frac{\Delta p T_o}{T_G}} \quad (10\text{-}2)$$

where:

T_o = calibration temperature

T_G = operating gas inlet temperature

Q = flow rate

K = overall calibration for specific flow meter

a = general constant for ratio d/D and geometry (from Fig. 10-4)

Δp = pressure drop across flow meter

D = pipe inside diameter

d = orifice diameter

It is convenient to plot Q versus Δp for a handy reference. Dimensions and calibration constants are shown for several orifice meter sizes in Fig. 10-5.

Rotational flow caused by pipe fittings can cause serious metering errors. It is good practice to provide at least eight pipe diameters upstream and three pipe diameters downstream of the orifice to minimize rotational errors.

Venturi Meters. The Herschel-type Venturi flow meter shown in Fig. 10-6 provides a higher pressure signal with a minimum pressure drop across the meter, because the divergent downstream section of the meter conserves gas momentum by converting velocity back into pressure. In addition, the Venturi meter does not present abrupt surfaces that otherwise would be subject to impact by tars and particles, so contaminant buildup is minimal and the maintenance required for a permanent installation is reduced. The cost of a Venturi meter is the highest of any of the differential pressure methods of velocity measurements presented here

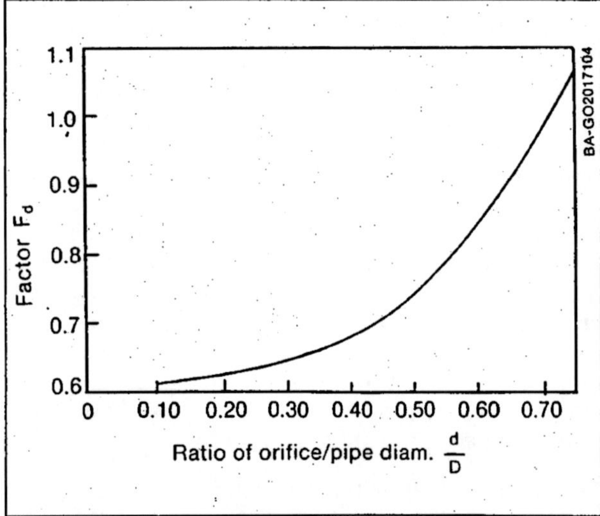

Fig. 10-4. Orifice meter disc factor (Source: Haaland 1968)

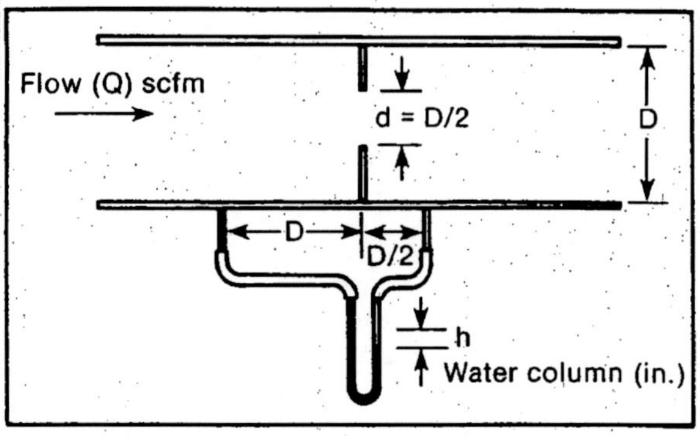

Pipe ID (D) in.	Orifice (d) in.	K* scfm/in.	a*
5/8	5/16	1.75	4.47
1-1/2	3/4	8.53	3.79
2	1	13.65	3.41

*K and a are the calibration factors in Eq. (10-2)

Fig. 10-5. Orifice meter calibration factors

because of the amount of precise machining required for its production.

10.2.2.3 Positive-Displacement Meters

The positive-displacement dry gas meter is a primary gas volume standard. It measures the accumulated total flow of a gas and is very accurate over a very wide range of flows. It is relatively inexpensive because of its use by the gas industry. Numerous moving parts are exposed to the gas, so the gas must be clean and dry. Its main applications are in measuring sampled dry gas in a contaminant sampling and for calibrating other flow meters.

10.2.2.4 Tracer-Gas Flow Measurement

Total gas flow can be measured by injecting a small, accurately measured flow of tracer gas into either the air or producer gas stream. The tracer gas must be one not normally present in producer gas, such as helium or n-butane, and capable of being measured with high precision. Alternatively, a gas that is normally present, such as nitrogen, can be injected intermittently into the stream, and the flow can be calculated from the changes of concentration in the product gas. Both of these techniques of tracer-gas flow measurement require accurate gas chromatographic analysis of the sample to determine the flow rate.

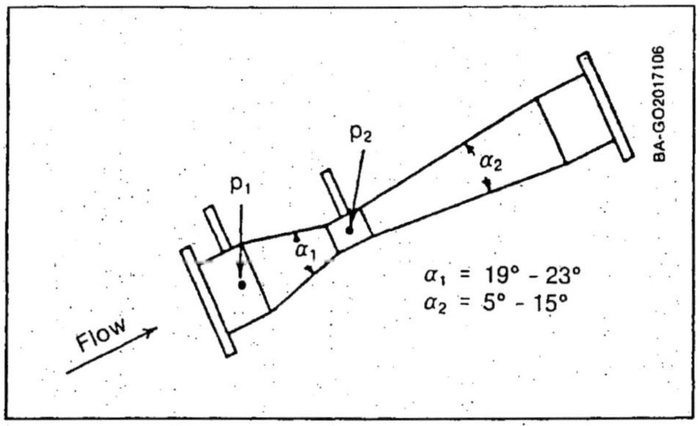

Fig. 10-6. Herschel-type Venturi flowmeter (Source: Perry 1973, Fig. 5-13. © 1973. Used with permission of McGraw Hill Book Co.)

Optical pyrometers can be used from about 300°C (infrared) or 700°C (visible) to 4000°C. They require a viewport and are generally hand-operated. They are useful for spot-checking surface temperatures.

Of these methods, thermocouples are best suited to most gasifier measurements. Thermocouples are available with either analog or digital readouts, and the signal can be recorded electronically. Analog thermocouple meters may cost as little as $50, but they require a fixed lead resistance. They are used widely in pottery and foundry trades. Digital pyrometer readouts start at $200. They read to greater precision than analog, and due to their high input impedance, lead length is not critical.

10.2.3 Solid Flow Measurement

Most World War II gasifiers were batch-fed, so it was very easy to weigh the biomass fuel as it was fed in order to record biomass consumption. However, this technique results in only a long time average measurement of the actual feed rate.

Many gasifier systems will be fed automatically, using level controls to determine the feed rate, as shown in Fig. 9-1. If the change in weight between levels is calibrated, then recording the number of feed cycles allows one to estimate biomass consumption. A more accurate technique involves a weighing means that is inserted along the feed train to measure the amount of feed delivered to the gasifier.

Alternatively, the gasifier itself may be weighed to record both the entry and consumption rates of fuel. Flexible connections on the gasifier are needed, so that the gasifier can be weighed free from the variable forces of the connections.

10.2.4 Temperature Measurements

Low temperatures (up to 300°C, 572°F) can be visually indicated with mercury thermometers or bimetallic-stem dial thermometers. Alternatively, thermistor sensors can be used in this temperature range to provide an electrical signal that can be used for automatic recordkeeping and/or control purposes.

Chromel-alumel (type K) thermocouples can be used continuously to 1000°C (1800°F) and intermittently to 1200°C, and they provide an almost linear electrical signal of 40 µV/°C. Sheathed thermocouples should always be used in connection with producer gas applications, because thermocouple alloys will react with hot reducing gases, H_2 and CO, changing the output voltage calibration. Thermocouples are used widely in industry for temperature measurements and are available from a number of sources.

10.3 Controls

For commercial application, gasifiers must be safe, dependable, and convenient. Automatic and unattended operation, an eventual necessity, will require suitable control and warning sensors and mechanisms.

10.3.1 Fuel-Level Controls

If a gasifier continues to operate after the biomass has been consumed, there is a danger of damaging the high-temperature region of the gasifier because of the extraordinarily high temperatures generated during char gasification and combustion. Therefore, level controls should be installed in the various fuel hoppers and the gasifier itself to warn when the biomass level is getting low.

A number of level indicators are available on the market for signaling the level of solids and liquids in containers. These operate on the basis of light or sound signals, bin wall pressure, resistance to vibration or rotation, and absorption of radiation. Suitability depends on vulnerability to failure from clogging or tarring the device. After installation, the control should be repeatedly tested to ensure that it is functioning, since we have seen these devices clog or not function for many reasons not anticipated by either the user or manufacturer.

10.3.2 Pressure Controls

Simple electromechanical switches that sense absolute or differential pressure are available. These switches provide on/off switch closure signal. An especially attractive, inexpensive, low pressure differential switch is the Dwyer Minitactor®.

Special pressure or differential pressure transducers are needed to produce analog electrical signals for an analog programmable controller, a control console, or data acquisition.

10.3.3 Temperature Controls

Thermistor, bimetallic, and thermocouple switches and controllers activate a switch closure in response to high- or low-temperature conditions to either take corrective action or give warning signals.

Additionally, thermocouples and thermistors generate analog electrical signals, which can be used by suitable proportional controllers.

10.4 Computer Data Logging and Control

Today, low-cost computers can exercise very sophisticated control over most processes in response to suitable signals, as can be seen under the hood of any modern car. Computer data recording and control should be considered in any research and development program and for any commercial gasifier.

Chapter 11
Engine Adaptation and Operation

11.1 Introduction

Producer gas technology is useful to provide clean combustion heat, shaft power, and electricity from widely available biomass fuels. Historically, engine shaft power generation has been the major application of small gas producers. Large producers, on the other hand, made "town gas," which was piped for cooking, lighting, and heating.

The major use of small producers during World War II was for transportation, especially trucks and buses. Although vehicles could become a major use of producer gas again, especially during a liquid fuel emergency, the added difficulty of using solid fuels for transportation does not make producer gas competitive with gasoline today. If programs for making synthetic liquid fuels (methanol, gasoline, or diesel) from biomass or coal are successful, producer gas may never be needed for transportation applications.

11.2 Producer Gas for Transportation

Transportation applications place a very heavy demand on producer gas systems: the system must be small, lightweight, and compact; operate at widely varying load conditions; have fast response times; be low in tar; be low in cost; be safe; and be convenient to use and service.

Currently, some vehicles are being powered regularly by gasifiers in Brazil and the Philippines (Mahin, June 1983). Sweden has maintained a stockpile of vehicle-ready gasifiers as part of its national preparedness strategy since the Suez oil crisis. A handful of groups offers plans for vehicle gasifiers (Mother 1982; Nunnikhoven 1984; Skov 1974). Various American vehicles have been converted and operated for demonstration purposes; however, regular use of these vehicles is rare (NAS 1983).

11.3 Producer Gas for Electric Power and Irrigation

Compactness and low weight are not as important for power and irrigation purposes as they are for transportation. Electric power production and irrigation place minimal requirements on turndown and response time of gas producers. The gas flow required for a synchronous generator varies over only a 3:1 ratio as the electric output goes from full load to no load.

11.4 Gasifier Types Suitable for Shaft-Power Generation

Updraft and fluidized-bed gasifiers have the slowest response times of the gasifier types and cannot be expected to follow changing loads with favorable results. The gas from both updraft and fluidized-bed gasifiers also contains large quantities of tars, making these gasifier types unsuitable for engine applications.

The fastest response time is obtained from crossdraft gas producers, but they are suitable only for low-tar fuels such as charcoal. Although low-tar operation has been observed in crossdraft gasifiers, nozzle spacing is very critical and load variations can cause tarring.

Downdraft gas producers provide a low-tar gas product from biomass and also have a rapid response time, so they are suited for powering engines with either varying or fixed loads. Low-tar gasifiers now under development incorporate design changes that recycle some of the gas with additional air to give even lower tar values (on the order of 50 mg/Nm^3) than conventional downdraft gasifiers (see Chapter 5 and especially Section 5.9).

11.5 Sizing the Gas Producer to the Engine

A common problem among gasifiers is the use of an oversized gasifier. An oversized gasifier produces excessive tars because lower flow rates do not develop the high temperatures necessary for good tar destruction. An undersized gasifier has excessive pressure drop, weak gas, and excessive raw gas temperature, and may be prone to burning out the grate. (See Sections 5.7.3 and 5.7.4.)

Begin sizing by evaluating the actual horsepower needed. For the task at hand, do not be misled by engine specifications. Vehicle applications seldom use full power output. A better indication of operating horsepower is the vehicle's measured fuel economy.

The average vehicle engine power required from a gasifier may be figured from the gasoline mileage at cruising speeds:

$$\text{Power (hp)} = \frac{\text{Cruising speed (mph)} \times \text{Conversion (hp-h/gal)}}{\text{Gasoline mileage at cruising speed (mpg)}} \qquad (11\text{-}1)$$

For a 55-mph cruising speed and a fuel-conversion factor of 10 hp-h/gal (a typical value for spark engines), the fuel consumption rates for various sizes of liquid-fueled engines are shown in Table 11-1.

Gasifiers typically require about 10 kg (22 lb) of wood or 5 kg (11 lb) of charcoal to replace one gallon of gasoline. Thus, it is possible to calculate wood consumption rates from the data in Table 11-1. In addition, the maximum gas production from Imbert gasifiers is determined by the maximum hearth load—typically 0.9 Nm^3/h of gas is produced per square centimeter of hearth area (see Section 5.8). In this way, the power requirements given above determine the hearth size, as shown in Table 11-2. See Table 5-1 for sizing of other gasifier types.

Retaining the liquid-fuel carburetor in conjunction with a gas mixer for dual-fuel operation realizes the best advantages of both fuels, saving liquid fuel for occasional full-power bursts while making economical use of producer gas during normal operation. A gas producer sized to provide cruising power as shown in Table 11-2 should cover at least 90% of expected driving conditions.

11.6 Engine Selection

Because engines are mass produced for vehicles, the vehicle engine is the most likely candidate for building small systems. Larger systems can use natural-gas engines, which in many ways are more suitable but also are more expensive.

11.6.1 Large-Vehicle Engines — Truck Engines up to 50 kW

The largest widely available spark-ignition vehicle engines are Big Lung V8 460-in.^{3} engines. The Ford truck engine of this size is used industrially for stationary applications and is highly regarded because its extra-heavy block readily and repeatedly can be rebuilt to its full original specifications. Lighter weight engines of this displacement from passenger vehicles (e.g., Pontiac and Buick engines) have lighter blocks. In addition to a shorter service life for these lighter engines, the rebuilding costs and overhaul parts can be more expensive than for the heavy-block engine. Also, the cylinder-wall thickness is inadequate to accommodate cylinder-wall sleeves, limiting the number of times that the light-block engines can be rebuilt.

Table 11-1. Fuel Consumption of Liquid-Fueled Engines

Fuel Economy (mpg)	Fuel Use Rate (gal/h)	Engine Output (hp)[a]
5.5	10.0	100
11.0	5.0	50
22.0	2.5	25
55.0	1.0	10

[a]For fuel consumption 10 hp-h/gal gasoline.

11.6.2 Small Engines

Engines suitable for long-life operation (2000 to 20,000 h) must have an air cleaner, oil-pump pressure-lubrication, an oil filter, and thermostatic engine-temperature regulation. Smaller engines of the air-cooled splash-lubrication types (e.g., many engines manufactured by firms such as Wisconsin, Briggs, Honda, Tecumseh, and Kohler) have very short service periods between oil changes (one small-engine mechanic recommends an oil change every 8 h of operation). These engines are designed for total life spans of around 1000 h (Onan 1981). Some of the above manufacturers have certain models that do meet long life requirements.

11.6.3 Natural-Gas Engines

Larger displacement spark-ignition engines (larger than 460 in.^{3}) are used for natural-gas-fueled stationary engines, many of which are used for irrigation applications. Like producer gas, natural gas has a low flame-velocity and a relatively high octane rating. These engines operate at a low rpm. The engines also operate at a high compression ratio and have a relatively high thermal efficiency. In addition, they have a very long service life and are particularly suited to large (greater than 250 kW) installations.

Natural-gas engines are built by Caterpillar, Waukesha, Minneapolis Moline, and others. These engines are available with standard options for industrial operation on gaseous fuel. An optional top-oiler is recommended, which supplies oil mist with the fuel to lubricate the piston rings. The top-oiler may greatly reduce the wear experienced with dry-gas fuels, and may be helpful with producer gas on smaller engines as well.

Large, natural-gas, industrial engines exhibit a quantum leap in equipment and overhaul costs over the cost of vehicular engines. Service parts are often available only through the manufacturer and may involve long downtimes as well as high costs.

11.6.4 Diesel Engines

Diesel engines also are suitable for large installations but have special problems, discussed in Section 11.10.

11.7 Cogeneration

Cogeneration involves using part of the greater than 70% of fuel energy that is otherwise lost as exhaust heat and engine heat during engine operation. If one has a use for this heat, then a water-cooled engine is both convenient and safe. The hot air from air-cooled

Table 11-2. Gasifier Size versus Engine Output

Power P (hp)	Wood Fuel[a] Consumption (lb/h)	Hearth Capacity Q (kBtu/h)	Imbert Throat Diameter[b] (cm)	Imbert Throat Diameter[b] (in.)	Stratified Hearth Diameter[c] (cm)	Stratified Hearth Diameter[c] (in.)
100	200	1400	16	6.3	40.5	16.0
50	100	700	11.3	4.5	29	11.4
25	50	350	8	3	20.5	8.0
10	20	140	5	2.0	13	5.1
5	10	70	3.6	1.5	9	3.6

[a] 1 kg (2.2 lb) of wood is required per hp-h for a 7000-Btu/lb average fuel heating value (Gengas 1950).
[b] Imbert throat constriction diameter is given by $D = 1.6\sqrt{P}$ cm.
[c] A maximum stratified bed hearth load of $h_{max} = 1$ MBtu/h-ft^2 is typical for wood chips. Stratified hearth diameter is given by $D = 1.6\sqrt{P}$ in.

engines requires safety precautions for heating because it may contain dangerously high levels of carbon monoxide so it should be vented. Heat from the exhaust gas can be recovered with a compact water-cooled heat exchanger or put to direct contact uses.

A particularly attractive application for the waste heat from engines is the drying of biomass feedstock before using it in the gasifier. The 50% moisture contained in fresh wood chips can be removed effectively by drying the chips with the engine's exhaust gas. There is drying capacity to spare if high-efficiency drying methods are used.

Because the engine's exhaust gas is hot enough (600°-700°C) to decompose biomass, it must be tempered (substantially diluted with cooler gases) before being used. One of the best ways to dilute is to take the warm, humid gas leaving the outlet of the biomass dryer, add about 10% by volume of engine exhaust gas, and recycle the mixture to the gas entry port of the dryer. Reusing the outlet gas from the dryer instead of using ambient air to temper engine exhaust gas offers several advantages: (1) Because the gas coming from the dryer is already warm, more of it must be used to dilute (and cool) the engine exhaust gas to a given tempered temperature. Therefore, more gas is available for drying the feedstock; more gas increases the capacity of the feedstock drying process. (2) The humidity in the dryer flume gas reduces the chance for pyrolysis of the feedstock when the gas recycles to the dryer. (3) Lastly, the gas from the dryer is essentially engine exhaust gas; as such, it is nonoxidizing and tends to quench any incipient fires in the dryer.

11.8 Spark-Ignition Engine Conversion

11.8.1 Engine System

A typical producer gas system for spark-ignition engines is shown in Fig. 11-1. The system consists of a gas producer (described in Chapter 5), a gas cleanup and cooling system (described in Chapter 8), a starting blower (described in Chapter 9), a carburetor (gas mixer), and an engine. During operation, suction from the engine draws air into the gas producer, through the cleaning system, and into the gas mixer where air is mixed with the gas.

11.8.2 Gas Mixers

Maximum power is achieved with a producer gas air mixture just slightly lean of the stoichiometric combustion ratio,* as shown in Fig. 11-2. On the other hand, gasoline delivers maximum power with mixtures significantly rich over its stoichiometric combustion ratio.

A gas mixer behaves in some ways like a liquid-fuel carburetor, but in reality it is much simpler. A carburetor must mix the correct weight of air with the liquid fuel (normally a air-fuel ratio of about 15 for gasoline and about 6.5 for methanol). When the throttle is opened suddenly, this mixture is momentarily enriched over the normal idling mixture by the accelerator pump, which supplies a squirt of fuel. This mixture is additionally enriched at high loads by the power circuit in order to boost power and to protect the engine from excessive temperatures. The mixture is enriched for starting by the choke when the engine is cold.

The producer gas mixer must mix a proper ratio of air with the producer gas, approximately a 1:1 ratio of fuel to air by volume. Variations in the producer gas mixture cause sharper changes in engine power than do similar variations in a gasoline mixture, as shown in Fig. 11-2. For this reason, the gas mixture must be controlled by the operator during producer gas operation. The gas quality can vary during operation and needs periodic adjustment; the gas mixer must allow for independent control of the gas, air, and mixed-gas flow to accommodate such adjustment.

A number of mixers used in the past are shown in Figs. 11-3 and 11-4. Note that, just like a carburetor, each mixer has one control (the throttle) to meter the total flow and another control (the choke) to adjust the ratio of gas to air.

*The stoichiometric combustion ratio is that fuel-air mixture that allows the gas to burn completely, with no surplus air remaining after the fuel supply has been exhausted.

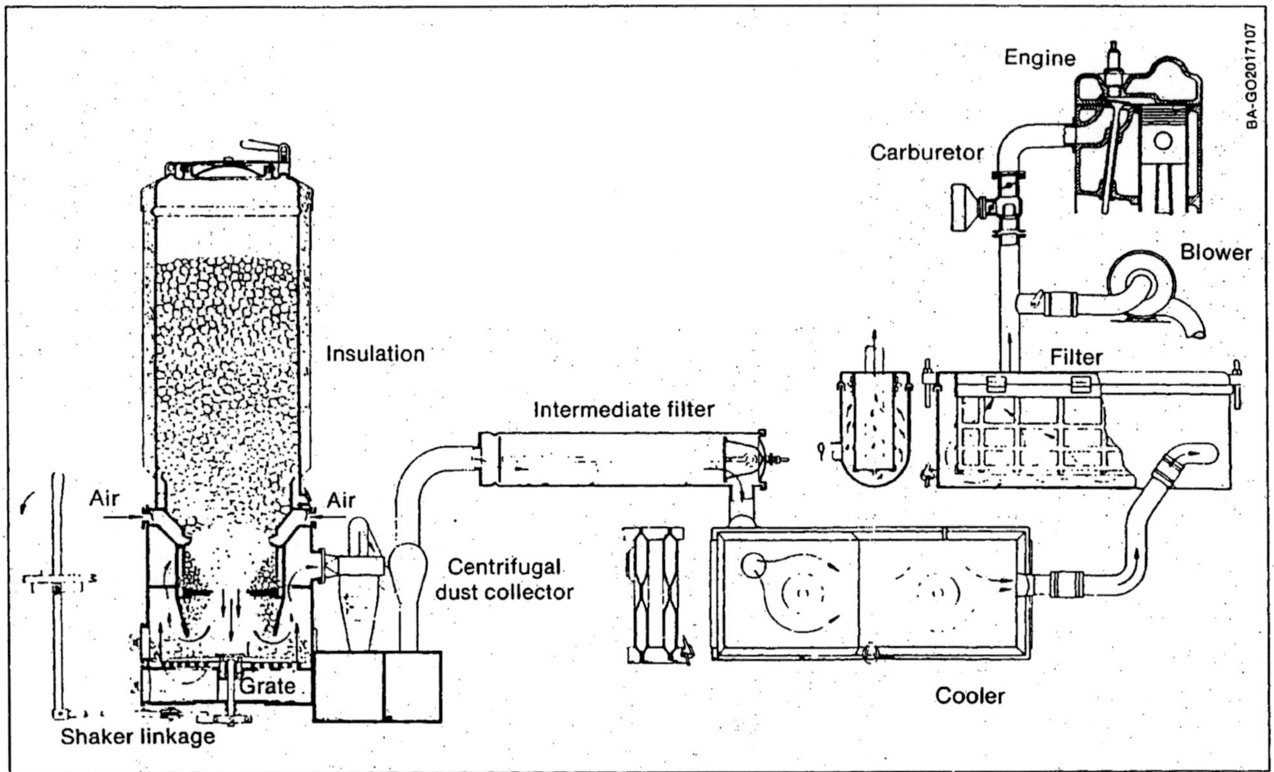

Fig. 11-1. Producer gas engine for spark ignition system (Source: Skov 1974, Fig. 46. © 1974. Used with permission of Biomass Energy Foundation, Inc.)

The butterfly valve often used in throttles and chokes has been found to respond to adjustment very nonlinearly and, therefore, can be a troublesome method for controlling the gas-air mixture. A gate valve provides much finer control of the air inlet than does a butterfly valve, so the narrow power peak is broadened out over one-half to one turn of the gate valve. Conversely, the entire range from too rich to too lean occupies only a few degrees of arc on a butterfly valve and is easily missed altogether. However, a butterfly valve can be used satisfactorily for throttle and gas-inlet control.

If it is essential to prevent gas from being released into the engine compartment after shutdown, then an additional gate valve should be used for positive gas shutoff. If idling is desired, a metering idle valve leading from the gas inlet around the throttle may be used to enrich the mixture during idle, without changing the running mixture, as shown in Fig. 11-4.

11.8.3 Power Time Lag

Lean gas-air ratios are normally produced for a short time after the throttle is opened, and richer gas is produced just after the throttle is closed. This change in gas quality alters the gas mixture and weakens engine output to a point at which the engine may even stall. The magnitude of this momentary power loss can be reduced by adjusting the fuel-air ratio to maintain a peak-power mixture, but this requires operator experience. One improvement that today's technology might bring to gasifier designs is automatic mixture control, employing a feedback signal from the engine exhaust. This type of control is already in use on feedback fuel-injection systems, and it should be adaptable to producer gas systems.

Gas-fueled buses, which necessarily make frequent stops, overcome the fuel-air problem with a bag inflated with gas by a blower, which maintains gas flow during stops. The surge of power needed for acceleration is then provided mostly by gas from the bag.

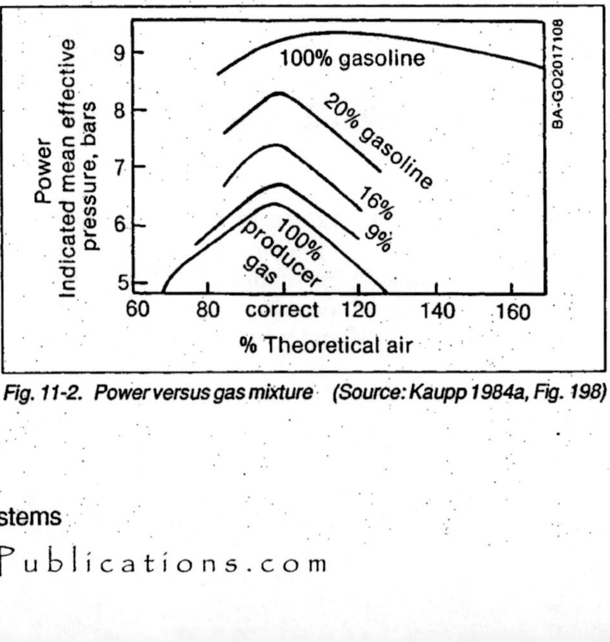

Fig. 11-2. Power versus gas mixture (Source: Kaupp 1984a, Fig. 198)

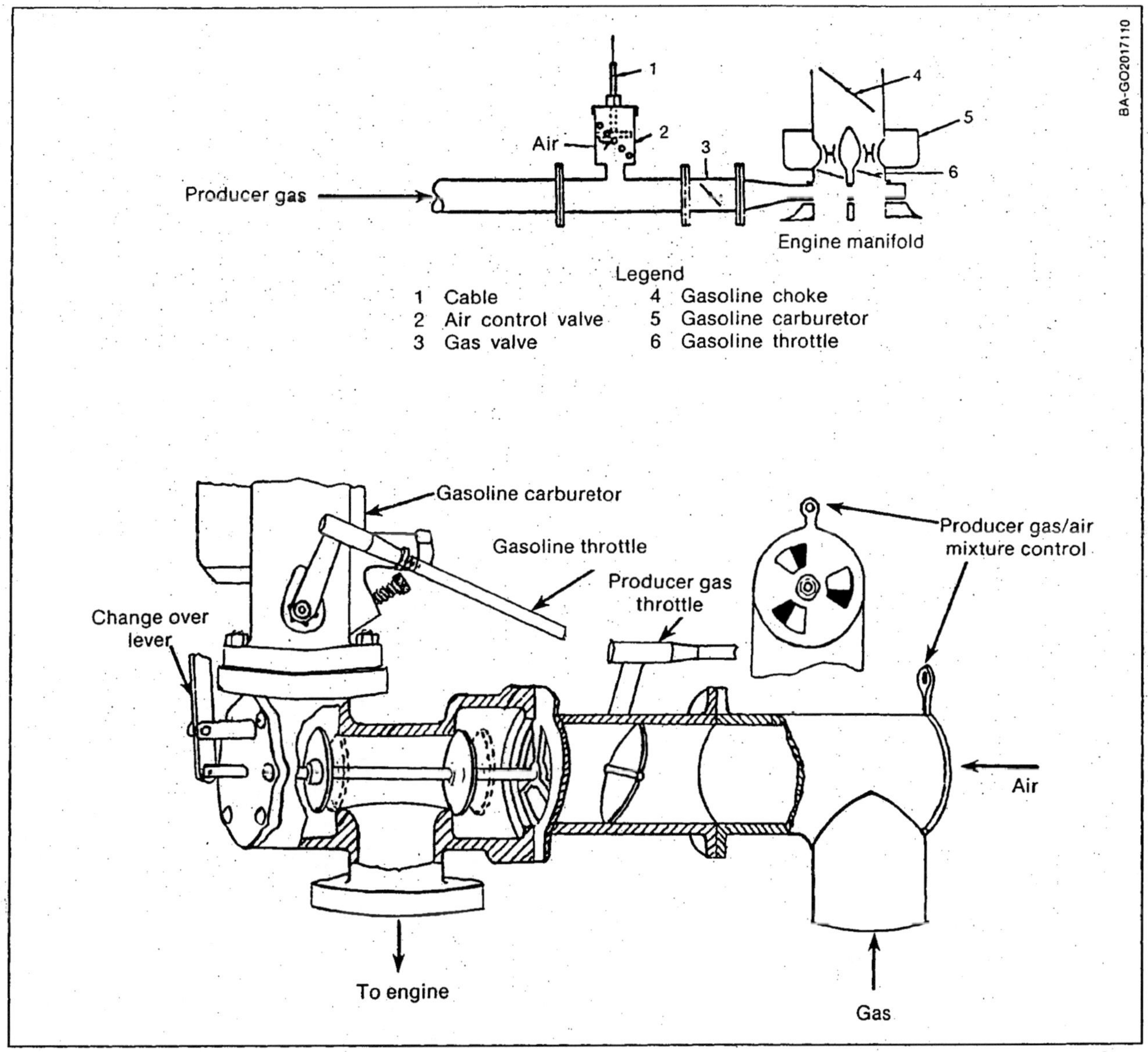

Fig. 11-3. Gas air mixers (Source: Adapted from Cash 1942 and Anon 1943)

Driving through hilly country in a producer-gas-fueled vehicle requires special skills. Often the driver races the engine while going downhill in order to build up the gas quality, so that high-quality gas is available when climbing is resumed. Traffic accidents increased in Sweden during World War II because drivers of producer-gas-fueled vehicles operated those vehicles at high speeds while running at hills (Gengas 1950).

11.8.4 Engine Startup

If liquid fuel is available, the engine can be started on fossil fuel (gasoline or propane), and the gasifier started separately with a fan or blower (which may be hand- or battery-operated) while flaring the gas. The blower outlet and flare are located upstream from the cleanup system to keep tars out of the cleanup stream during starting procedures. When the gas is less tarry and can support combustion on its own, the blower is turned off, and the gas valve is opened. Then, the producer gas air/throttle is slowly opened; simultaneously, the gas-air mixture is adjusted to obtain maximum power output. At the same time, the fossil fuel throttle is slowly closed until the changeover is complete. Once the mixture has been initially adjusted, repeated changeovers can be made quickly. The liquid fuel supply should then be securely closed to prevent liquid-fuel consumption. This type of dual-fuel capability permits rapid changeover between solid and liquid fuels.

One novel and simple alternative to a rotary blower for startup is to use the engine exhaust gas as the propelling gas for an aspirator ejector, as shown in Fig. 11-5.

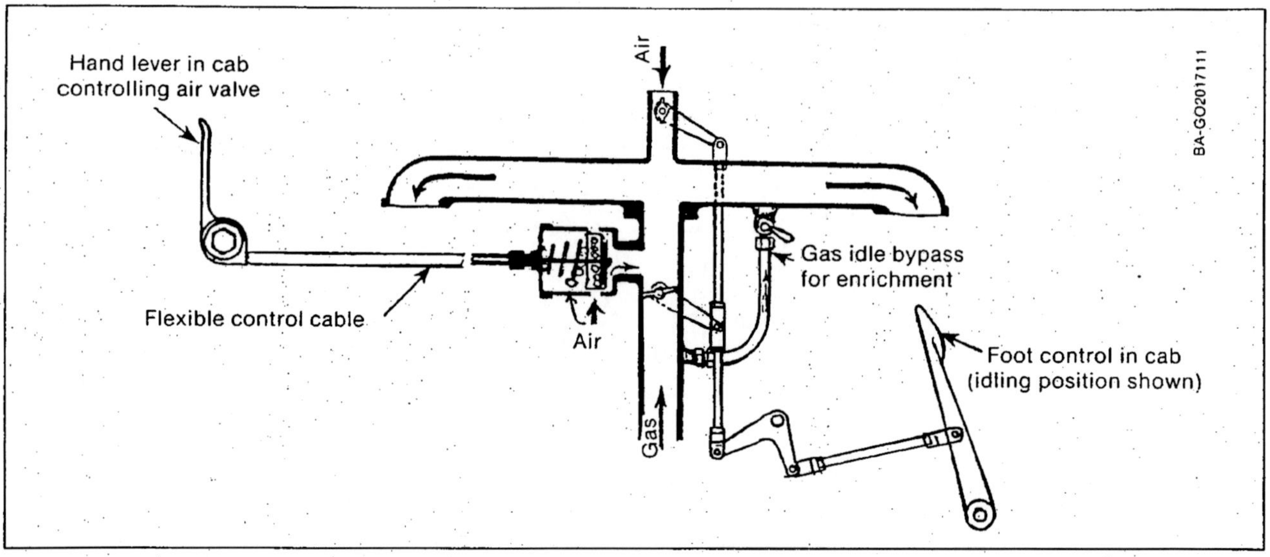

Fig. 11-4. Gas mixer showing idle enrichment (Source: Adapted from Heywood 1944, Fig. 22)

This technique allows a considerable saving in battery power at the expense of only a little extra liquid fuel.

11.8.5 Ignition Timing

Producer gas burns slowly, as shown in Fig. 11-6, giving it a high octane rating. It is usually found that optimum engine operation is obtained by advancing the ignition timing 5 to 15 degrees more than the advance used with gasoline, as shown in Fig. 11-7.

The "ping-sensor" and "electronic ignition" timing controls used in many late-model automobiles may be helpful in automatically providing the ideal spark advance for producer gas, while permitting dual-fuel operation.

11.8.6 Spark Plugs

The porcelain insulator of a spark plug may be glazed with feldspar or quartz for generator gas operation, to permit easier deposit removal. Spark plugs for gasoline engines are normally not glazed because lead anti-knock compounds form a conductive glass with the glaze. Therefore, lead-free gas should always be used for start-up procedures with glazed plugs. The heat range of the spark plug should be high enough to permit the plugs to self-clean, but not so high as to cause preignition. If fouling is encountered, use a lower heat number (hotter electrode).

11.9 Two-Cycle Engine Conversion

Two-cycle engines present a special challenge for conversion to producer gas. The crankcase must be lubricated and protected from contact with gas impurities. A number of the most novel innovations from the entire World War II era appeared on the two-stroke engines of that time.

In the Bolinder Munktell system shown in Fig. 11-8, the lower crankcase served as an air pump to supply air to pressurize the gasifier and supply pressurized air to the gas mixer. In the two-cycle conversion, either the head must be fitted with both intake and exhaust valves, or the piston-porting system must be isolated from the crankcase to prevent gas deposits from interfering with lubrication.

Similarly, the Pulsator prevents producer gas from passing through the crankcase. A side-chamber is provided, as shown in Fig. 11-9, and connected to the crankcase through a butterfly valve and to the gas supply and intake manifold through check valves. The pulsating movement of air in the crankcase alternately sucks in and mixes air and gas, and then delivers the mixture to the combustion chamber. A check valve

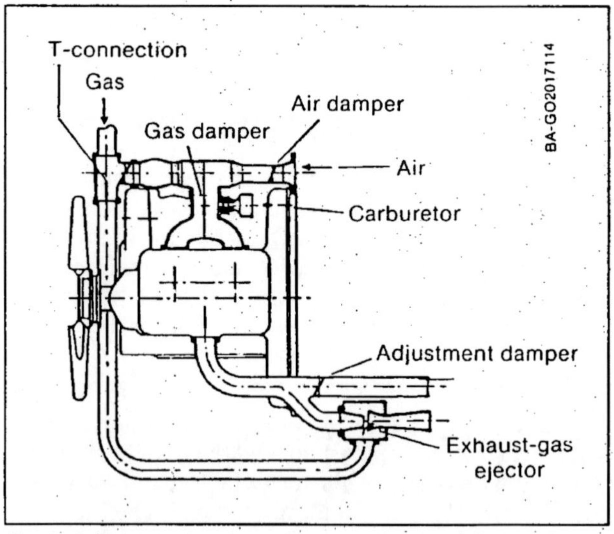

Fig. 11-5. Exhaust pipe ejector for start-up fanning (Source: Gengas 1950, Fig. 151)

leading from the air control into the crankcase permits air flow through the engine. The volume of the pulsator mixing chamber necessarily prevents gas from being sucked all the way into the crankcase.

The third method of two-cycle conversion, the two-cycle crank case system shown in Fig. 11-10 is the simplest. Mixed gas is drawn through check valves into the crankcase, where the gas enters the combustion chamber through the piston ports. Engines popular for use on today's motorcycles use this same principle.

11.10 Diesel Engine Conversion

11.10.1 Diesel Operation with Producer Gas

Diesel engines are operated on producer gas in the "aspirated" mode. The producer gas is mixed with the intake air stream. A small quantity of diesel fuel is introduced continuously through regular injectors to ignite the gas mixture and provide timing. This so-called pilot-injection mode or diesel dual-fuel mode is also used occasionally with natural gas or alcohol fuels. Direct-injection diesel engines are better suited for conversion to producer gas than antechamber diesel engines.

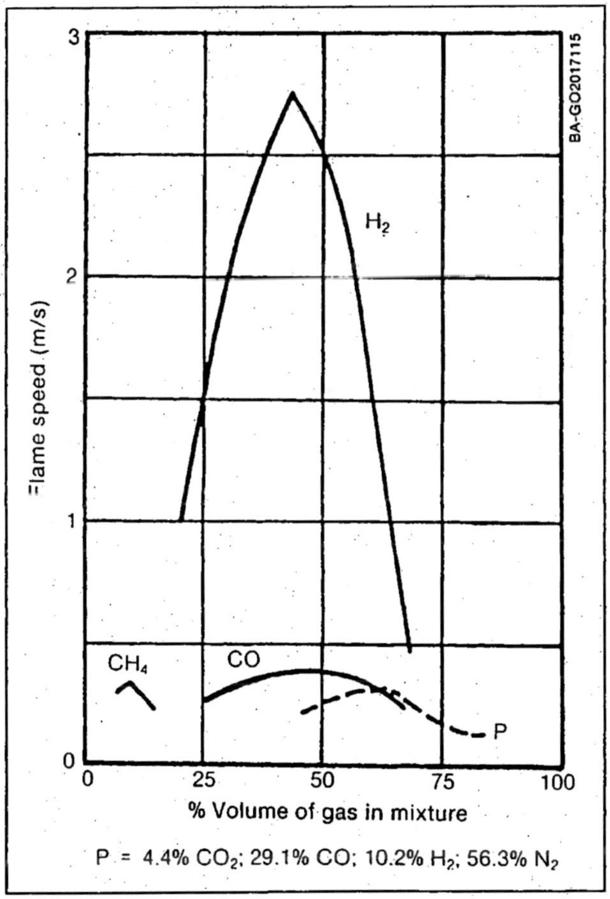

Fig. 11-6. Flame speed of various gases versus mixture (Source: Kaupp 1984a, Fig. 182)

The power lost when diesel engines are converted to operate on producer gas is less than that lost by spark-ignition engines for three reasons. First, some diesel fuel is introduced; second, diesel engines operate at a higher compression ratio; and third, diesel engines operate with a large excess of air, so the energy per

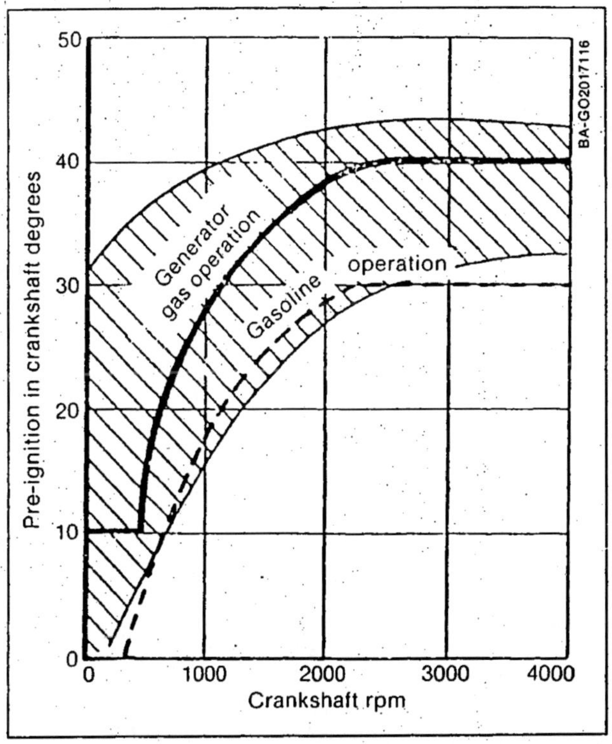

Fig. 11-7. Ignition advance for producer gas operation. Shaded area is range of producer gas operation; solid line is typical operating condition (Source: Gengas 1950, Fig. 198)

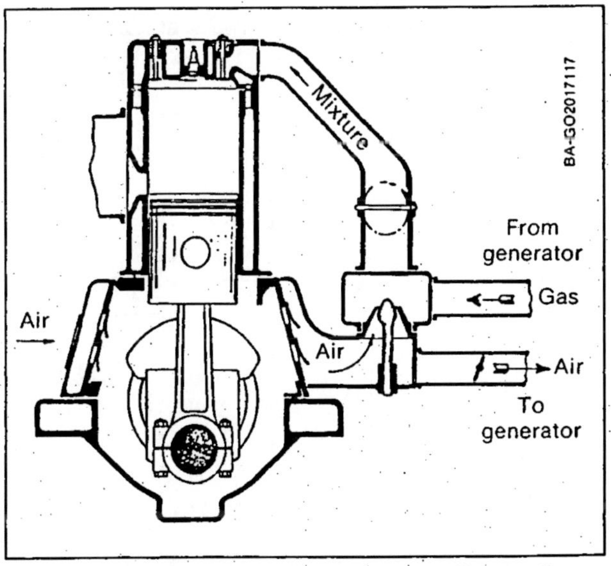

Fig. 11-8. Bolinder Munktell two-stroke engine (Source: Gengas 1950, Fig. 191)

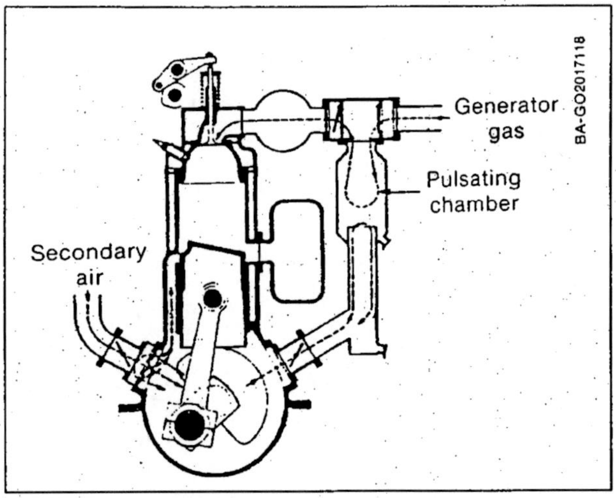

Fig. 11-9. Pulsator two-stroke engine (Source: Gengas 1950, Fig. 187)

stroke is very close to the heating value of producer gas-air mixtures. Furthermore, full diesel operation can be achieved readily by simply increasing the quantity of diesel fuel to normal levels when needed. This can be accomplished easily by using the existing governor on many diesel engines. The brake thermal efficiency is higher for dual-fuel operation than it is for straight diesel fuel operation, as shown in Fig. 11-11.

Standard diesel engines inject the liquid fuel into the compressed-air charge in order to start combustion at a time preset relative to the crank position. When using producer gas, it is necessary to advance the injection timing angle with engine speed especially at high rpm. The optimum injection timing for one engine is shown in Fig. 11-12. An engine with an improperly set timing angle will run roughly at low speed and expel smoky exhaust at high speeds.

It has been reported that when operating conditions allowed producer gas-air mixtures to self-ignite without engine knock, a single-cylinder, direct-injection, slow-speed (1000-1500 rpm) diesel engine was able to run on 100% producer gas for extended periods (Cruz 1984).

Ignition reliability is better with pilot injection than with spark ignition. The minimum injector quantity for stable pilot ignition and efficient combustion is given by a diesel fuel-mixed gas ratio of 10 mg/kg. Increasing this ratio will boost power output, up to a ratio of 20 mg/kg; above this injection rate, thermal efficiency falls with no significant power increase. Partial load operation is complicated by the fact that a full inlet charge is required to develop ignition temperature. Throttling for no-load operation must lean out the gas mixture and reduce the injection quantity to as low as 5 mg/kg (Anon 1944).

Diesel engines also can be converted completely to producer gas by reducing the compression ratio, adding a spark-ignition system, and replacing the injectors with spark plugs. The modification expense is hardly justified, considering the successes of pilot injection, except in situations where complete replacement of diesel fuel is required. The compression ratio should be reduced to 9.5-10:1. For antechamber diesel engines, the heads must be replaced.

Pilot injection helps the producer gas-air mixture to ignite. Very stable and reliable ignition has been observed using pilot injection; in fact, it has been shown to perform better than spark ignition. The power output from weak gas (e.g., 118 Btu/scf) can be 90% of that from straight diesel fuel with 40% excess air (Anon 1944). Minimum pilot injection is limited by its ability to provide stable ignition at roughly 10% to 20% of the regular idle injection.

11.10.2 Starting Diesel Engines

Diesel engines are started by first igniting and fanning the gasifier with the blower until clean, burnable gas is

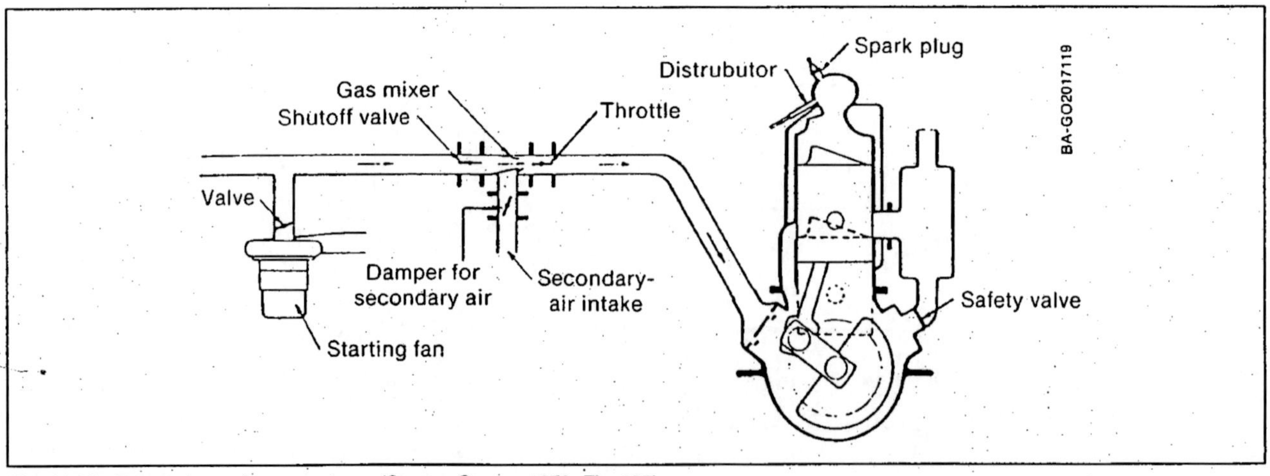

Fig. 11-10. Two-cycle crankcase system (Source: Gengas 1950, Fig. 185)

produced. The engine itself is started and warmed up on diesel fuel. As the producer gas-air mixture is admitted, the governor automatically reduces the liquid-fuel feed. Fuel savings up to 90% over straight diesel operation have been reported (Kjellstrom 1983). The governor can automatically boost the injection rate in the event of an increased power demand or a variation in the quality of the producer gas.

11.10.3 Throttling at Partial Load

High compression is essential for ignition, so throttling should be applied only at the gas inlet. Throttling at the air inlet results in excessive suction, which can cause bad combustion and smoky exhaust. However, at medium-to-high loads, where the gas-inlet throttle is fully open, the air inlet throttle may be partially closed to increase gas suction.

11.11 Increasing Power from Producer-Gas-Fueled Engines

11.11.1 Mechanisms of Power Loss

A given volume of producer gas, when mixed with the correct quantity of air for combustion, contains 70% of the energy of an equal volume of gasoline or propane mixed with air. The relative power at any given low rpm is proportional to the mixed gas energy, as shown in Fig. 11-13. Full-throttle power increases in direct proportion to the rpm rate, up to the point where pressure drops in the intake and exhaust valves limit the power. Beyond that point, power drops off rapidly with an increasing rpm rate.

The volume of intake gases for producer gas use is roughly doubled over gasoline or propane use, and increasing pressure drops reduce the peak-power rpm by up to 30%. The combined effects of a reduced mixed-gas heating value and a reduced peak-power rpm result in an overall reduction in the maximum engine power up to 50%.

11.11.2 Engine Breathing

A producer gas engine must handle roughly twice the volume of intake gases as gasoline engines. The engine's breathing ability therefore begins to limit engine power at lower rpm. The peak power output, shown in Fig. 11-14, operating on producer gas, occurs at roughly two-thirds of the rpm for gasoline fuel. Improvements in the intake induction system can both raise the speed attained during the peak-power output and decrease the overall power loss at each rpm rate.

One might ask how producer gas containing 150 Btu/scf can produce more than 150/1000ths of the power of methane containing 1000 Btu/scf. The reason is that engine power depends on the mixed-gas/air

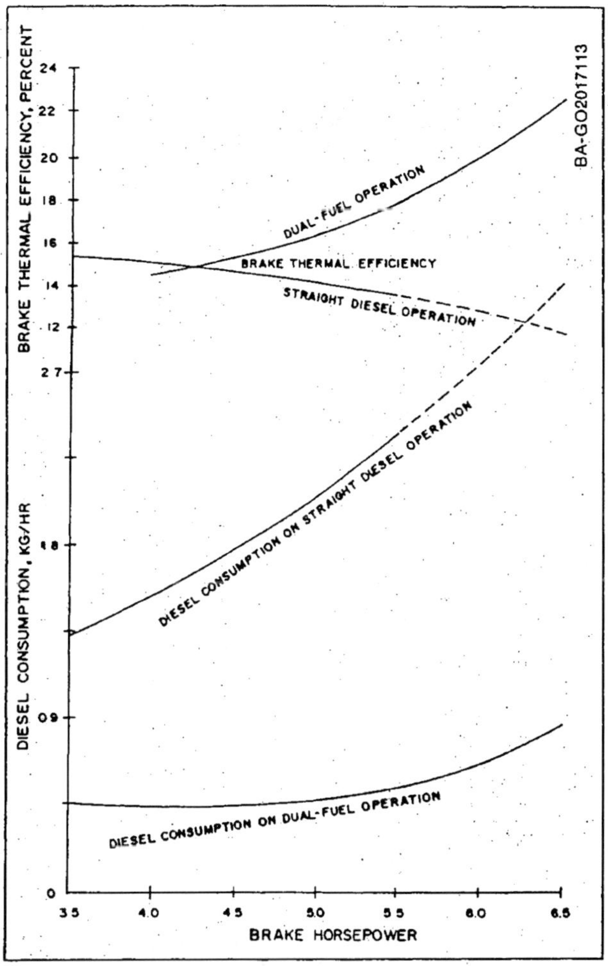

Fig. 11-11. Diesel dual-fuel efficiency and fuel savings (Source: Cruz 1983, Fig. 14. © 1973. Used with permission of the Beijer Institute)

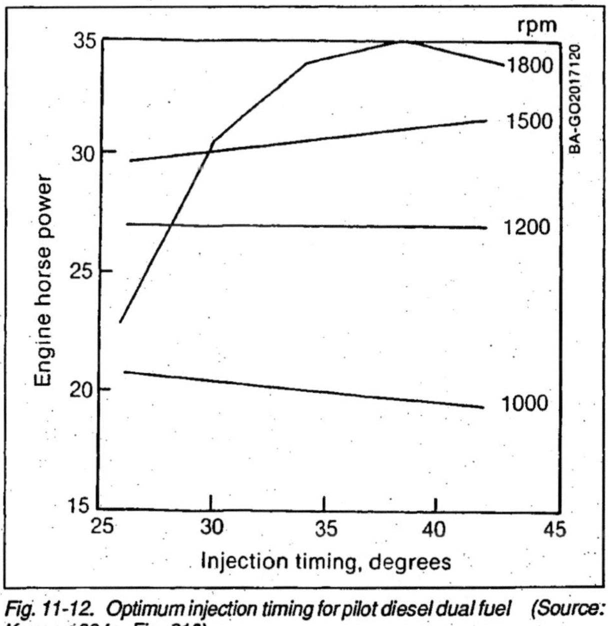

Fig. 11-12. Optimum injection timing for pilot diesel dual fuel (Source: Kaupp 1984a, Fig. 210)

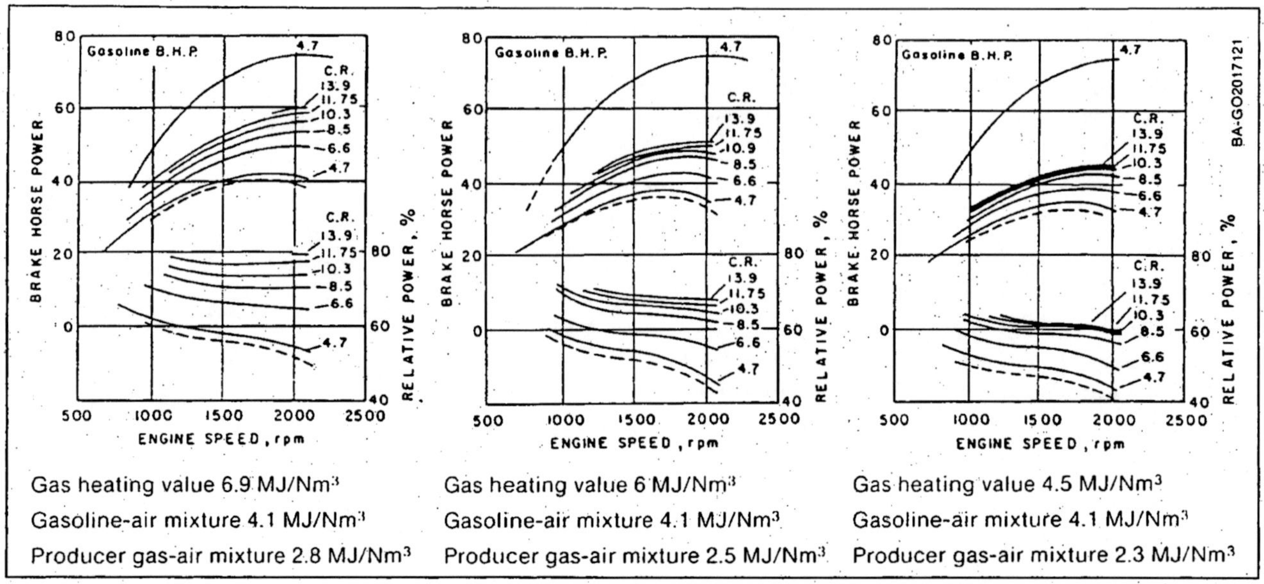

Fig. 11-13. Power versus rpm compared with gasoline (Source: Kaupp 1984a, Figs. 186, 187, 188)

heating value, not on the pure-gas heating value. The mixed-gas heating values are shown in Fig. 11-15.

11.11.3 Efficiency and Power Loss

The spark engine operating on gasoline achieves a thermal efficiency of 25%-30%. The same engine operating on producer gas may achieve 15%-25% thermal efficiency, depending on how well the engine is converted to producer gas. A diesel engine using diesel typically achieves 30%-35% thermal efficiency. Operating on 90% producer gas, it can be expected to give 25%-30% thermal efficiency. The overall efficiency of the system must be computed from engine efficiency and gasifier efficiency.

Engine friction losses are primarily a function of engine speed. Partial-throttle operation increases the fraction of fuel consumption devoted to fixed engine losses. The highest shaft efficiency for any speed occurs with the highest power consistent with complete combustion. Therefore, it is primarily for the benefit of fuel economy that various techniques are used to increase the power output obtained from producer gas.

11.11.4 Blowers and Superchargers

The power from producer gas can be increased to that available from naturally aspirated gasoline or propane (or even higher) in several ways.

A Roots-type blower or turbosupercharger can be used to increase the pressure in the cylinders above atmospheric pressure, and thus to increase the air-fuel charge. A positive pressure of 6 psig at the intake manifold can recover full gasoline power.

The Roots blower uses engine shaft power for compression and necessarily reduces engine efficiency. The turbosupercharger uses the pressure of the exhaust gases to operate a turbine compressor, thereby recovering most of this lost energy.

Since engines are designed to withstand a particular combustion pressure (BMEP, or brake-mean-engine-pressure), power increases must be limited to restoring the original peak-combustion pressure.

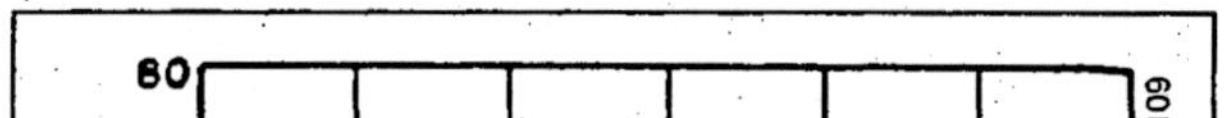

Fig. 11-14. Power output as a function of engine speed (Source: Kaupp 1984a, Fig. 199)

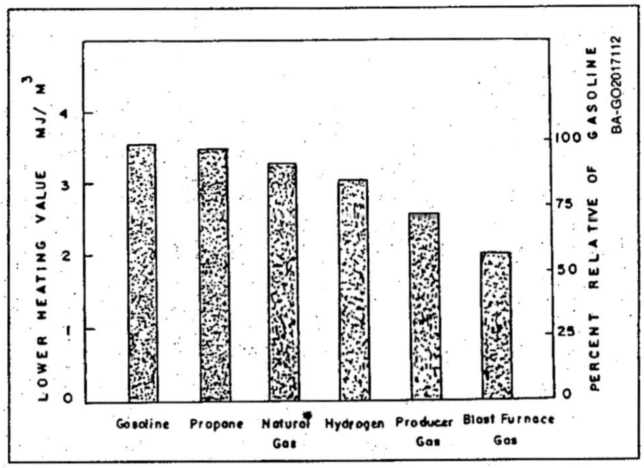

Fig. 11-15. Heat values for stoichiometric fuel air mixtures (Source: Kaupp 1984a, Fig. 177)

11.11.5 Other Methods for Increasing Producer Gas Power

A portion of the power loss can be recovered even for a naturally aspirated engine by increasing the compression ratio, as shown in Fig. 11-16, advancing the spark, and improving the intake systems. Producer gas has a higher octane rating than gasoline, so the power can be increased by increasing the compression ratio to between 11:1 and 14:1. Increasing the compression ratio, however, may prevent dual-fuel operation on lower octane fuels because of excessive knocking and detonation. One method for adapting high-compression-ratio-modified engines to operate at a partial load on lower octane fuels is to recycle some exhaust gas back to the intake manifold. The mixture of gasoline, air, and exhaust gas burns more slowly, minimizing knock and permitting partial power operation.

Cam shaft design may be optimized for producer gas by faster valve opening, higher lift, and no overlap, with valve timing-duration optimized for producer gas. Synchronous generators require engine speeds of 1800, 1200, or 3600 rpm. Cam shaft design has been overlooked so far in the lower rpm region.

Power levels also have been increased to near-supercharged performance by improving the intake system through installing larger inlet valves, or especially by using tuned ram induction and larger intake manifolds. Additionally, intake manifolds may be improved with flow straighteners after each bend, shell-casting techniques to give smoother inlet surfaces, and carefully tuned long ram intake manifolds to give a sharper peak pressure at the optimum rpm rate.

The decrease in power can be made more acceptable if one increases the engine speed by lowering the final drive-gear ratio. Full power cannot be recovered completely by increasing the rpm rate alone. Many interesting details for converting piston engines to operate on generator gas during World War II are given in *Generator Gas* (Gengas 1950).

Further work in engine adaptation would be beneficial. In particular, combustion chamber design for improved swirl and squish can lead to better combustion efficiencies, and microprocessor controls using signals from pressure, temperature, ping, and oxygen sensors can be applied to provide continuously optimized, unattended engine operation, to follow changing gas quality and variations in solid fuels.

11.12 Engine Life and Engine Wear

11.12.1 Engine Life Expectancy

There are reports from Sweden during the late World War II era (undocumented) and also from recent investigations (Kjellstrom 1981), of longer engine life experienced with producer-gas-fueled engines than with gasoline-fueled engines. Note in Table 8-3 that cylinder wear is less on producer gas with fabric filter than wear on straight diesel fuel for two out of three tractors tested, and that oil contaminants were also less on well-cleaned producer gas than on diesel oil.

Discussion with rebuilders of industrial engines indicates that life expectancy for fossil industrial engines is related to engine speed. High-speed, 3600-rpm operation is discouraged because it may cause premature failure. Lower speeds give commensurately longer engine life expectancies (e.g., engines operated at 1800 rpm may last 1-1/2 years, 1200-rpm-operation may allow 3-year lifetimes, and 900-rpm-operation

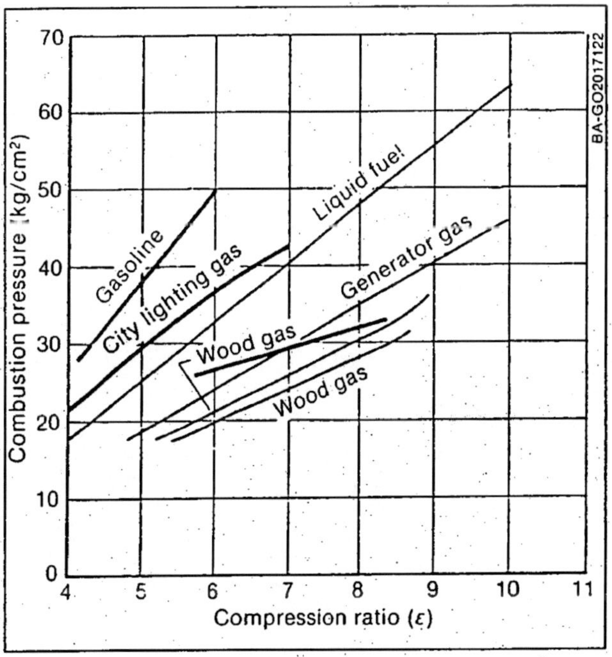

Fig. 11-16. Combustion pressure versus compression ratio for various fuels (Source: Gengas 1950, Fig. 174)

may allow 5 years of engine life). No known property of clean producer gas should shorten engine life. One truck, which has been driven 100,000 km on producer gas without any unusual maintenance problems, is shown in Fig. 11-17.

Many of the clean-up mechanisms outlined in Chapter 8 have yet to be applied to producer gas cleanup, and the gas in the bulk of today's test and production systems is far from clean enough for reliable operation, much less long-life operation. We will discuss the effects of dirty gas on engine life in the following sections.

11.12.2 Sticking Intake Valves

Tar mists and dusts form an asphalt-like deposit around the intake valve stems. Valve-sticking is more a problem of accumulated deposits than of engine wear. Sluggish valve closing decreases power until cylinder compression is lost. The valves may be reconditioned and cleaned, and the engine returned to service without serious damage to the rings and bearings. Intake valve sticking can be corrected without a complete engine overhaul. However, sticking valves clearly indicate the need for a more efficient cleanup system (Chapter 8) and possibly redesign of the gasifier for a higher hearth load to reduce tar production. (See Chapter 5 for information on gasifier sizing and design.)

Based on cylinder, bearing, and cam wear, a life-expectancy of 5000 hours was estimated for an engine after extended tests with gas so dirty that the valves required cleaning eight times in the first 1000 hours (Breag 1982).

11.12.3 Oil Thickening and Contamination

In gasoline engines, gasoline vapors from blowby tend to thin oil with time. With clean producer gas, the volatile motor oil fractions evaporate, causing the oil to thicken naturally. Thickening improves the oil's lubricating ability, but it also increases cranking requirements.

Fig. 11-17. Swedish logging truck that operated 100,000 km on producer gas without any unusual maintenance (Source: NAS 1983, p. 46)

Dirty gas can introduce tar particulates and corrosives into the oil through cylinder blowby. Particulates in the oil, and ash and char accumulation in crankcase oil can increase engine wear. Particles larger than the oil-film thickness may scratch bearing surfaces; however, char and ash tend to be easily crushed to smaller particles, minimizing this problem. If ash slagging occurs, harder abrasive particles are produced that could cause severe wear. Small amounts of sodium from ash in either the lubricating oil or the fuel gas can lead to engine valve corrosion. It becomes obvious that particulate removal is very important to engine life at many levels.

11.12.4 Tar/Oil Accumulations

Tar/oil mists occur as very finely divided droplets much smaller than char and ash. Therefore, they are much more difficult to remove from the gas and are more likely to cause engine problems. Tar accumulation does not cause wear; rather, it causes moving parts to stick, and plugs passages, intake valves, throttle valves, and piston rings. Tar deposits become a viscous fluid at engine operating temperature; on cooling and drying, they can become a hard varnish that will prevent or hinder engine startup, or bend a pushrod in the event of full valve seizure. The best cure is prevention through the cleaning methods of Chapter 8 and verification of gas cleanliness regularly, using the tests of Chapter 7.

Mild valve-sticking can sometimes be reduced or prevented by operating the engine on liquid fuels (especially methanol) for a short time before shutdown. This tends to rinse tar accumulations into the crankcase oil, where they do less damage. This technique may merely prolong the time lag between repair, and delay use of proper cleanup techniques.

Oil should be changed when it is visibly dirty or excessively thick. Crankcase oil analyses are more widely available today than in the past and can help diagnose causes of engine wear.

11.12.5 Engine Corrosion

Corrosive engine cyclinder wear increases with low operating temperatures as shown in Fig. 11-18. Below 120°F, bearing corrosion, cylinder wear, and ring wear rise rapidly with increased water condensation and corrosive hydrolysis products due to carbonic acid. Below 185°F, cylinder and ring wear depends on oil consumption and oil-film thickness. Above 185°F, wear is minimal and independent of oil-film thickness (Mahin, June 1983). Wear is maximum after a cold-engine startup; preheating the engine block to 120°F before startup can minimize this wear for maximum engine life.

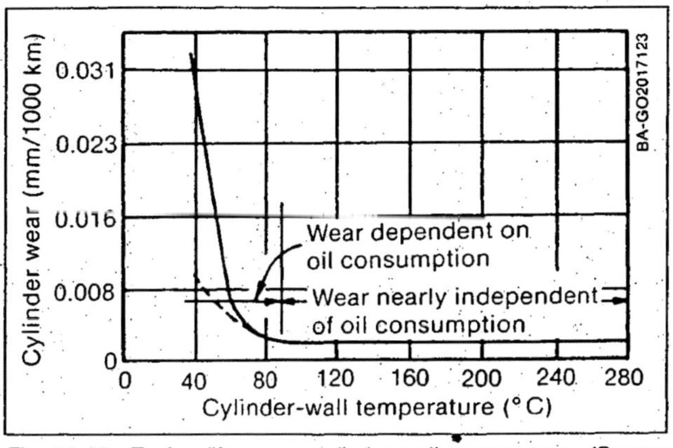

Fig. 11-18. Engine life versus cylinder wall temperature (Source: Kaupp 1984a, Fig. 211)

11.12.6 Engine Warranty

Reputable sellers of gasifier engine systems should offer an engine warranty, backed up with service contracts, operator and maintenance training, and continuous monitoring of the performance of installed units. In this way the maker will be the first to know of the need for design improvements. Minneapolis Moline has written a 6000-h warranty on a producer-gas-powered engine (Mahin, June 1983).

11.13 Exhaust Emissions

Carbon monoxide exhaust emissions from a properly running producer-gas-fueled engine can be expected to be inherently lower than emissions from gasoline-fueled engines because maximum power is achieved from a mixture that is lean of the stoichiometric combustion mixture, whereas the maximum power from gasoline is achieved by burning a rich mixture. Hydrocarbon emissions also can be expected to be lower than those from gasoline because of the relative absence of hydrocarbons in the producer gas. The lower flame temperature of the producer gas, along with the excellent antiknock characteristics and low prompt nitrogen content of producer gas, suggests that NO_x emissions should be lower as well.

The major pollutant source from producer gas is the necessary disposal of cleaning condensate, which may be high in tars, especially phenols. This source can be minimized by using the driest fuels possible and using the best low tar gasifier design available to prevent tar and condensate production. If the fuel is dry enough to render condensate removal unnecessary, then the gas can be used above its dew point, eliminating condensate collection altogether. This scheme has been used in a few recent designs. Another approach to condensate removal is evaporation, coupled with recycling the residue to the gasifier. Still another approach applied to an especially tarry system is incineration of the condensate tars outside of the gas producer. However, this approach consumes additional fuel. Prevention remains the most attractive cure. (See the discussion on low tar designs and drying in Chapter 5.)

11.14 Other Devices for Producer-Gas Power Generation

11.14.1 Gas Turbines

Historically, producer gas has been used in internal-combustion piston engines of both the spark-ignition and diesel-powered types. Accordingly, these engines remain the major area of interest at present. However, the gas turbine may be an attractive application for producer gas for electric generation, since it operates at a relatively high efficiency (25%-35%) in intermediate sized units (200-10,000 kW). Gas turbines can be started and stopped in a short time, so they are especially useful for producing peaking power.

Gas turbines were developed after producer gas use was abandoned, and turbines still have not been operated on producer gas. Producer gas could eventually become a major turbine fuel for electric power generation. Turbines offer simplicity, long life, and reliability. The simplicity of one moving part is unmatched. Long bearing life with minimal wear is achieved through an even, nonimpulsive bearing load rotating at high speeds to establish a stable and continuous lubricant film.

Turbines now in use operate with intake gas pressures of 75 to 150 psi. Converting existing turbines to producer-gas use would require operating the gas producer at pressures up to 150 psi in order to avoid compressing the gas for the turbine combustor.

Gas must be exceptionally clean and particularly free from alkali metal content for turbine operation, because the blades operate at high temperatures and velocities, and are easily eroded and coated. Based on tests of a 4250-hp turbine fired with powdered coal, erosion and blade deposits were eliminated using high-temperature cyclonic inertial cleanup, which captured 99% of 20-μm and 92% of 10-μm particles (Yellott 1955). The Aerospace Corporation has developed a wood-fired turbine system operating at 5 MW under a DOE contract. Initial results show that ash from the wood, though relatively bulky, has not presented problems that were expected (PNL 1985).

11.14.2 Fuel Cells

Another potentially interesting use of producer gas is to generate electric power using fuel cells. A fuel cell is an electrochemical device that converts chemical energy into electricity directly without moving parts. Fuel cell development is very active at present, and some fuel cells may soon be available for use with producer gas.

11.14.3 External-Combustion Devices

A variety of external-combustion devices can produce power from producer gas. The Stirling-cycle engine uses an external heat exchanger and external combustion. The main drawback is the lack of widespread availability of inexpensive Stirling-cycle engines. The steam engine also uses external combustion in a boiler to produce high-pressure steam for use in piston or turbine steam engines.

When clean producer gas is available, external-combustion devices are not necessary.

Chapter 12
Safety and Environmental Considerations

12.1 Introduction

During the emergency development of gasifiers during World War II, various dangers were discovered in conjunction with gasifier operation. These dangerous areas were divided into toxic hazards, and fire and skin burn hazards. In addition, we have since recognized the potentially damaging effects that our activities can have on the environment, and we shall point out areas that may be affected by widespread gasifier use.

12.2 Toxic Hazards

12.2.1 Carbon Monoxide

The principal fuel component of generator gas is carbon monoxide (CO), a deadly poison that ties up hemoglobin in the blood and prevents the transport of oxygen to the tissues. Death from CO is death by suffocation. Lesser exposures cause headaches, nausea, dizziness, and irritability. CO is an insidious poison because it is odorless and tasteless.

Exposure during pregnancy, even at levels too low to show symptoms in the mother, may affect development of the fetus, lower its birth weight, and increase the risk of abortion and stillbirth. There are no indications that CO causes mutations or cancer (Kjellstrom 1981).

Seventeen people were killed in Sweden between December 1939 and March 1941 because of careless gasifier operation (Foley 1983). More recently, two researchers at a Midwestern university died from CO inhalation when they climbed inside of a gasifier fuel bin.

CO occurs widely in our industrial civilization in small quantities. Smokers typically inhale concentrations of several thousand parts per million, and some smokers can have as high as 20% hemoglobin blood saturation (Kjellstrom 1981). Gasoline-fueled automobile exhausts used to contain as much as 5% CO, and after a decade of improved pollution controls on cars, CO is still a major contributor to pollution in our cities. However, until the advent of natural-gas pipelines, CO was the primary fuel component of manufactured city gas, coal, and blue water gas. These were widely distributed and used around the world and are still used in many countries, evidence that CO can be handled safely when proper procedures are followed.

12.2.1.1 Acute Carbon Monoxide Poisoning

The symptoms of carbon monoxide poisoning are shown in Table 12-1.

First aid treatments for CO poisoning follow.

- Move the poisoned person to the open air or a room free of CO. Prevent the victim from exerting himself.

Table 12-1. Symptoms of Carbon Monoxide Poisoning

% Saturation of the Blood with Carbon Monoxide	Symptoms: At Rest	Symptoms: During Physical Exertion
0-10	None	None
10-20	None	During exertion, dizziness, heart pounding, and difficulty in breathing may occur.
20-30	Headache may occur.	In case of exertion, pressure at the forehead. Mild headache.
30-40	Headache in the forehead or back of the head, pulse increase, heartbeat, nausea.	In case of exertion, dizziness,fainting, possibly unconsciousness are added.
40-50	All symptoms more pronounced, nausea, vomiting, dizziness,increased tendency for unconsciousness.	
50-60	Deep unconsciousness with increased breathing and pulse rate.	
60-70	Deep unconsciousness with slow pulse and low breathing rate; possible death.	
70-80	Respiratory failure and death.	

Source: Gengas 1950.

Loosen clothing around the neck and throat, and check for the presence of an adequate airway in the throat.

- If the person is unconscious, administer artificial respiration immediately.
- Administer oxygen or mixed resuscitation gas (7% CO_2 in oxygen) as soon as possible; see Fig. 12-1.
- Summon medical aid but not at the expense of leaving the victim unattended.
- Keep the victim warm.
- Keep the victim under surveillance, as relapses often occur.

Exertion presents two profoundly dangerous effects on CO poisoning victims. First, increased exertion increases the speed of CO absorption into the blood stream, as shown in Fig. 12-2. Second, physical exertion increases the impact of CO that has already been absorbed, as shown in Table 12-1. Removing a person suffering from CO poisoning to fresh air should be done as quickly as possible, but preferably without the victim's active participation. If possible, the person should be carried. Over-exertion on the part of the rescuers should also be avoided.

Every second is valuable, and the difference may have a life-saving effect. Any producer gas installation should include two bottles of mixed resuscitation (Karbogen) gas (O_2 + 7% CO_2) or pure oxygen, with two face masks and shoulder harnesses to free the hands of the rescuers. The Karbogen gas removes CO as much as three times as fast as pure oxygen alone since the CO_2 in Karbogen gas stimulates the vagus nerve, causing more rapid breathing and, hence, faster CO removal.

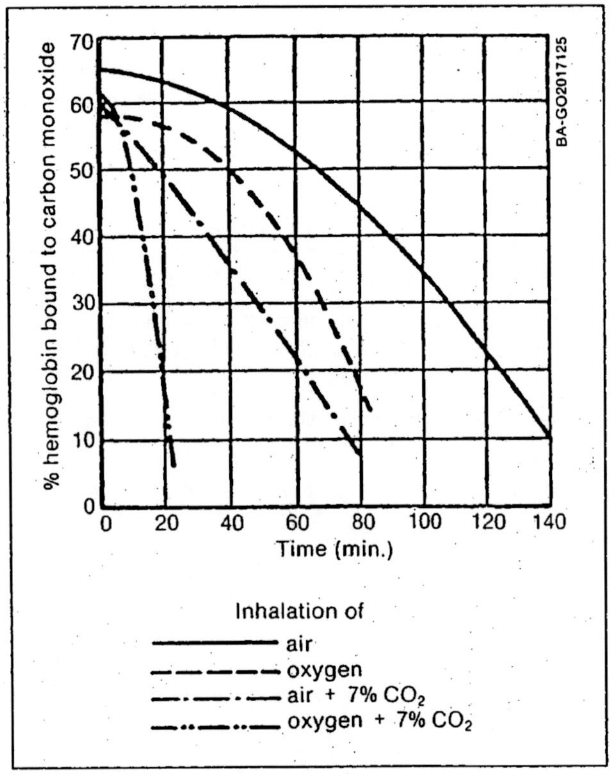

Fig. 12-1. Treatment of carbon monoxide poisoning (Source: Gengas 1950, Fig. 265)

Chronic CO poisoning symptoms, which often "sneak up" on the victim, cause him to become tired, uncomfortable, and irritable, and also induce sleeping difficulty. Sex drive virtually disappears, and urinary and heart problems have been common. Memory and eyesight may be temporarily impaired. More extensive mental symptoms may occur, including impaired memory, reduced concentration and perseverance, and possibly brain damage. The effects of chronic CO poisoning may go undiagnosed or be attributed to other causes by the uninformed victim.

12.2.1.2 Carbon Monoxide Exposure Standards

It is curious that no safe threshold CO exposure limit was recognized during the development and use of Swedish World War II-era gasifiers. Such a threshold limit would recognize CO levels below which one could expect to be free from the effects of chronic CO poisoning. The current threshold limit value (TLV) in the United States is 50 ppm CO in the work place time weighted average for an 8-h work day. Short-term exposure limit (STEL) is 400 ppm (MSA and ACGIH [OSHA 1982]). The international standards are shown in Table 12-2. Atmospheric concentrations above 1% (10,000 ppm) CO may cause unconsciousness in only a few minutes.

12.2.1.3 Safe Operating Procedures

From a safety standpoint, the best gasifier systems operate at negative pressure (suction gasifiers), so that leaks result in air being drawn into the system (possibly causing gas explosions, which the equipment should be designed to handle without harm) rather than CO being expelled into the workshop. If the gasifier is pressurized or if there is an outlet blower, then the gas will be under positive pressure and leaks will release CO. The importance of achieving a leak-tight system cannot be overemphasized.

The entire system should be leak-tested upon installation and regularly with engine service thereafter. The pipes carrying producer gas should be provided with fittings to close off and pressurize the system to 4 in. water gauge, and the entire system should be checked with soap bubbles, especially at fittings, valves, welds, lids, and seals.

Where possible, generators should be used outside, with adequate ventilation. Indoor installations should provide adequate ventilation that effectively changes the inside air supply every 1 to 2 minutes. Alarms are available for measuring and signaling excessive CO

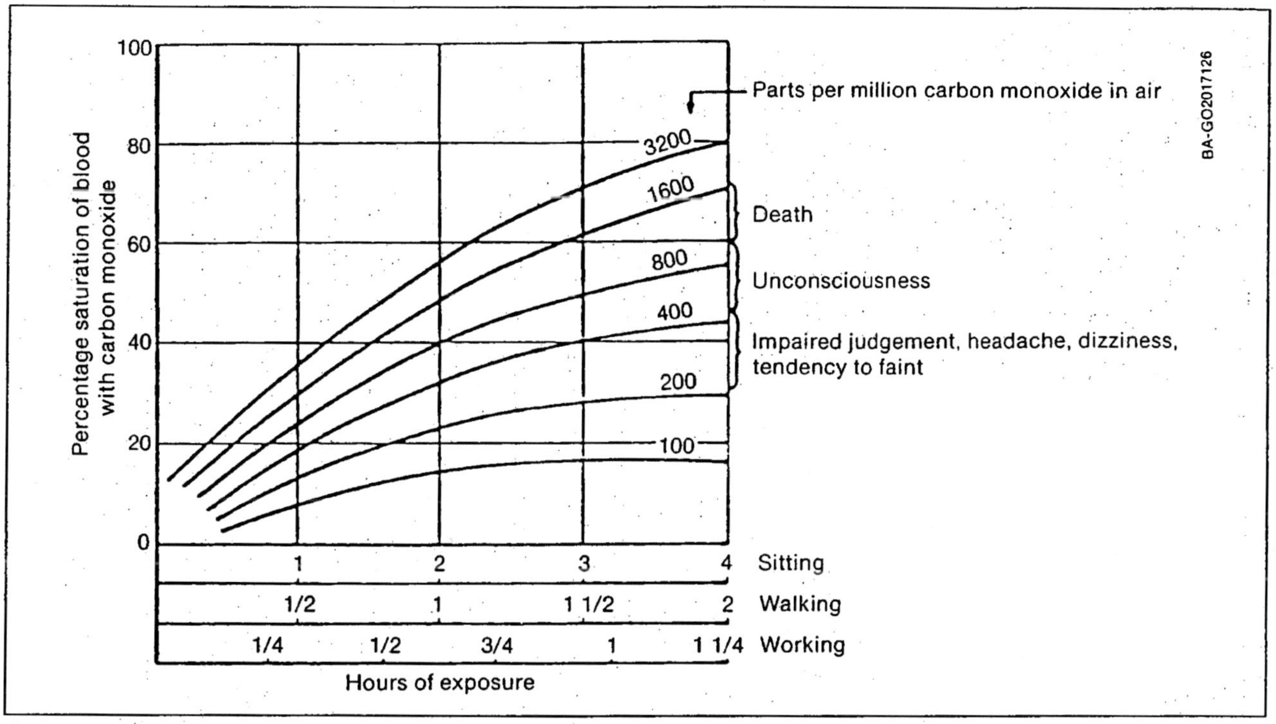

Fig. 12-2. Absorption of carbon monoxide in the blood (Source: Gengas 1950, Fig. 264)

levels in ambient air. Mini (Peeper) CO models I, II, and III (MSA catalog nos. 465398, 465530, and 466523, with ranges of 0-100, 0-500, and 0-250 ppm, respectively) indicate CO levels, while model IV (MSA catalog no. 468572) indicates the CO level and sounds an alarm (MSA). These precautions should be observed and enforced for the protection of all who run the risk of inhaling gas.

12.2.1.4 Startup and Shutdown

Carbon monoxide gas release also occurs during startup and shutdown. Gas is released when the gasifier is fanned or when the engine is started on producer gas. These two CO releases can be eliminated by installing a propane-enriched burner at the fan outlet and by starting the engine on liquid fuel.

When a gas generator is shut off, CO continues to be evolved, sometimes for hours. In this case, it is particularly important to have adequate ventilation. Since suction is no longer present within the gasifier, pressure builds up, and CO and smoke can leak out. Under no circumstances should a generator mounted on a vehicle be driven into a closed garage and shut off.

These concerns were the impetus for vehicle gasifier regulations between 1939 and 1946 that required the gas-mixer air inlet to be extended out of the engine compartment. In this way, the opening was a safe distance downwind from doors and windows. Similar regulations applied to the gasifier air inlet.

12.2.2 Creosote

Although downdraft gasifiers usually convert less than 0.1% of the input into tars and oils, these heavy products still must be scrubbed from the gas and disposed of. In earlier days, they were probably flushed down the sewer or buried. Today, such practices have come under close scrutiny, and neither can nor should be tolerated.

Man has always lived with the smoke and tars from fires, and we know that these materials can be tolerated in reasonable quantities. We also know now that the smoke from wood cooking can cause cataracts and that some tars contain dangerous carcinogens, so they should be handled with care.

Furthermore, the tars contain phenols that are potent bacteriocides. Relatively small amounts can interfere with the proper operation of septic tanks and municipal sewage systems. In Sweden during World War II, the maximum permissible phenol content of water released to sewers was 10 g/m^3 (10 mg/L), approximately 10 ppm. The phenol content of typical gasifier or gas-cooler condensate is from 1500 to 3000 mg/L. Dumping these condensates onto the ground or into the sewers or waterways is not acceptable. It should be mandatory that we determine safe disposal for these materials.

The best way to solve this disposal problem is prevention at the source; in other words, gasifiers must be designed to convert the maximum amount of tars to gas.

Table 12-2. Maximum Carbon Monoxide Concentrations in Work Places — International Standards

Country	mg/m³	ppm	Year	Fraction of Hemoglobin (%)
Australia	55	50	1973	8
Belgium	55	50	1974	8
Bulgaria	20	17	1971	3
BRD	55	50	1978	8
Czechoslovakia	30	26	1976	4.5
Denmark	40	35	1978	5.5
DDR	55	50	1978	8
Finland	55	50	1972	8
Hungary	30	26	1974	4.5
Italy	55	50	1975	8
Japan	55	50	1975	8
Netherlands	55	50	1973	8
Norway	40	35	1978	5.5
Poland	30	26	1976	4.5
Romania	30	26	1976	4.5
Soviet Union	20	17	1977	3
Sweden	40	35	1978	5.5
USA	55	50	1979	8
Yugoslavia	58	51	1971	8

Source: Kjellstrom 1981

(See Section 5.9.) Some gasifiers now incorporate catalysts that claim to eliminate the problem entirely (Ekstrom 1985). Others employ a high degree of tar recycle to eventually crack the tar thermally (Susanto 1983; Groeneveld 1980b; Kaupp 1984a).

Other methods for disposing of these materials depend on the method of gas cleanup. In some cases, wood chips or burnable filter materials are used to filter the tars. The chips used in this type of process can be dried and used as a fuel; if the tar is collected on solid filters, the filters can be incinerated. If the tars are collected in water, waste heat can be used to concentrate the tars and they can then be gasified with the gasifier fuel. Well-dried biomass fuel minimizes condensate production and eliminates the need for condensate removal for the purpose of improving gas quality. In this case, the gas can be used above its dew point, thus largely eliminating condensate collection.

Finally, it should be noted that there are neither accounts nor any evidence of damage occurring from draining the nearly one million vehicle gasifiers that were in operation worldwide between 1939 and 1945 (Kjellstrom 1983).

12.3 Fire Hazards

In the early days of gas generators, there were a large number of garage fires. These hazards were reduced through education and regulation to discourage indoor gasifier filling. Gases inside the hopper can flash when the lid is opened during filling, which in turn can ignite nearby flammable materials.

Flammable and explosive gas mixtures are usually present inside a cold gasifier. When a flame is introduced to start the gasifier, care should be exercised to prevent explosions. Proper precautions include the following:

- Always fan a cold gasifier before igniting it, to remove residual producer gas; in this way, one ensures that the gasifier contains only fresh air.
- Always stand to one side when igniting the gasifier; never look into the ignition opening while applying a flame.
- If the gasifier has a tight-fitting lid, it should be equipped with a safety release valve that will harmlessly lower the pressure resulting from a gas explosion.
- A flame arrestor, as shown in Fig. 8-12, should be placed at the gas mixer to prevent explosions. The flame arrestor can also serve as a safety filter, since it will plug rapidly when the cleanup system fails.

Openings through which fuel is loaded should be designed to provide spill shields that will prevent spilled fuel from falling onto hot surfaces and possibly causing a fire. During installation, one should install insulation, heat shields, or warning signals for workers around the hot surfaces of a gasifier. Hot metal surfaces can cause nasty skin burns at temperatures well below those that will cause the metal to glow.

Widespread gasifier use would require insurance and local fire inspectors to advise users on proper precautions that might not be evident to the inexperienced.

12.4 Environmental Hazards

The environmental problems of gasifier tars have been discussed in previous sections. As gasifiers come into wide use again, there will be an increased demand for wood and agricultural residue fuels. This can have either negative or positive environmental effects, depending upon the approach to the problem. Selective wood cutting and use of some residues can improve the forests and fields. Indiscriminate cutting and use of all residues can lead to poor wood lot stands, erosion, and soil depletion.

"It is essential that a careful survey of available sources of biomass fuel, and analysis of alternative uses and existing competing markets of these fuels, be carried out prior to large-scale introduction of producer gas plants." (Kjellstrom 1983). Permanent deforestation and desertification, which already threaten much of our planet, can be initiated by indiscriminate use of producer gas (Giono 1976). The use of wood for gasifier fuel should be introduced only in regions of fast-growing vegetation, accompanied by effective programs of forest management and reforestation (Kaupp 1984a).

Removing excessive quantities of biomass from agricultural land depletes the soil, removing not only nutrients but also reducing tilth, water permeability, and storage ability, and leaving the bare soil exposed to erosion by wind and water. Good soil conservation practice requires careful regional determination of maximum acceptable biomass removal rates, coupled with cultivation methods that make best use of the biomass left on the field (Lowdermilk 1975).

Erosion is more destructive than drought because erosion destroys the soil's water permeability. Typically, 99% of rainfall soaks into healthy soil, but only 50% of rainfall may soak into deteriorated soil. The other 50% becomes runoff, thus making the soil twice as arid with the same rainfall and simultaneously accelerating the erosion process (Carter 1974).

Chapter 13
Decision Making

13.1 Introduction

Gasifiers are technically practicable. But other criteria also factor into decisions about their use. These criteria include gasifier application, the availability of suitable equipment, biomass fuel availability and fuel-source reliability, regulations, operator availability, and of course cost and financing.

13.2 Logistics Assessment

The decision to install, invest in, or finance a gasifier electric project in a particular location requires a favorable assessment of the factors that affect the practicability of the concept over the operational life of the project.

13.2.1 Gasifier Application

Gasifier system application can range from a fuel or chemical source for in-house or external use to a vast complex that produces heat and electric energy as well as inert byproducts. Generally, the system becomes more cost-effective as the range of products increases. Gasifier system planning at least should consider systems that are larger than needed for internal use (with sell off of surplus product) and systems providing cogeneration.

13.2.2 Equipment Selection Factors

Equipment needs vary with application (e.g., heat only, fuel only) as well as the gas and location of the gasifier. All gasifier installations, however, require control systems (however rudimentary), feedstock storage, feedstock feeding mechanisms, and, obviously, a gasifier. Devices for making use of the gas are also needed. Combined with internal-combustion engines, for example, equipment add ons for gasifiers are engines and gas-cleanup equipment. In addition, electric generators and electric power conditioners will be needed where electricity is to be produced.

The larger issues in equipment selection are project scope (i.e., number of uses), size, and composition (i.e., equipment origins and mix). Decisions depend on available financing, on whether the equipment eventually will be used for purposes exceeding immediate needs, and on the talent of available labor.

Various factors control gasifier system size. Since this document covers only small gasifiers, the largest project size of interest to readers of this text is limited to a few gasifiers, each producing a maximum of 20 MBtu/day—sufficient to meet the needs of a small housing development (fewer than 20 homes), a moderately sized industrial complex, or a good-sized farm.

The equipment may consist of all "home-made" components, all individual purchases of manufactured parts, a combination of "build and buy," or a turnkey purchase. Money, the availability of suitable equipment, use of the equipment, and local talent enter into decisions affecting equipment makeup and mix. Where there's need for an uninterrupted gas supply, choices may include multiple small gasifiers or fewer large ones with attendant gas storage.

13.2.3 Feedstock Supply

More than price is involved in assessing biofuels availability. Adequate resources should be available within a 50-mile radius to minimize concerns for an uninterrupted supply. The types of biofuels available are also important, and there should be alternates compatible with the gasifier design. Since the resource may ultimately attract other customers, one or more long-term contracts guaranteeing a supply is prerequisite. Also the reputation of the suppliers to meet commitments should be thoroughly verified. (As the reliability of the supply diminishes, the size of the biofuels storage increases.)

13.2.4 Regulations

Local and federal regulations may influence decisions regarding the type, form, and size of a gasifier installation. The federal government's PURPA legislation should cause planners to recognize the potential to be earned by selling energy to utilities and then balance advantages against liabilities. Other regulations (e.g., environmental) may mandate use of particular (e.g., ash-disposal) equipment.

13.2.5 Labor Needs

Most individuals, given minimal training, can operate a batch-fed gasifier that is not tied into other equipment. Mechanically or electrically competent labor is required where the gas is used to produce electric power. Where automatic control is extensive, there will be additional need for competence with controllers and computers.

13.2.6 Final Logistics Considerations

Before the first piece of equipment is ordered or financed, the success of the project should be secured by the following:

- All equipment should be checked as being of proven design, which can be underwritten by long-term service contracts and warranties for prime power application.
- Systems engineering should be certified by a reputable engineer.
- A long-term fuel purchase agreement at a firm price with assured supply (multiple sourcing is preferred) should be secured as critical to the project's success.
- All zoning questions should be resolved, and all permits, licenses, and approvals acquired.
- Liability insurance should be available and assured.
- A long-term power sales contract should be in place, with a levelized power-price or fuel-cost escalator, if the project includes utility buyback.

13.3 Economics

Logistics is only part of the equation determining the practicality of gasifier implementation. The other part is economics.

Determining the economic feasibility of a gasifier project in a specific situation involves realistic and site-specific estimates of capital, feedstock, labor, and maintenance costs; the value of the electricity and heat produced; net fossil-fuel and energy savings; and experience with similar systems. Naturally, costs can be measured by using low-cost and no-cost burnable wastes (provided they are available on a reliable basis), low-cost equipment options, renewable energy tax credits, reduced-rate financing, automatic operation, and reduced maintenance and longer overhaul cycles.

The intent of the information that follows is to provide a basis for understanding the elements entering into an economic decision about investing in gasification facilities. Potential users of gasifiers are referred to a number of excellent references for economic assessment of gasifiers (Hodam 1983; NYSERDA 1980; EIA 1983a; EIA 1983b). The circumstances affecting economics, however, are in constant flux, and only general rules are noted here.

13.3.1 Costs

An assessment of the overall economic feasibility of using biomass to generate electric power should consider the following factors when considering a specific project:

- the cost of biomass fuel
- the cost of the gasifier system, including fuel storage bins, fuel feeding devices, gas cleanup systems, a utility connection, switchgear, and installation costs
- the cost of money at prevailing interest rates
- the cost of operating labor
- overhaul and replacement costs
- design life of the equipment between overhauls or replacement
- the value of the power produced
- the cost of ash and tar disposal after cleanup
- retail cost of buying electricity
- benefits of using renewable energy.

13.3.2 Calculating Energy Costs

The cost of energy from biomass should be compared with the costs for all other fuels with which it might compete. The convenient fuels (electricity, gas, and oil) are more expensive than the solid fuels (coal and biomass). Furthermore, a reliable infrastructure is available for transport and delivery of electricity, gas, and oil, and reliable, inexpensive equipment is available for their use. Biomass use in the absence of a well-developed infrastructure is highly dependent on the specific situation. The fact that biomass use has doubled in the past 10 years, however, suggests that the infrastructure and equipment for biomass use are growing steadily.

The cost of various forms of biomass should be compared using a common base. Since neither moisture nor ash contributes to fuel value, biomass cost is mostly quoted in dollars per ton MAF ($/ton MAF), where MAF denotes moisture-and ash-free basis (i.e., AS IF the biomass had its moisture and ash removed).

$$\text{Biomass Cost (\$/MAF ton)} = \frac{\text{Actual Cost}}{(1 - M - A)} \qquad (13\text{-}1)$$

where M is the fraction of moisture and A is the fraction of ash in the biomass. Freshly harvested biomass often contains a moisture fraction of 0.5. The internal ash content of most wood is less than 0.01 (1%), but as-delivered it may contain extraneous matter in fractions up to 0.05 (5%). Agricultural residues contain 0.05 to 0.20 ash fractions, so the normalization equation clearly is important in calculating biomass costs.

The actual costs of various forms of biomass, shown in Table 13-1, vary from a negative $20/dry ton (depending on landfill tipping fees) for landfilled burnable residues and municipal waste to more than $100/ton for firewood delivered in a city. Costs, in fact, depend on many factors, which include

- quantities available
- whether the biomass is a byproduct or principal product
- distance that the biomass must be hauled
- amount of pretreatment, sizing, drying, and storage needed
- tax benefits for biomass use.

Table 13-1. Typical Fuel Costs for Various Forms of Biomass (1986 $)

Form	Shape	Cost ($/ton)	Cost ($/MBtu)
Wet chips from tree service	Chips with twigs and bark	$0-20	$0-1.25
Demolition hammer milled dry	Chips with slivers	10-20	
Dry chips - whole tree	Chips, uniform size, no twigs or slivers	20-40	1.25-2.50
Densified biomass (pellets, cubes)	Uniform cubes	40-60	2.50-3.75
Cord wood	Large 2 ft long irregular diameter	100-150	6.25-10.00
Municipal solid waste	Very irregular shape with high-ash, mixed composition	Credit -(0-20)	Credit -(0-1.25)
Refuse-derived fuel	Irregular shape, composition more more predictable than MSW	0-20	(0-1.25)

MSW or refuse-derived fuels (RDF) appear to be the most inexpensive biomass fuels. However, they are technically the most difficult fuels to gasify (or burn) because of their high ash, heavy metal, and plastic content.

13.3.3 Equipment Cost

A gasifier is basically only an empty can, so the gasifier by itself can be a very low-capital-cost device, ranging from $2000 to $10,000/MBtu ($40-$200/kWh). On the other hand, a gasifier system for generating process heat or power may cost two to eight times this amount, depending on the auxiliary equipment required, including

- fuel storage bins
- fuel drying, screening, and pretreatment systems
- devices to deliver and meter fuel fed to the gasifier
- the gasifier itself
- ash removal
- a gas cleanup system
- gasifier operational controls.

These factors are specific to the application, and they must be evaluated economically and technically for each application.

In many cases, a gasifier might be considered simply as a retrofit to provide low-cost gas in place of more expensive fossil fuels because a boiler or other fuel-requiring device already exists. In these cases, the economic analysis is greatly simplified because the operation and economics of existing equipment are already well understood.

If the gas is to be burned directly, then an equipment comparison should be based on the cost per million Btu per hour. A compilation of local energy costs similar to those in Table 13-2 gives a good idea of the economic attractiveness of a gasifier-for-heat project. Many of the factors discussed for power generation will also apply to heat generation.

The cost of a gas producer and the cleanup portion of the gasifier system primarily should be the cost of fabricating these units from sheet, plate, bar, and tube stock. As the production volume for a particular design increases, there may be some benefit from using custom stampings, spun domes and cones, and custom castings. However, tooling costs are significant for specialized, single-purpose components, so the effect of mass production methods can be considered minor.

Fabrication costs can be estimated by tabulating the costs of material (e.g., pounds of sheet metal, as well as accessory hardware and fittings), fabrication (based on factors such as the length of cuts, number of welds and bending operations, and amount of assembly), and overhead expenses (Perry 1973). In addition, there will be extra costs for completely automatic systems and for

Table 13-2. Typical Costs of Various Fuels in Colorado (1983)

Fuel	Energy Equivalence	Typical Cost	Typical Cost ($/MBtu)
Natural Gas	0.1 MBtu/therm	$0.50/therm	$5.00
Oil	6.2 MBtu/bbl	$30/bbl	4.80
Electricity	3412 Btu/kWh	$0.05/kWh	14.60
Coal	24 MBtu/ton	$40/ton	1.66
Biomass	16 MBtu/ton	$32/ton	2.00

the engineering and development costs of the initial design.

13.3.4 Conversion Efficiency and Fuel Consumption

The efficiency with which the gasifier converts biomass fuel to a final product is an important factor for calculating operating costs. Overall efficiency is the product of a several efficiencies, including those for drying the biomass, operating the gasifier, and use of the product gas.

Drying efficiency varies widely, depending on the equipment design and heat source (see Chapters 3 and 8). A typical fuel consumption for wood with 20% moisture is roughly 2 lb/hp-h (3 lb/kWh) (Gengas 1950).

13.3.5 The Cost of Operating Labor

Batch-fed gasifiers can be used in the lowest cost system designs. Batch-fed gasifiers are suitable for many situations, especially in the context of the workplace where change of shifts, lunch, and breaks serve as natural intervals for fueling and ash removal. The cost of operator labor for refueling can be determined from the equation in Table 13-3 or from the graph in Fig. 13-1. If the gasifier requires continual monitoring by a skilled attendant, this cost introduces a significant economy-of-scale factor against very small systems.

It should be possible in many situations to operate batch-fed gasifiers with minimal attention. However, if round-the-clock operation or minimal attendant labor is desired, then automatic operation may be more suitable.

Automatic fuel feeding and ash removal require additional equipment. The necessary materials and equipment and reliable controls can add to the equipment costs on a one-time basis. The costs of automatic material handling can be compared with the expected savings in operator labor costs using Fig. 13-1 to evaluate whether to install the automatic equipment.

13.3.6 Maintenance Costs

Maintenance costs (as well as original-equipment costs) increase rapidly as engine size increases. For instance, a major overhaul on Caterpillar and Onan 100-kW natural-gas engines costs from $6000 to $9000, representing more than a threefold increase in overall cost per kilowatt hour over a 50-kW system.

Table 13-3. Sample Calculation of Electric Production Costs

For example at $1,000/kW equipment cost, 15% interest, operating at 80% duty cycle

$$\text{Interest} = C_{int} = \frac{\text{Equipment Cost \$/kW)(Loan Interest \%/yr)}}{\text{(Duty Cycle \%)(365 days/yr)(24 h/day)}} = \frac{(1000)(15)}{(80)(365)(24)} = 2.14¢\text{/kWh}$$

For wood chips at $24/ton, 10% H2O + Specific Fuel Consumption of 3 lb/kWh, we get Fuel Cost

$$C_{fuel} = \frac{\text{(Fuel Price \$/ton)(Specific Fuel Consumption lb/kWh)}}{\text{(2000 lb/ton)[1-(moisture \%/100)]}} = \frac{(24)(3)}{(2000)[1-(10/100)]} = 4.0¢\text{/kWh}$$

The cost of engine wear for a 50 kW engine with 2000 h engine life and $1000 rebuild cost

$$C_{wear} = \frac{\text{Rebuild Cost \$}}{\text{(kW)(engine life)}} = \frac{\$1000}{50(2000)} = 1.0¢\text{/kWh}$$

Labor Cost for 1/2 h per 8 h shift at $5/h wage rate

$$C_{labor} = \frac{\text{(Wage Rate \$/h)(Attention Hours/Shift)}}{\text{(Capacity kW)(Hours/Shift)}} = \frac{(5)(0.5)}{(50)(8)} = 0.625¢\text{/kWh}$$

Normal Maintenance $6.25 for 5 qt oil capacity, $15 oil analysis, $15 plugs, $8 points, 1 h labor, 200 h maintenance interval

$$C_{maint} = \frac{\text{Parts + Labor + Oil Analysis}}{\text{(kW)(maintenance interval)}} = \frac{(6.25 + 15 + 8) + (5) + (15)}{(50)(200)} = 0.5¢\text{/kWh}$$

Total Cost of Electric Generation is the sum of the above components of production cost.

$$C_{total} = C_{int} + C_{fuel} + C_{wear} + C_{labor} + C_{maint}$$

$$C_{total} = 2.14 + 4.0 + 1.0 + 0.625 + 0.5 = 8.525\text{ /kWh}$$

Furthermore, the supply of parts and service is more limited, usually only through dealers.

13.4 Cost Benefits

Using a gasifier may be justified on the basis of a single-purpose, in-house need. If, additional uses can be found, the gasifier may be more attractive. Some factors to consider are sale of excess electricity and cogeneration.

13.4.1 Value of Power Produced

PURPA requires utilities to buy back power generated from biomass at a price equal to the utility's full, avoided cost for generating power. Plants with excess capacity may be worth considering if the extra capacity can be sold at a profit. Unfortunately, the electric utility climate for PURPA buyback is one of confusion, misinformation, and frequent changes. The PURPA law does not clearly define avoided costs and leaves interpretation to the individual states, with consequent

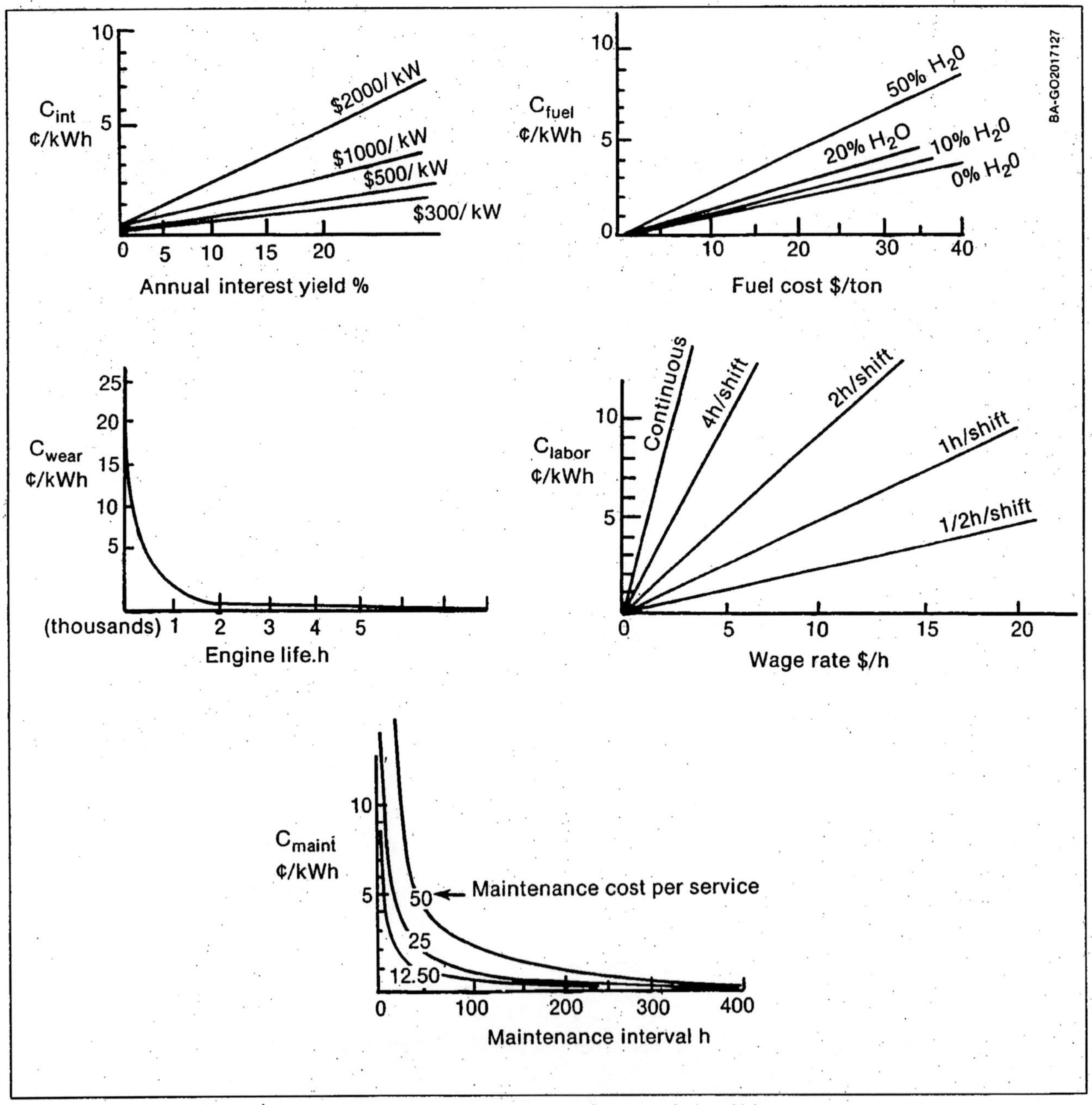

Fig. 13-1. Electricity generation costs: graphic calculators for interest, wear, maintenance, fuel, and labor

disparities between states. It is prudent to seek long-term power purchase agreements only when and where buyback rates are favorable.

If sale of power to utilities is being considered, it is important to determine the optimum size for a gasifier electric-generation project. If the buyback rate is low and the retail rate is high, then a prospective gasifier should be downsized. If, however, the buyback rate is high enough, the decision may be to choose a much larger project with expectations of making money by selling power. Certain system sizes may be allowed more favorable buyback rates; this therefore should be explored. For instance, some states offer full retail value to renewable electricity projects under certain sizes.

Thus, it is necessary to investigate thoroughly the PURPA and state power-generation structure before making binding decisions on an electric power project. The state utility commissions are listed in Table 13-4.

Optimum-size considerations for a biomass gasifier electric project also should include available fuel supply, fuel transport equipment, and the PURPA climate.

13.4.2 Cogeneration Possibilities

The "waste" engine heat in the exhaust gases and engine block coolant represents from 2/3 to 3/4 of the energy supplied by the fuel. Using this waste heat from an engine (cogeneration) allows a much higher degree of energy utilization and is sometimes eligible for additional tax credits.

The size of a cogeneration system depends, of course, on the size of the heat load. Each kilowatt from an engine generator yields arund 15,000 Btu/h in waste heat. This heat can be applied for such applications as space heating, greenhouse heating, grain drying, and absorption cycle refrigeration. Figure 13-2 indicates a heat budget for cogeneration heat recovery.

13.5 Financing

Several potential funding sources exist for implementing a gasifier project.

13.5.1 Government Subsidies in the Form of Tax Incentives

The U.S. government in the past has provided tax incentives for using renewable-energy power. So have some states. The situation is in flux and existing situations should be checked with an accountant or lawyer specializing in energy issues. More general (energy) tax breaks (e.g., accelerated depreciation) may exist even where no special tax privileges exist for renewable-energy projects, and these possibilities, too, should be explored.

13.5.2 Financial Institutions

If a gasifier/generator system is being financed, investors and the financial community will require a complete assessment of the project with performance guarantees and benefits and risks clearly identified.

Interest rates in the 8% to 20% range place a high demand on the economic yield of an installation. For instance, 15% interest on a $1000/kW installation costs 2.5¢/kWh in interest for continuous operation. Actual interest costs can be determined from either the equations in Table 13-3 or the graphs in Fig. 13-1. The sensitivity of the generating cost to engine life, fuel cost, and labor is indicated in graphic form in Fig. 13-1.

13.6 Other Considerations

The factors considered in this chapter directly influence the viability of the project. Other factors, although peripheral, are worth noting because such important considerations as bank-loan approvals can hinge on their perceived value. These factors include job creation and economic benefit to the community. New jobs can be expected because of a biomass gasifier electric project. Also most of the operating expenses of biomass gasifier electricity generation involve payment directly to individuals in the local community through fuel purchases (including collection, preparation, handling, and transport) and

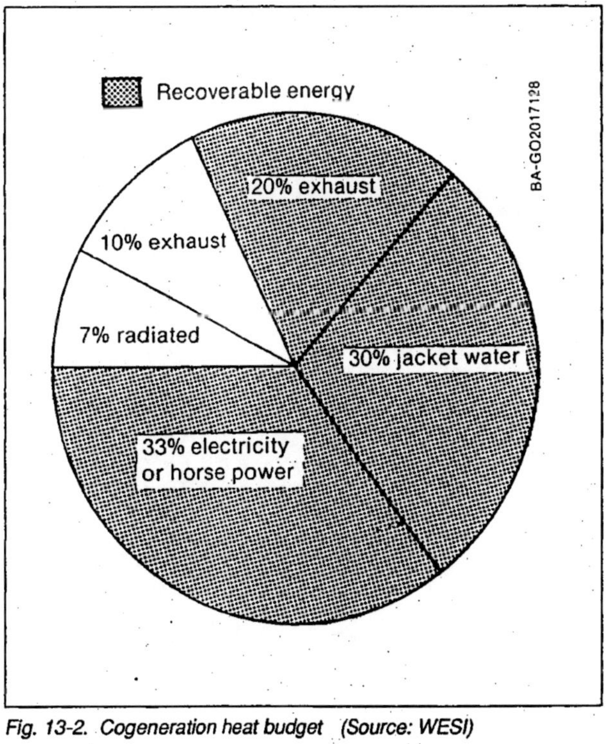

Fig. 13-2. Cogeneration heat budget (Source: WESI)

equipment operation and maintenance. Engine overhaul and investment earnings also may be retained in some local economies. In such cases, virtually all the expenses of electricity generation may generate local wages and income to the benefit of the local economy. From 30% to 70% of the value of electricity sales could result in local wages in addition to earning a 15% return on equipment investment.

Table 13-4. State Utility Commissions

State	Agency	Phone	Contact
Alabama	PSC	205-832-3421	Wallace Tidmore
Alaska	PUC	907-276-6222	Judy White
Arizona	ACC	602-255-4251	Jim Apperson
Arkansas	PSC	501-371-1792	Dana Nixon
California	PUC	415-557-1159	John Quinley
Colorado	PUC	303-866-4300	Mike Homeac
Connecticut	DPUC	203-827-1553	Research Division
Delaware	PSC	302-736-3233	Leon Ryan
District of Columbia	PSC	202-727-3062	
Florida	PSC	904-487-2740	Bonnie Davis
Georgia	PSC	404-656-4141	Sam Weaver
Hawaii	PSC	808-548-3990	Leroy Yuen
Idaho	PUC	208-343-3456	William Drummond
Illinois	ICC	217-785-0326	Joseph Gillan
Indiana	PSC	317-232-2711	William Boyd
Iowa	ISSC	515-281-5701	Robert Latham
Kansas	KCC	913-296-5468	Eva Powers
Kentucky	KURC	502-564-3940	Richard Heman
Louisiana	PSC	504-342-1403	Arnold Chauviere
Maine	PUC	207-289-3831	
Maryland	PSC	301-659-6021	Paul Daniel
Massachusetts	DPU	617-727-9748	Fuel Charge Div.
Michigan	PSC	517-373-8171	Donald Johns
Minnesota	PSC	612-296-8662	Stuart Mitchell
Mississippi	PSC	601-354-7265	Keith Howle
Missouri	PSC	314-751-3234	
Montana	PSC	406-449-2649	Ted Otis
Nebraska		No Authority	
Nevada	PSC	702-885-3409	
New Hampshire	PUC	603-271-2437	Sarah Voll
New Jersey	BPU	201-648-3448	Steve Gable
New Mexico	PSC	505-827-3361	Tom Halpin
New York	PSC	518-474-6515	Craig Indyke
North Carolina	UC	919-733-2267	Tim Carrere
North Dakota	PSC	701-224-4078	Steven Kahl
Ohio	PUC	614-466-7750	Alan Pound
Oklahoma	OCC	405-521-2335	Jim Winters
Oregon	PUC	503-378-7998	Leon Hagen
Pennsylvania	PUC	717-783-1373	Tim Clift
Rhode Island	PUC	401-277-3500	Doug Hartley
South Carolina	PSC	803-758-5632	Randy Watts
South Dakota	PUC	605-773-3201	Walter Washington
Tennessee	PSC	615-741-2125	
Texas	PUC	512-458-0202	Mike Williams
Utah	PSC	801-533-3247	Douglas Kirk
Vermont	PSB	802-828-2880	Perer Zamore
Virginia	SCC	804-786-4932	Bill Stevens
Washington	UTC	206-753-1096	Dick Bostwick
West Virginia	PSC	304-348-2174	Rich Hitt
Wisconsin	PSC	606-266-5620	Jennifer Fagen
Wyoming	PSC	307-777-7472	Dave Walker
Institute for Local Self Reliance		202-232-4108	David Morris

References

(Aarsen 1985) van den Aarsen, F.G. et al., "Wood Pyrolysis and CO_2 Char Gasification in a Fluidized Bed," in *Fundamentals of Thermochemical Biomass Conversion: An International Conference*, Estes Park, Colo., October 18-22, 1982. Proceedings edited by R.P. Overend, T.A. Milne, and L. Mudge, Elsevier, 1985.

(ACGIH 1982) *TLV's for Substances and Physical Agents in Work Environments 1982.* The American Conference of Government Industrial Hygenists, P.O. Box 1937, Cincinnati, Ohio, 45211, ISBN 9-936712-39-2.

(Anon 1943) "Air-Gas Mixing Valves," *Automobile Engineer*, Sept. 1943, p. 377.

(Anon 1944) "Conversion of Compression Engines to Producer Gas Operation," *Gas and Oil Power*, Sept. 1944, pp. 244-249.

(Antal 1979) Antal, M.J., "The Effects of Residence Time, Temperature and Pressure on the Steam Gasification of Biomass," in ACS Symposium *Biomass as a Nonfossil Fuel Source*, Honolulu Meeting, 1 April 1979.

(Arthayukti 1984) Arthayukti, W., *Biomass Gasification in Thailand*, National Energy Administration Ministry of Science, Technology and Energy, Bangkok, Thailand, 1984.

(ASHRAE 1981) *ASHRAE Handbook of Fundamentals*, American Society of Heating, Refrigerating, and Air-Conditioning Engineers, Atlanta, Ga., 1981.

(ASME 1965) *Determining the Properties of Fine Particulate Matter*, The American Society of Mechanical Engineers, ASME PTC 28, 1965.

(ASME 1969) *Gaseous Fuels: Performance Test Codes*, The American Society of Mechanical Engineers, ASME PTC 3.3, 1969.

(ASME 1980) *Determining the Concentration of Particulate Matter in a Gas Stream*, The American Society of Mechanical Engineers, ANSI/ASME PTC 38, 1980.

(ASTM 1977) *Calorific Value of Gases in Natural Gas Range by Continuous Recording Calorimeter*, ASTM D 1826, 1977. Appears in *Gaseous Fuels: Performance Test Codes*, The American Society of Mechanical Engineers, ASME PTC 3.3, 1969.

(Bailie 1979) Bailie, R., *Hesselman Gas Generator Testing*, The Environmental Energy Engineering, Inc., Report AH-8-1077-1 to Solar Energy Research Institute, 1979.

(Beenackers 1982) Beenackers, A.A.C.M. and van Swaaij, W. P. M., in Palz, W. and Grassi, G., (Eds), *Energy from Biomass*, Vol. 3, p. 201, Reidel, Dordrecht, 1982.

(Bioenergy 1985) *BioEnergy '84, Proceedings of a World Conference, Goteborg, Sweden, June 21, 1984*, edited by H. Egneus and A. Ellegard, Elsevier, London, 1985.

(Bowen 1978) Bowen, M.P. et al., "A Vertical Bed Pyrolysis System," Solid Wastes and Residues ACS Symposium Series, No. 76, p. 94, 1978.

(Breag 1982) Breag, G.R., Harker, A.C., Hollingdale, A.C., and Pearce, D., *Experimental Trials on a Spark Ignition Engine Fueled by Producer Gas from Charcoal to Provide up to 4 kW Shaft Power*, Report of the Tropical Products Institute, 56162 Gray's Inn Road, London WCIX8LU, England, Publication L50, ISBN 0 85954-154-1.

(Bridgwater 1984) Bridgwater, A.V. ed., *Thermochemical Processing of Biomass*, Butterworths, London, 1984.

(Buekens 1985) Buekens, A.G. and Schoeters, J.G., "Modelling of Biomass Gasification," in *Fundamentals of Thermochemical Biomass Conversion: An International Conference*, Estes Park, Colo., October 18-22, 1982, Proceedings edited by R.P. Overend, T.A. Milne, and L. Mudge, Elsevier, 1985.

(Calvert 1972) Calvert, S., Goldshmid, J., Leith, D., and Mehta, D., *Wet Scrubber System Study: Vol. I, Scrubber Handbook*, The U.S. Environmental Protection Agency, EPA-R2-72-118a, 1972. Also available from NTIS, PB 213016, 213017, July 1972.

(Carré 1985) Carré, J. et al., "Densified Products to be Used in Gas Producers," in *Symposium on Forest Products Research International* — 22 April, Pretoria, South Africa, Vol. 5, National Timber Research Institute.

(Carter 1974) Carter, V.G. and Dale, T. *Topsoil and Civilization*, University of Oklahoma Press, Norman, Okla., 1980.

(Cash 1942) Cash, J.D. and Cash, M.G., *Producer Gas for Motor Vehicles*, Angus Robertson Ltd., Sydney and London, 1942.

(CEC 1980) "Energy from Biomass," papers from conference, Commission of the European Communities, Brighton, England, 1 Nov. 1980, London: Department of Energy.

(CEC 1983) *Energy from Biomass: 2nd EC Conference.* Edited by A. Strub, P. Chartier, and G. Schleser, Applied Science Publishers, New York, 1983.

(Chern 1985) Chern, Shyh-Ming, *The Gasification of Biomass in Commercial Downdraft Gasifiers*, Masters Thesis, Kansas State University, 1985.

(Chittick 1983) Chittick, D.E., "Method for Converting Organic Material into Fuel," U.S. Patent 4,421,524 assigned to Pyrenco Inc., Prosser, Wash., 1983.

(Chrysostome 1985) Chrysostome, G. and Lemasle, J., "Syngas Production from Wood by Oxygen Gasification Under Pressure," in *BioEnergy '84*, edited by H. Egneus and A. Ellegard, Vol. III, Elsevier, London, 1985.

(Cousins 1983) "Some Properties of Carbon Black from Wood Gas," *New Zealand Journal of Science*, Vol. 26., pp. 277-281, 1983.

(Cruz 1983) Cruz, I.E., "Developments in the Design of Gas Producer Systems in the Philippines," In *Producer Gas 1982: A Collection of Papers on Producer Gas With Emphasis on Applications in Developing Countries*, Producer Gas Conference, Sri Lanka, ISBN:91-86618-00-8, The Beijer Institute, Stockholm, Sweden, 1983.

(Cruz 1984) Cruz, I.E., *Producer Gas Technology for Rural Applications*, Philippines National Oil Company Energy Research and Development Center, Don Mariano Marcos Ave., Diliman, Quezon City, Philippines, 1984.

(Das 1985) Das, A., *Contaminant Testing Method for Biomass Gasifier Engine Systems*, TIPI Workshop, P.O. Box 84, Allenspark, Colo., 80510, 1985.

(Das 1986) Das, A., The authors have had more than 10 years' experience in gasification, much of which is unpublished. This reference is to such unpublished results.

(Davidson 1978) Davidson, P.E. and Lucaw, T.W., "The Andco-Torax High Temperature Slagging Pyrolysis System," in *Solid Wastes and Residue Conversion by Advanced Thermal Processes*, ACS Symposium Series 76, American Chemical Society, Washington, D.C., 1978. ISSN 0097-6156. Also see (IGT 1984).

(Desrosiers 1982) Desrosiers, R., *Fundamental Air Gasification Engineering Parameters*, SERI/PR-234-1470, (February 1982). Also *Gasification Engineering. Quarterly Report, SERI/PR-622-1348, (December 1981) and SERI/PR-622-1153 (April 1981).*

(Diebold 1984) Diebold, J.P. and Scahill, J.W., "Ablative, Entrained-Flow Fast Pyrolysis of Biomass," in *Proceedings of the 16th Biomass Thermochemical Contractors' Meeting*, Portland, Oreg., May 8, 1984, Pacific Northwest Laboratory, PNL-SA-12403, 1984.

(Diebold 1985) Diebold, J.P., *The Cracking Kinetics of Depolymerized Biomass Vapors in a Continuous Tubular Reactor*, Master's thesis, Colorado School of Mines, Golden, Colo., 1985.

(Diebold 1986) Diebold, J.P., Scahill, J.W., and Evans, R.J., "The Thermal and Catalytic Upgrading of Oxygenated, Primary Biomass Pyrolysis Oil Vapors," in *Proceedings of the 17th Biomass Thermochemical Contractors' Meeting*, Pacific Northwest Laboratory, PNL-SA-1357, 1986.

(DOE 1982) *Estimates of U. S. Wood Energy Consumption from 1949 to 1981*, U.S. Department of Energy, DOE/EIA-0341, 1982.

(Domalski 1986) Domalski, E.S., Jobe, T.L., Jr., and Milne, T.A., *Thermodynamic Data for Biomass Conversion and Waste Incineration*, SERI/SP-271-2839, Solar Energy Research Institute, Golden, Colo., 1986.

(Donnot 1985) Donnot, A., Reingovolo, J., Magne, P., and D'Eglise, X., "Flash Pyrolysis of Tar from the Pyrolysis of Pine Bark," *Journal of Analytical and Applied Pyrolysis*, Elsevier Science Publishers, B. V. Amsterdam-Netherlands, 1985, p.14.

(Dwyer 1960) F.W. Dwyer Mfg. Co., Michigan City, Ind., Bulletin AA-2, 1960.

(Easterling 1985) Easterling, J.C. et al., *Identifying the Barriers to Commercialization of Low-Btu Gasifier: Proceedings of a Workshop*, (Proceedings of the Low-Btu Gasifier Workshop, Atlanta, Ga., Nov. 27, 1984), Pacific Northwest Laboratory, CONF-8411156, PNL-SA-13123, April 1985.

(Edrich 1985) Edrich, R., Bradley, T., and Graboski, M., "The Gasification of Ponderosa Pine Charcoal," in *Fundamentals of Thermochemical Biomass Conversion: An International Conference*, Estes Park, Colo., October 18-22, 1982. Proceedings edited by R.P. Overend, T.A. Milne, and L. Mudge, Elsevier, 1985.

(Egloff 1941) Egloff, G. and Van Arsdell, P., *Journal Inst. Petroleum Tech.*, Vol. 27, p. 121. 1941.

(Egloff 1943) Egloff, G. and Van Arsdell, P., "Motor Vehicles Propelled by Producer Gas," *Petroleum Engineer*, Vol. 15, p. 645, 1943.

(EIA 1983a) Energy Information Administration, *Cogeneration: Regulation Economics and Capacity Volume 1: Regulation*, Energy Information Administration, Office of Coal, Nuclear, Electric, and Alternate Fuels, U.S. DOE, Washington, D.C., 20585, DOE/NBB-0031, April 1983.

(EIA 1983b) Energy Information Administration, *Cogeneration: Regulation Economics and Capacity Volume 2: Economics and Capacity*, Energy Information Administration, Office of Coal, Nuclear, Electric, and Alternate Fuels, U.S. DOE, Washington, D.C., 20585, DOE/EIA-0421, Sept. 1983.

(Ekstrom 1985) Ekstrom, C., Lindman, N., and Peterson, R., "Catalytic Conversion of Tars, Carbon Black, and Methane," *Fundamentals of Thermochemical*

Biomass Conversion: An International Conference, Estes Park, Colo., October 18-22, 1982. Proceedings edited by R.P. Overend, T.A. Milne, and L. Mudge, Elsevier, 1985.

(Emrich 1985) Emrich, W., "Handbook of Charcoal Making," Series E, Vol. 7 of *Solar Energy in the European Community*, D. Reidel, Boston, Mass., 1985.

(Evans 1984) Evans, R.J., Milne, T.A., and Soltys, M.N., "Fundamental Analysis Studies," in *Proceedings of the 16th Biomass Thermochemical Conversion Contractors Review Meeting*, Portland, Oreg., May 1984.

(Fisher 1983) Fisher Scientific Company, *1983 Supply Catalog*, 711 Forbes Ave., Pittsburgh, Pa., 15219.

(Foley 1983) Foley, G. and Barnard, G., *Biomass Gasification in Developing Countries*, Earthscan Techn. Rept. No. 1, Earthscan, London, 1983.

(FPRS 1983) Forest Products Research Society, *Proceedings of the 6th Industrial Wood Energy Forum, Forum '82*, 1983. See also previous and subsequent FPRS symposia (available from FPRS, Madison, Wis.).

(Freeth 1939) Freeth, E.E., "Producer Gas for Agricultural Purposes," *Journal of the Department of Agriculture of Western Australia*, Vol. 16, No. 4, December 1939, pp. 371-415.

(Gengas 1950) *Gengas: Svenska Erfarenheter Fran Aren 1939-1945*, Stockholm, 1950; Translated as *Generator Gas: The Swedish Experience from 1939-1945* by M. Geuther (T. Reed and D. Jantzen, eds.) by the Solar Energy Research Institute, SERI/SP-33-140, 1979; reissued as *Gengas* with index added by A. Das, TIPI Workshop Books, P.O. Box 84, Allenspark, Colo., 1982 (see TIPI 1986).

(Giono 1976) Giono, J., "Trees," *Coevolution Quarterly*, Summer 1976, pp. 54-59.

(Goldman 1939) Goldman, B. and Jones, C., "The Modern Portable Gas Producer," *Journal of the Institute of Fuel*, Vol. 12, No. 63, Feb. 1939, pp. 103-140.

(Goss 1979) Goss, J.R., *An Investigation of the Down-Draft Gasification Characteristics of Agricultural and Forestry Residues*, Interim Report, California Energy Commission P 500-79-0017, 1979.

(Graboski 1985) Graboski, M. et al., *Oxygen Blown Downdraft Gasification of Wood*, report to the Solar Energy Research Institute, contract ZK-5-05058-1, Syngas Systems Inc., 1985.

(Graboski 1986) Personal Communication, Mike Graboski, Syngas Systems Inc., Golden, Colo., 80401.

(Graham 1983) Graham, R.G. and Huffman, D.R., *Gasification of Wood in Commercial-Scale Downdraft Gasifiers*, Research Paper (no number) from Forintek Canada Corp., 800 Montreal Rd., Ottawa, Canada KIC 325; in (IGT 1984), Vol. 5.

(Groeneveld 1980a) Groeneveld, M.J., *The Co-Current Moving Bed Gasifier*, PhD thesis, Twente Univ. of Technology, The Netherlands, 1980. Published by Krips Repro Meppel.

(Groeneveld 1980b) Groeneveld, M.J. and van Swaaij, W.P.M. "The Design of Co-Current, Moving Bed Gasifiers Fueled by Biomass," in Jones, J.L. and Radding, S.B., *Thermal Conversion of Solid Biomass Wastes*, ACS Symposium Series #130, Am Chem. Soc., Washington, D.C., 1980.

(Guzdar 1982) Guzdar, A. R. and Harvey, A. C., "Feeding Coal and Biomass into High Pressure Reactors," in *Biomass-to-Methanol Specialists Workshop, Tamarron, CO, 3-5 March 1982*, edited by T. B. Reed and M. Graboski, SERI CP-234-1590, 1982.

(Gumz 1950) Gumz, W., *Gas Producers and Blast Furnaces*, J. Wiley and Sons, New York, 1950.

(Haaland 1968) Haaland, H.H., *Methods for Determination of Velocity, Volume, Dust, and Mist Content of Gases*, Bulletin WP-50, Western Precipitation-Joy and Cox Div. Anderson Samplers, Inc., 4215-C Wendell Drive, Atlanta, Ga., 30336.

(Heywood 1941) Heywood, H., "Loss of Power in Petrol Engines Running on Producer Gas," *Engineering*, Jan. 24, 1941, pp. 61-63.

(Heywood 1944) Heywood, H., *Tests on Transport Producer Gas Units* (with Discussions), Institution of Mechanical Engineers, Journal and Proceedings, Vol. 151, Jan.-Dec. 1944, pp. 192-208.

(Hodam 1983) Hodam, R.H., Williams, R.O., and Lesser, M.C., *Engineering and Economic Characteristics of Commercial Wood Gasifiers in North America*, SERI/TR-231-1459, Solar Energy Research Institute, Golden, Colo., 1982.

(Hubis 1983) Hubis, M.J., *The Mechanical Behavior of Biomass Pellets During Gasification*, Masters thesis, Colorado School of Mines, 1983.

(Humphries 1985) Humphries, J. P., "Biomass Gasification," in *Symposium on Forest Products Research International*, 22 April 1985.

(IGT) This reference is to the Proceedings of the 11 technical symposia on biomass energy organized and presented by the Institute of Gas Technology, Chicago, Ill. They are "Clean Fuels From Biomass, Sewage, Urban Refuse, Agricultural Wastes," January 27-30, 1976, Orlando, Fla.; "Clean Fuels From Biomass and Wastes," January 25-28, 1977, Orlando, Fla.; "Energy from Biomass and Wastes," August 14-18, 1978, Washington, D.C.; "Energy from Biomass and Wastes

IV," January 21-25, 1980, Lake Buena Vista, Fla.; "Energy from Biomass and Wastes V," January 26-30, 1981, Lake Buena Vista, Fla.; "Energy from Biomass and Wastes VI," January 25-29, 1982, Lake Buena Vista, Fla.; "Energy from Biomass and Wastes VII," January 24-28, 1983, Lake Buena Vista, Fla.; "Energy from Biomass and Wastes VIII," January 30-February 3, 1984, Lake Buena Vista, Fla.; "Energy from Biomass and Wastes IX," January 28-February 1, 1985, "Energy from Biomass and Wastes X," April 7-11, 1986, Washington, D.C., and "Energy from Biomass and Wastes XI," March 16-20, 1987, Orlando, Fla.

(Jasas 1982) Jasas, G. and Kajpak, J. "Gas Turbine Demonstration of Pyrolysis-Derived Fuels," in *Proceedings of the 14th Biomass Thermochemical Contractors Review Meeting*, Oct. 1982, Arlington, Va., Pacific Northwest Laboratory, PNL-SA-10646, CONF-820685.

(Johansson 1980) Johansson, E., *Swedish Tests of Otto and Diesel Engines Operated on Producer Gas*, National Machinery Testing Institute, Upsala, Sweden, 1980.

(Johansson 1985) Johansson, K. G., "Wood Gas Engine Applications in Southern Africa," in *Symposium on Forest Products Research International*, 22 April Pretoria, South Africa, National Timber Research Institute, 1985.

(Kadyszewski 1986) Personal Communication. J. Kadyszewski, U.S. Agency for International Development, Arlington, Va., 1986.

(Kaupp 1984a) Kaupp, A. and Goss, J.R., *State-of-the-Art for Small (2-50 kW) Gas Producer-Engine Systems*, Final Report to USDA, Forest Service, Contract No. 53-39R-0-141, (March 1981). A bibliography of the 800 papers used by authors has been prepared and can be computer accessed by title, subject, or author. Published as *Small Scale Gas Producer Engine Systems*, by Deutsches Zentrum fur Entwicklungstechnologien - GATE, Vieweg, Wiesbaden, 1984 (see TIPI 1986).

(Kaupp 1984b) Kaupp, A., *Gasification of Rice Hulls: Theory and Praxis*, published by Deutsches Zentrum fur Entwicklungstechnologien - GATE. Vieweg, Wiesbaden, 1984 (see TIPI 1986).

(Kishimoto 1985) Kishimoto, S. and Sugiwa, G., "Charcoal as a Soil Conditioner" Article 12-23, Vol. 5 in *Symposium on Forest Products Research International* — 22 April, Pretoria, South Africa, National Timber Research Institute, 1985.

(Kjellstrom 1981) Kjellstrom, B., *Producer Gas 1980: Local Electricity Generation from Wood and Agricultural Residues*, FV-80-0035/01, The Beijer Institute, Stockholm, Sweden, 1981.

(Kjellstrom 1983) Kjellstrom, B., Stassen, H., and Beenackers, A.A.C.M., *Producer Gas 1982: A Collection of Papers on Producer Gas With Emphasis on Applications in Developing Countries*, Producer Gas Conference, Sri Lanka, ISBN:91-86618-00-8, The Beijer Institute, Stockholm, Sweden, 1983.

(Kjellstrom 1985) *2nd International Producer Gas Conference*, 19 March 1985, Bandung, Indonesia, Edited by B. Kjellstrom, (To be published)

(LaFontaine 1984) Personal Communication, H. LaFontaine, The Biomass Energy Institute, 1995 Keystone Blvd., Miami, Fla., 33181, 1984.

(LaFontaine 1987) FEMA/DOE/Martin Marietta Contract No. 98x-6446-7V to H. LaFontaine, Miami, Fla.

(LePori 1983) LePori, W. A. and Parnell, C. B., *Cleanup of Gases Produced from Gasification of Agricultural Biomass*, prepared for the USDA, Grant No. 59-2481-0-2-092-9 by Department of Agricultural Engineering, Texas A&M University, College Station, Tex., 1983.

(Lowdermilk 1975) Lowdermilk, J.C., *Conquest of the Land Through Seven Thousand Years*, Agriculture Information Bulletin No. 99, USDA Soil Conservation Service, U.S. Supt. of Documents, Stock No. 001-000-03446-4, 1975.

(Lutz 1940) Lutz, H., "Die Verbesserung des Fahrzeug-Holzgaserzeugers durch Wärmetichnische Massnahmen," *ATZ Automobiltechnische Zeitschrift*, No. 23, 1940, pp. 589-595.

(Mahin) *Bioenergy Systems Report*, Published quarterly by D. Mahin, International Energy Projects, Box 591, Front Royal, Va., 22630.

(Makray 1984) Makray, Z.T., *Gasification of Biomass in a Downdraft Gasifier*, Post Graduate Thesis at the Faculty of Engineering of Campina, Department of Mechanical Engineering, Unicamp, Brazil. (See also, Termoquip, Alternate Energy Ltd., Campinas, Brazil.)

(Manurung 1985) Manurung, R. and Beenackers, A., *An Open Core Rice Husk Gasifier for Small-Scale Application*, Twente University of Technology, Box 217, 7500 AE Enschede, The Netherlands, 1985.

(MASEC) *Woodstove Reliability Standard*. Mid-America Solar Energy Complex, Minneapolis, Minn., n.d.

(Masuda 1980) Masuda, T. and Fisher, T.F., "Purox System Demonstration Test on Simulated Japanese Refuse," in Jones, J.L. and Radding, S.B., *Thermal Conversion of Solid Biomass Wastes*, ACS Symposium Series #130, Am Chem. Soc., Washington, D.C., 1980.

(McGowan 1980) McGowan, T.F., Bryson, F., and Leverette, J.T., *A State Demonstration Program in*

Wood Energy, Volume III Wood Fuel Processing, Georgia Institute of Technology, Atlanta, Ga., 1980.

(McGowan 1984) Personal Communication, T. McGowan, Western American Combustion Co., Atlanta, Ga.

(Miles 1982) Miles, T.R., "The Preparation of Biomass for Gasification," in *Biomass-to-Methanol Specialists Workshop, Tamarron, CO, 3-5 March 1982*, edited by T.B. Reed and M. Graboski, SERI/CP-234-1590, Solar Energy Research Institute, Golden, Colo., 1982.

(Miles 1983) Miles, T. R. and Miles, T. R. Jr., "Biomass Preparation for Thermochemical Conversion," in *1st European Workshop on Thermochemical Processing of Biomass*, U. of Aston, Birmingham, U. K., April 12, 1983.

(Miller 1983) *State of the Art Survey of Wood Gasification Technology*, prepared by Fred C. Hart Associates, Inc. (B. Miller, Principal Investigator) for the Electric Power Research Institute, AP-3101, Project 986-9, 1983.

(Mother 1982) Personal Communication, R. Freudenbeger, *Mother Earth News*, Box 70, Hendersonville, N.C.

(MSA) Mine Safety Appliance Catalogs 465398, 465530, 466523, 468572, Data Sheet 08-01-02-TLV Standards, MSA Inc., 600 Pen Center Blvd., Pittsburgh, Pa., 15235, n.d.

(Nandi 1985) Nandi, S.P. and Onischak, M., "Gasification of Chars Obtained from Maple and Jack Pine Woods," in *Fundamentals of Thermochemical Biomass Conversion: An International Conference*, Estes Park, Colo., October 18-22, 1982. Proceedings edited by R.P. Overend, T.A. Milne, and L. Mudge, Elsevier, 1985

(NAS 1983) *Producer Gas: Another Fuel for Motor Transport*, National Academy Press, Washington, D.C., 1983. A companion bibliography with 421 references is also available from NAS.

(NTRI 1985) Symposium on Forest Products Research International, 22 April 1985, Organized by the National Timber Research Institute (NTRI), CSIR Conference Centre, Pretoria, Republic of South Africa, 11 Volumes. Volume 5 is titled *Carbon from Biomass* and *Woodgas in Practice*.

(Nunnikhoven 1984) Plans for Vehicle Gasifier, Nunnikhoven Industries, Oakville, Iowa.

(Nygards 1979) *Wood Gas Generator for Vehicles*, translated from Swedish by Nils Nygards, available from PURWACO International, Minneapolis, Minn., 1979.

(NYSERDA 1980) *Site Owners Manual for Small Hydro*, New York State Energy Research and Development Authority Report 79-3, 1980.

(Onan 1981) *Manual for Professional Engineers, Electric Generating Set Specifications 1.75 to 1000 kW*, The Onan Corp., 1400 73rd Ave. NE, Minneapolis, Minn., 55432, 1981.

(OSHA 1982) *Standards for Working Environments Maximum Exposure Limits 1982*, Occupational Safety and Health Administration, Washington, D.C., 1982.

(OTA 1980) *Energy from Biological Processes*, OTA-E-124, Office of Technology Assessment, Washington, D.C., 1980.

(OTA 1984) *U.S. Vulnerability to an Oil Import Curtailment*, OTA-E-243, Office of Technology Assessment, Washington, D.C., 1984.

(Overend 1982) Overend, R. *Wood Gasification — Review of Recent Canadian Experience*, National Research Council (Canada) Report NRCC 20094, March 1982.

(Parikh 1985) Parikh, P.P., *State of Art Report on Gasification of Biomass*, Interim Report of DNES (201/20/84-BM), Dept. of Mech. Eng., Indian Inst. of Technology, Bombay 400 076, India. Vol. I in print; Vols. II and III in press.

(Perry 1973) Perry, R.H. and Chilton, C.H., Eds. *Chemical Engineers' Handbook*, 5th Edition, McGraw-Hill, New York, 1973.

(Peterson 1965) Peterson, C.M. and Whitby, K.T., "Fractional Efficiency Characteristics of Unit Type Cyclones," *ASHRAE Journal* Vol. 7, p. 42, 1965.

(PNL) *Annual Proceedings of the Biomass Thermochemical Conversion Contractor's Review Meeting*; Edited by Pacific Northwest Laboratory, Richland, Wash. These proceedings, describing work supported by the U.S. Department of Energy, are available from the National Technical Information Service (NTIS), U.S. Department of Commerce, 5285 Port Royal Rd., Springfield, Va., 22161.

(Pyrenco) Company Brochure, The Pyrenco Co., Box 903, Prosser, Wash. n.d. See also U.S. Patent 4,421,524, Dec. 1983.

(RAC) RAC Staksamplr Bulletin 2343-R3, Sales Bulletin of Research Appliance Company, 4215 Wendell Dr., Atlanta, Ga., 30336. n.d.

(Rambush 1923) Rambush, N.E., *Modern Gas Producers*, Benn Bros. Ltd., London, 1923. Microfiche copy available at Energy Branch Library, CISTI, National Research Council of Canada, Ottawa, Ontario, KlA OR6.

(Reed 1978) Reed, T.B., "Biomass Energy - A Two-Edged Sword," presented at the Annual Meeting of the International Solar Energy Society, Denver, Colo., Aug. 28, 1978, SERI DDS-011, Solar Energy Research Institute, Golden, Colo., 1978.

(Reed 1979) *Retrofit 79: Proceedings of a Workshop on Air Gasification*, Seattle, WA; 2 Feb. 1979, SERI/TP-49-183, edited by T. B. Reed and D. Jantzen, Solar Energy Research Institute, Golden, Colo., 1979.

(Reed 1981) *A Survey of Biomass Gasification, Vol. I-III*, SERI/TR-33-239, edited by Solar Energy Research Institute, Golden, Colo., 1979. Published in hard cover as T. Reed, editor, *Biomass Gasification: Principles and Technology*, Energy Technology Review No. 67, Noyes Data Corp, Park Ridge, N.J. 1981.

(Reed 1982) *Biomass-to-Methanol Specialists Workshop, Tamarron, CO, 3-5 March 1982*, edited by T.B. Reed and M. Graboski, SERI/CP-234-1590 Solar Energy Research Institute, Golden, Colo., 1982.

(Reed 1983a) Reed, T.B. and Markson, M., "A Predictive Model for Stratified Downdraft Gasification," in *Progress in Biomass Conversion*, edited by D. Tillman and E.C. Jahn, Vol. 4, pp. 217-254, Academic Press, New York, 1983.

(Reed 1983b) Reed, T.B., Levie, B., Markson, M.L., and Graboski, M.S., "A Mathematical Model for Stratified Downdraft Gasifiers," in *Symposium on Mathematical Modeling of Biomass Pyrolysis Phenomena, August 1983, Washington, D.C.*, Preprints of the American Chemical Society, Division of Fuel Chemistry, 28, 5, 1983.

(Reed 1984) Reed, T.B. and Levie, B., "A Simplified Model of the Stratified Downdraft Gasifier," in *The International Bio-Energy Directory and Handbook 1984*, edited by Paul F. Bente, Jr. The Bio-Energy Council, 1625 Eye St. NW, Suite 825A, Washington, D.C., 1984, pp. 379-389.

(Reed 1985a) Reed, T. and Markson, M. "Reaction Velocities in a Downdraft Gasifier," in *Fundamentals of Thermochemical Biomass Conversion: An International Conference*, Estes Park, Colo., October 18-22, 1982. Proceedings edited by R.P. Overend, T.A. Milne, and L. Mudge, Elsevier, 1985.

(Reed 1985b) Reed, T.B., "Principles and Technology of Biomass Gasification," in *Advances in Solar Energy, Vol. 2*, K. Böer and J. Duffie, eds, Plenum, 1985.

(Reed 1985c) Reed, T.B., "Role of Carbon Saturation in Thermal Conversion Processes," in *Proceedings of 16th Biomass Thermochemical Conversion Contractors Review Meeting*, Pacific Northwest Laboratory, PNL-SA-12403, CONF-8405157, 1985.

(Reed 1985d) Reed, T.B., Levie, B., and Das, A. "Understanding, Operating, and Testing Fixed Bed Gasifiers," in *Bioenergy '84, Proceedings of a World Conference, Goteborg, Sweden, June 21, 1984*, Elsevier, 1985.

(Reed 1986a) Reed, T., The authors have had more than 10 years' experience in gasification, much of which is unpublished. This reference is to such unpublished results.

(Reed 1986b) Reed, T.B., "Moving Bed Air and Oxygen Biomass," in *Proceedings of the 1985 Biomass Thermochemical Conversion Contractors' Meeting*, Pacific Northwest Laboratory, PNL-SA-13571, CONF-8510167. 1986. A more complete report is presently being prepared with a similar title, covering the six years of work on oxygen gasifiers at SERI and SGI.

(Reines 1983) Personal Communication, R. Reines, Open University, London, England, 1983.

(Rissler 1984) Personal Communication, R. Rissler, Missouri Gasification, Inc., R.D. 3, Box 198, California, Mo., 65018, 1984.

(Rogers 1985) Personal Communication, The Buck Rogers, Co., 150 Industrial Parkway, Industrial Airport, Kans., 66031, 1985.

(Roy 1983) Roy, C. et al., "Design et Construction d'un Reacteur de Pyrolyse sous Vide," in *Comptes Rendu de L'Aatelier de Travail sur la Liquefaction de la Biomasse*, Proceedings of the 3rd Liquefaction Experts Workshop, Sept. 28, 1983, Sherbrooke University, Quebec, Canada, NRCC 23130.

(Schläpfer 1937) Schläpfer, P. and Tobler, J., "Theoretical and Practical Studies of Operation of Motorcars on Wood Gas," originally published in 1937, translated to English by H. Stassen in *2nd International Producer Gas Conference*, 19 March 1985, Bandung, Indonesia, B. Kjellstrom, ed., to be published.

(Schroeder 1985) *Biomass Gasification for Electric Utility Diesel Generating Units*, prepared by Burns & McDonnel Engineering Company, Inc. (Schroeder, R.J., Ege, H.D., and Hunt, F.E., Principal Investigators) for the Electric Power Research Institute, AP-3865, Project 1438-16, 1985.

(Scott 1983) Scott, D.S. and Piskorz, J. "The Continuous Flash Pyrolysis of Biomass," in *Comptes Rendu de L'Aatelier de Travail sur la Liquefaction de la Biomasse.* Proceedings of the 3rd Liquefaction Experts Workshop, Sept. 28, 1983, Sherbrooke University, Quebec, Canada, NRCC 23130.

(Singer 1958) Singer, C.J., *History of Technology*, Vol. IV, p. 252, Oxford, 1958.

(Skov 1974) Skov, Niels A. and Papworth, Mark L. *The Pegasus Unit: The Lost Art of Driving Without Gasoline*, Pegasus, 1974; now available from H. LaFontaine, Biomass Energy Foundation, Inc., 1995 Keystone Blvd., Miami, Fla., 33181.

(Stevenson 1982) Stevenson, W.A., "Large Wood to Methanol Plants," in *Biomass-to-Methanol Specialists*

Workshop, Tamarron, CO, 3-5 March 1982, edited by T.B. Reed and M. Graboski, SERI/CP-234-1590. Solar Energy Research Institute, Golden, Colo., 1982.

(Strauss 1975) Strauss, W., *Industrial Gas Cleaning*, Pergamon Press, Oxford, 1975.

(Ström 1985) Ström, E., Liinanki, L., Sjöström, K., Rensfelt, E., Waldheim, L., and Blackadder, W., "Gasification of Biomass in the Mino-Process," in *Bio-Energy '84*, edited by H. Egneus and A. Ellegard, Vol. III, Elsevier, London, 1985.

(Susanto 1983) Susanto, H., Beenackers, A.A.C.M., and van Swaaij, W.P.M., "Moving Bed Gasifier with Internal Recycle of Pyrolysis Gas," in *Producer Gas 1982: A Collection of Papers on Producer Gas With Emphasis on Applications in Developing Countries*, Producer Gas Conference, Sri Lanka, ISBN:91-86618-00-8, The Beijer Institute, Stockholm, Sweden, 1983.

(Taylor 1986) Taylor, L., Personal Communication, Gentronix, Inc., Albuquerque, New Mex., 1986.

(Thompson 1981) Thompson, S.P., *Fuel Preparation Systems Using a Rotary Dryer*, Bulletin 8006, Rader Companies Inc., Box 20128, Portland, Oreg., 20128, 1981.

(TIPI 1986) The TIPI Workshop Books (A. Das, Director) publishes a number of books dealing with gasification. Write for list of current publications to TIPI Workshop Books, P.O. Box 84, Allenspark, Colo., 80510.

(Walawender 1985) Walawender, W.P., Chern, F.M., and Fan, L.T. in *Fundamentals of Thermochemical Biomass Conversion: An International Conference*, Estes Park, Colorado, October 18-22, 1982. Proceedings edited by R.P. Overend, T.A. Milne, and L. Mudge, Elsevier, 1985.

(WESI) WESI Engineering Bulletin WE12832M, Waukesha, 17803 S. Santa Fe Ave., Compton, Calif., n.d.

(Work 1958) Work, L.T., "Size Reduction," *Industrial and Engineering Chemistry*, Vol. 50, No. 3, Part 2, p. 483, 1958.

(Yellott 1955) Yellott, J.I., Broadley, P.R., Meyer, W.M., and Rotzler, P.M., "Development of Pressurizing, Combustion and Ash Separation Equipment for A Direct Fired Coal Burning Gas Turbine Locomotive," ASME Paper 54A 201, American Society of Mechanical Engineers, 1955.

Appendix
Units and Conversions

Although it was our intention to present the collected data within a consistent framework of acceptable metric units, this goal could be only partially achieved. The still widespread use of English units did not in all cases allow transfer of the reported data to metric units. The internationally established gram (g), meter (m), second (s), and joule (J) system (SI) is therefore occasionally replaced by units that may be more familiar and more convenient to the reader. Conversion factors are from the *Handbook of Chemistry and Physics*, 67th edition, CRC Press.

Weight

1 kg = 2.204 lb = 15,432 grains = 32.105 oz (troy)
= 0.984×10^{-3} ton (long) = 1.1023×10^{-3} ton (short)

Pressure

1 atm = 1.0133 bar = 101.33 kPa = 14.7 psia
= 29.921 in. Hg = 1419 in. H_2O = 760 mm Hg

Velocity

1 m/s = 3.281 ft/s = 3.6 km/h = 2.237 mph

Energy

1 Btu = 1.055 kJ = 252 Cal = 778.2 foot-pound-force

1 kWh = 3.6 MJ = 3413 Btu

1 Cal = 4.187 J

Density

1 g/cm^3 = 1000 kg/m^3 = 62.43 lb/ft^3

1 lb/ft^3 = 0.01602 g/cm^3 = 16.02 kg/m^3

Power

1 watt = 1 J/s = 3.43 Btu/h = 0.2389 Cal/s = 3.6 kJ/h
= 1.341×10^{-3} hp

Temperature

K = °C + 273.15

°R = °F + 459.67 = 1.80 K

°F = 1.8°C + 32

Concentration

1 grain/ft^3 = 2.571 g/m^3

Length

1 in. = 2.54 cm

1 micron (micrometer) = 1 μm = 10^{-6} m

Volume compressible gas (1 atm dry)

1 Nm^3 (0°C) = 38.55 scf (77°F) = 37.32 scf (60°F)
= 37.90 scf (68°F)

Volume noncompressible

1 m^3 = 35.315 ft^3 = 1000 liters

1 ft^3 = 0.02831 m^3 = 7.48 gal

1 gal (U.S.) = 3.785 liters = 0.1336 ft^3 = 231 in.^{3}

1 liter = 0.353 ft^3 = 0.2642 gal = 33.82 fl oz = 61.02 in.^{3}

Flow gas

1 Nm^3/h = 0.632 scfm (68°F)

Area

1 m^2 = 10.76 ft^2 = 1550 in.^{2} = 1.30 yd^2

Hearth Load (for 130 Btu/scf gas)

0.9 Nm^3/h-cm^2 = 537 scfm/ft^2 = 3.73 scfm/in.^{2}
= 4.2 MBtu/h-ft^3

Gas Energy Content

1 Btu/scf (68°F) = 9.549 kCal/Nm^3 (0°C)
= 39.98 kJ/Nm^3 (0°C)

Fuel Energy

1 Btu/lb = 0.5555 Cal/g = 2.326 J/g

1 Cal/g = 1.8 Btu/lb = 4.187 J/g

Gasification Rules of Thumb — approximately true

Fuel Consumption

1 hp-h ≅ 2 lb biomass = 2 scfm gas

1 kWh ≅ 3 lb biomass

1 ppm ≅ 1 mg/m^3

Standard conditions for gas measurements vary widely depending on the organization setting the standards. Since gas properties can be measured to a precision considerably finer than the error introduced by a misunderstanding in standard conditions, expressions of gas volume and heating value must specify the measurement conditions.

Two often-used references (Kaupp 1984a; Gengas 1950) use gas measurement conventions different from common gas practice. The following table is provided to indicate the sources of gas measurement conventions. However, the reader should exercise extreme caution in all conversions using the term "scf.". If the gas measurement conditions are not specified or cannot be safely presumed, then gas heating value is subject to 3% uncertainty.

We have used 1 atm = 101.32 kPa = 760 mm = 29.92 in. Hg for a standard pressure and a temperature of 20°C = 68°F in our data reporting.

Standard Measuring Conditions for Gases

Unit	Where Used	Standard Conditions		Reference
		Pressure	Temperature	
scm/scf	EPA Method 5	Dry gas 760 mm = 1 atm	68°F = 20°C	(1)
scf	American Gas Association	762 mm = 30 in. Hg dry gas	60°F = 15 5/9°C	(2)
scf	Compressed Gas Institute	760 mm - 1 atm 29.92 in. Hg	68°F = 20°C	
Nm^3	NTP = STP	760 mm = 1 atm dry gas	0°C	
Nm^3	Standards Council of Canada	1 atm dry	15°C	(3,4)
ft^3	Environment Canada Air Pollution Control Directorate	29.92 in. Hg	537°R 78°F	(5)
scf	ANSI Standard Z132.1 for gas calorific value	Saturated with water 14.73 psia	60°F	(6)
	Molecular weight standard conditions	29.92 in. Hg = 101.325 kPa 760 mm = 1 atm dry gas	0°C = 32°F	(7)
scf	Kaupp and Goss	1 atm	77°F = 25°C	(8)
Nm^3 scf	*Generator Gas*	1 atm dry 1 atm dry	0°C dry 70°F = 21.11°C	(9)

(1) Environmental Protection Agency, "Determination of Particulate Emissionsfrom Stationary Sources," Codified Federal Register 40, Pt. 60, Appendix A,Method 5.

(2) *Dictionary of Scientific and Technical Terms*, 3rd Ed. New York: McGraw-Hill, 1984.

(3) International Standards Organization, *Measurement Conditions for GaseousFuels*, ISO STD 5024-TC28, 1979.

(4) ENFOR Project C172, DDS File No. 41SS, KL229-1-4117, 1979.

(5) Environment Canada Air Pollution Directorate, EPS 1 AP.74-1, 1974.

(6) *Calorific Value of Gases in Natural Gas Range by Continuous RecordingCalorimeter*, ASTM D 1826, 1977.

(7) *Determining the Properties of Fine Particulate Matter*, ASME PTC-28, 1965.

(8) Kaupp, A. and Goss, J.R. *State-of-the-Art for Small (2-50 kW) Gas Producer-Engine Systems.* Final Report to USDA, Forest Service, 1981.

(9) *Generator Gas: The Swedish Experience from 1939-1945.* SERI/SP-33-140, 1979 (Solar Energy Research Institute, Golden, Colo).

SAVE HARMLESS AGREEMENT

Because use of the information, instructions and materials discussed and shown in this book, document, electronic publication or other form of media is beyond our control, the purchaser or user agrees, without reservation to save Knowledge Publications Corporation, its agents, distributors, resellers, consultants, promoters, advisers or employees harmless from any and all claims, demands, actions, debts, liabilities, judgments, costs and attorney's fees arising out of, claimed on account of, or in any manner predicated upon loss of or damage to property, and injuries to or the death of any and all persons whatsoever, occurring in connection with or in any way incidental to or arising out of the purchase, sale, viewing, transfer of information and/or use of any and all property or information or in any manner caused or contributed to by the purchaser or the user or the viewer, their agents, servants, pets or employees, while in, upon, or about the sale or viewing or transfer of knowledge or use site on which the property is sold, offered for sale, or where the property or materials are used or viewed, discussed, communicated, or while going to or departing from such areas.

Laboratory work, scientific experiment, working with hydrogen, high temperatures, combustion gases as well as general chemistry with acids, bases and reactions and/or pressure vessels can be EXTREMELY DANGEROUS to use and possess or to be in the general vicinity of. To experiment with such methods and materials should be done ONLY by qualified and knowledgeable persons well versed and equipped to fabricate, handle, use and or store such materials. Inexperienced persons should first enlist the help of an experienced chemist, scientist or engineer before any activity thereof with such chemicals, methods and knowledge discussed in this media and other material distributed by KnowledgePublications Corporation or its agents. Be sure you know the laws, regulations and codes, local, county, state and federal, regarding the experimentation, construction and or use and storage of any equipment and or chemicals BEFORE you start. Safety must be practiced at all times. Users accept full responsibility and any and all liabilities associated in any way with the purchase and or use and viewing and communications of knowledge, information, methods and materials in this media.